NANOTECHNOLOGY: SOCIETAL IMPLICATIONS II

About the cover

Protein-templated assembly image, courtesy of Andrew McMillan, NASA Ames Research Center (ARC). The computer-generated central image models heat shock proteins that have self-assembled into a double ring structure, 17 nanometers in diameter, called a chaperonin. The researchers can tailor both the chemical functionality of the chaperonin—in this case, the proteins have been genetically modified to bind to nanoscale gold particles—and its structural features. They can coax groups of chaperonins into a variety of 1-, 2-, or 3-dimensional structures, which serve as templates for creating ordered nanoparticle arrays. NASA researchers are exploring the use of protein-templated arrays as sensors and electronic devices.

This is Volume II of a two-volume set resulting from a workshop held under the auspices of the U.S. National Science Foundation and the Nanoscale Science, Engineering, and Technology (NSET) Subcommittee of the U.S. National Science and Technology Council on Dec. 3-5, 2003. The primary purpose of the workshop was to examine trends and opportunities in nanoscience and nanotechnology toward maximizing benefit to humanity, and also potential risks in nanotechnology development. Volume II contains essays contributed by the workshop participants. The companion Volume I is a summary of the findings of and discussions at the workshop.

National Science Foundation

NANOTECHNOLOGY: SOCIETAL IMPLICATIONS II

Individual Perspectives

Edited by

Mihail C. Roco

and

William Sims Bainbridge

*National Science Foundation,**
Arlington, VA, U.S.A.

Sponsored by

U.S. National Science Foundation

*Any opinions, findings, and conclusions or recommendations expressed here are those of the authors and do not necessarily reflect the views of National Science Foundation.

Springer

A C.I.P. Catalogue record for this book is available from the Library of Congress.

ISBN-13 978-90-481-7165-1
ISBN-10 1-4020-5432-7 (e-book)
ISBN-13 978-1-4020-5432-7 (e-book)

Published by Springer,
P.O. Box 17, 3300 AA Dordrecht, The Netherlands.

www.springer.com

 Publication of this work was supported in part by grant (0423742) from the National Science Foundation to The World Technology Evaluation Center (WTEC), Inc.

Printed on acid-free paper

Table of Contents

INTRODUCTION

This volume contains the 48 essays contributed to the most significant single effort to chart the societal implications of nanoscience and nanotechnology. The authors include natural scientists, social and behavioral scientists, engineers, philosophers, and legal experts. They work in industry, universities, government, and private practice. All have thought deeply about the profound implications of the newest realm of science and technology.

The first major examination of the human future of nanotechnology was a workshop held in September 2000 at the National Science Foundation that resulted in an influential book-length report [1]. More than three years later, in December 2003, knowledgeable science and policy leaders met again at NSF to consider the great wealth of information and ideas that had accumulated during that period, and to envision the opportunities and challenges that nanoscience and nanotechnology would present in the coming years. The meeting was sponsored by the Nanoscale Science, Engineering, and Technology Subcommittee (NSET) of the U.S. National Science and Technology Council's Committee on Technology, and it has recently published the major findings and recommendations of the 10 topical task forces that were organized in this large and complex meeting: Productivity and Equity; Future Economic Scenarios; The Quality of Life; Future Social Scenarios; Converging Technologies; National Security and Space Exploration; Ethics, Governance, Risk and Uncertainty; Public Policy, Legal and International Aspects; Interaction with the Public; Education and Human Development [2] (also included as Vol. I of this book).

The participants prepared drafts of their contributions before the meeting, but then revised them in response to what they learned from the rich exchange of information and insights, with the constant encouragement of the editors. As a result, the 48 essays have stronger intellectual connections than conference papers usually do, as well as greater depth. They range from rigorously factual reports of recent developments, to well-informed projections of trends, to visionary outlines of what the technology could mean in future years.

These individual contributions support the conclusions and recommendations of the 10 task forces [2]. These include the need for clear, scientifically grounded statements of the principles of nanotechnology to enable non-scientists to participate effectively in public policy discussions. Participants strongly recommended supporting a broad range of research studies at the nanoscale, selected primarily from peer-reviewed investigator-initiated proposals and evaluated with public involvement. They judged that research will be needed on the human health and environmental consequences of nanostructured materials and that government will need to review the adequacy of the current regulatory environment for nanomaterials, given the existence of size-dependent properties.

Another point of consensus was that a careful and rigorous analysis of the adequacy of current NNI funding levels and of future investment priorities is

M.C. Roco and W.S. Bainbridge (eds.),
Nanotechnology: Societal Implications — Individual Perspectives, 1–5.

necessary to optimize societal benefit. Participants were especially concerned about the need for educational reforms not only to teach principles of nanoscience and nanotechnology to students at many levels, but also to support cross-disciplinary training, to equip underutilized scientists and engineers with nanotechnology-related skills, and to train social scientists and other scholars to conduct science and technology studies related to the field. Enthusiasm was also expressed for research and development partnerships linking industry, academia, national laboratories, U.S. funding agencies, and corresponding international organizations.

The essays in this volume do far more than merely support such recommendations. They also outline a complex agenda for future research, offer creative ideas for policy initiatives or industrial entrepreneurship, and inform the reader about vast possibilities for future human progress. Although the 10 task force areas were an excellent rubric for organizing input to the formal recommendations, we found that the individual contributions best fit a seven-part categorization: economic impacts including commercialization, social scenarios charting alternate paths for the future, converging technologies, ethical or legal issues, appropriate governance mechanisms, public perceptions of the science and technology, and educational issues such as curriculum development and multidisciplinarity. Brief descriptions of the seven sections of this volume follow.

1. Economic Impacts and Commercialization of Nanotechnology

A good beginning has been made on developing methods and databases to chart the growth of nanotechnology-based industries. Pressing challenges include the need to identify the best innovation models for management, the processes that will provide essential human resources, and the most informative measures of success. Some participants believed that the value of nanoscale and nano-enabled innovations would be great enough to support a distinctive nanoeconomy. More specifically, the economic growth and improvement of human life attributable to information technology depends on continued improvement in microelectronics, which has recently entered the nanoscale and must become nanoelectronics. Larger questions concern the sustainability of economic growth stimulated by nanotechnology and its benefit for the wages of workers.

2. Social Scenarios

Although one cannot accurately predict the future of any technology, it will be necessary to develop ways of identifying likely possibilities in order to anticipate ethical issues, based on multiple points of view, metaphors, contexts, and timeframes. For example, nano-enabled, mobile information technology will allow people to record all their experiences, thereby increasing many opportunities for personal fulfillment while raising a privacy issue for the people with whom the individual interacts. The question often will be not how a given nanotechnology will affect otherwise stable conditions, but how it will interact with the chaotic forces that swirl in an already unstable world, such as the impending population collapse in postindustrial nations caused by insufficient fertility. Sometimes, a conceivable but unproven nanotechnology application might have tremendous impact, however; for example some analysts think it may be possible to create general-purpose fabricators capable of producing many valuable products from readily available materials locally at low cost. Many other issues deserve analysis, such as whether nanotechnology really will contribute to broader prosperity, whether it will encourage research universities to emulate commercial enterprises, what new possibilities it will open for artists, and how we can actually arrive at an appropriate vision of its potential.

3. Converging Technologies

Nanotechnology will have much of its effect on humans in partnership with other technologies, notably biotechnology, information technology, and new technologies based on cognitive science [3]. Examples include molecular machines comparable to the natural machinery inside living cells, medical devices and materials that might be implanted inside the human body, and the application of principles from computerized natural language processing to genomics and proteomics. The complexity of such multidisciplinary innovation will make it difficult to distinguish real risks from phantom risks in practical contexts such as product liability law. Convergence of historically distinct sciences and branches of engineering is not an automatic process, and it may be difficult to achieve because the communities involved are culturally heterogeneous and may unnecessarily defend their professional autonomy. At the same time, convergence of technical disciplines at the nanoscale will be essential for society, for example, by providing the knowledge and the tools for environmental protection.

4. Ethics and Law

There is general recognition of the importance of incorporating an ethical consciousness throughout the process of nanotechnology innovation, rather than waiting until an innovation has been deployed and reacting belatedly to the consequences. However, a vigorous debate concerns the proper roles of professional ethicists in this process, with some participants doubting their objectivity and technical competence. The conference itself illustrated the important roles that academic philosophers can play, raising issues that others had overlooked, stressing values like transparency in public decision-making, and sharpening conceptions of the quality of life. Among the challenges for law and government regulation are establishing consistent definitions of terms, finding ways to make sure that courts benefit from the necessary technical expertise, and ensuring that the patent system protects intellectual property rights in an era when the technical realities are changing rapidly.

5. Governance

In the modern world, government not only regulates the applications of technology but also invests in the fundamental scientific research that makes it possible in the first place. Comparison with past examples, such as nuclear power or genetically modified organisms, can alert one to issues that may arise, without providing perfect guidance for managing development to maximize human benefit. The multidisciplinary field of Science and Technology Studies needs to apply increasingly improved research methodologies to a growing range of questions about nanotechnology. Some of the implications of government action may be indirect, such as the way in which large-scale funding initiatives transform the goals and functioning of universities. Given the heightened concern for national security at the present time and the dual-use nature of nanotechnologies that have military applications, policy debates in the defense area will be both complex and crucial.

6. Public Perceptions

For the interests of the general public to be represented in the making of public policy and investment strategies concerning nanotechnology, public perceptions of the field could be decisive. It is difficult to communicate about risks, especially when the probability of harm and its degree of severity cannot be estimated, and when the general public has its own ideas about how frightening an imaginable but improbable accident might be. Research on the actual likely societal impact will be useful, as will studies of how the mass media are reporting nanotechnology, and of how relevant social movements muddy the waters with hyperbole. However, public involvement cannot be a one-way street, in which our only responsibility is to make sure that correct information flows to the public. Rather, the public must be fully engaged, with ample opportunity to express its concerns and state its goals for the technology.

7. Education

In significant measure, nanoscience and nanotechnology education is a human resources challenge, because an unknown but probably large number of scientists, engineers, and other skilled workers will be required for economic progress, national security, and new frontiers such as space exploration. Both by its own nature and through convergence with other fields, nanotechnology is inherently multidisciplinary, which means that many students must be trained to be interactional experts mediating between disciplines. New nanoengineering curriculum will also be needed, as well as new interactive learning environments to help students visualize reality at the nanoscale. Some knowledge about nature at the nanoscale must be incorporated in even the earliest school grades. Many non-technical advanced students will need to learn about nanotechnology, and future nanoengineers will need to study the ethical and societal implications of their work.

Research on the societal implications of nanoscience and nanotechnology has only just begun, but the 48 essays offered here demonstrate that already it has discovered some valuable findings, that rich theory to guide research already exists, and that a solid basis has been prepared for rapid progress. The insights collected in this volume will be of value for scientists, engineers, students, policy makers, journalists, investors, and anyone who wants a clear view of the remarkable future made possible by humanity's new power to understand and create at the nanoscale.

References

1. M. C. Roco, W. S. Bainbridge, eds., *Societal Implications of Nanoscience and Nanotechnology*, Dordrecht, Netherlands: Kluwer (2001).
2. M.C. Roco, W.S. Bainbridge, eds., *Nanotechnology: Societal Implications—Maximizing Benefit for Humanity, Nanoscale Science, Engineering and Technology Subcommittee*, Arlington, VA (2005).
3. M. C. Roco, W. S. Bainbridge, *Converging Technologies for Improving Human Performance*, Dordrecht, Netherlands: Kluwer (2003).

1. ECONOMIC IMPACTS AND COMMERCIALIZATION OF NANOTECHNOLOGY

SOCIO-ECONOMIC IMPACT OF NANOSCALE SCIENCE: INITIAL RESULTS AND NANOBANK

Lynne G. Zucker and Michael R. Darby, UCLA [1]

Research on the nanoscale has revolutionized areas of science and has begun to have an impact on, and be impacted by, society and economy. Early traces of these processes are already available to us, and we are capturing these data in NanoBank now before the ephemeral traces are lost to the social science and ethics research communities. NanoBank is a large- scale, multi-year project to provide a data resource for social scientists, ethicists, nanoscientists, government officials, and the public. NanoBank will hold data elements that document the socio-economic impact of nanoscience and nanotechnology, and institutional change that occurs either to support the development or as a response to it. The research of the discovering scientists, those who learn from them, the nonprofit organizations that assess risks and/or benefits of the new technology, and the process of industry formation will be documented. NanoBank traces the knowledge flows that underlie these changes, with special emphasis on cross-discipline flows and flows that transfer knowledge from discovering scientists to scientists working in firms. We begin the early part of the process of disseminating findings based on NanoBank in the figures included in this report.

The U.S. government has identified nanoscience and nanotechnology as a scientific and technological opportunity of immense potential, formally launching a National Nanotechnology Initiative (NNI) in January 2000. It is difficult to define simply the full range of nanoscience, but the NNI's steering committee settled on a definition of nanotechnology that incorporates the scale ("approximately 1–100 nanometer"), the understanding, creation, and use of novel properties and functions that occur at the nanoscale, and the integration into larger scale assemblies [1]. Roco, Williams, and Alivisatos; Siegel, Hu, and Roco; Roco; and Roco and Bainbridge [2, 3, 4, 5] provide a thorough review of the present state of nanoscience and technology, the implementation of the NNI, and an introduction to thinking about the implications of nanoscience and technology for our economy and society in the context of international developments in nanoscale research and commercialization.

---

[1] This research is supported by National Science Foundation Grant SES 0304727 (as a Nanoscale Interdisciplinary Research Team project), University of California's Industry-University Cooperative Research Program, UCLA's International Institute, and the Harold Price Center for Entrepreneurial Studies at the UCLA Anderson School.

M.C. Roco and W.S. Bainbridge (eds.),
Nanotechnology: Societal Implications — Individual Perspectives, 7–74.

Our Approach

"Technology transfer is the movement of ideas in people" (Donald Kennedy, Stanford University, March 18, 1994).

NanoBank is built on three insights into the processes of knowledge transfer, commercialization, and industry change. Turning first to knowledge transfer, scientific breakthroughs often yield new knowledge that is initially tacit—not yet codified. New codes and formulae describing breakthrough discoveries often develop slowly—with little incentive if value is low and many competing opportunities if high. This tends to keeps the knowledge tacit.

Second, those with the most information about breakthrough discoveries are the scientists actually making them, so there is initial scarcity: Scientists must learn the new knowledge from the discoverer or someone trained by the discoverer, limiting diffusion [6]. The combination of scarcity and tacitness yields natural excludability, a barrier to the diffusion of the valuable knowledge. Indeed, cooperation by the inventor is required for successful commercialization by the licensee for 71 percent of the inventions licensed at universities [7, 8, 9, and Table 7.1].

Third, commercialization of scientific breakthroughs requires access to naturally excludable knowledge, both tacit and scarce, that constitutes intellectual human capital retained by the discovering scientists. Thus, top scientists become the main resource around which firms are built or transformed in both biotechnology [10, 11, 12] and nanotechnology [13]. Top discovering scientists who collaborate with company scientists have strong positive effects on company success that increases as the extent of involvement goes up [14, 15].

Technological change at any given time is highly concentrated in a relatively few firms in a few industries [13, 16]. This metamorphic progress dramatically transforms existing industries, forms new industries, or both. It is misleading to concentrate on the many firms in many industries achieving perfective progress through gradual improvement or inching up. To understand or affect technological progress we must focus on the exceptions—the industries and firms achieving metamorphic progress.

The source of the driving innovations for metamorphic change may be internal or external to the industry, with external innovations using different technological bases the most threatening to existing firms in a transforming industry [17]. Biotechnology transformed the pharmaceutical industry, and nanotechnology also uses different technological bases likely to transform industries—but it is too early yet to identify which industries will experience the largest impacts. In both cases, natural excludability of breakthroughs gives discovering and other top scientists and engineers a key role and increases the likelihood of metamorphic change.

In this chapter, we report preliminary results based on core data files from an early pre-beta test form of NanoBank that focuses on nanoscale research and commercialization—an area with dramatic, recent breakthrough academic discoveries and evidence of likely metamorphic industry change. For purposes of comparison, we will refer to biotechnology, which is a well-studied recent and continuing case of the development of a science-driven industry. In the first section, we outline the central features of NanoBank and report on our current work identifying nanoscale search terms and phrases. We compare nanotechnology to biotechnology in the next section to motivate our approach and analyses. The third section explores the extent and geography of localization of nanoscience, including where and when firms are entering into nanotechnology and in what kinds of technologies. The fourth section provides some comparison of the U.S. nanoscale science base with that in Europe, Japan, and some interesting recent developments in China (PRC). The final section of the chapter presents a summary of the evidence and our conclusions.

NanoBank Under Construction

Theory-based databases are not merely lists of variables and their related data, but build theoretically important relationships among variables that are predicted to alter the socio-economic impact of nanoscale research, as well as variables predicted to alter the socio-economic feedback effects on both the science and its commercialization. Nanotechnology affects society, but society also affects nanotechnology. NanoBank is designed to provide the raw materials to conduct further research that can help understand and potentially guide the development and deployment of nanoscience and its commercialization, while simultaneously addressing basic processes of interest in social science.

NanoBank is designed as a data archive, an active site for exchanging papers and ideas for social scientists and ethicists, and a site for interdisciplinary learning across scientific disciplines through the construction of analogies and other methods. It will be located as one of the sites available from the NNIN portal, and also located as one of the resources available from the California NanoSystems Institute (CNSI at UCLA and UC Santa Barbara).[2] From design to launch in under four years requires rapid decisions and an active program of informing and engaging social scientists and ethicists likely to use this resource through professional organizations, national and local government agencies, and nonprofit policy advisory organizations. We are also drawing in scientists who are crossing interdisciplinary boundaries, via analogies and other tools to aid understanding and use of concepts from other disciplines.

The research on social and ethical impacts of nanotechnology will be facilitated by and, in many cases, enabled by NanoBank. NanoBank is an integrated database, which will be a public Web-deployed digital library (DL). NanoBank links currently disparate data sets such as articles, patents, firm financial reports and directory listings, and university data. Thus, a nanoscientist or social scientist will be able to focus, for example, on articles and patents by a particular scientist through implementation by and success of a company or companies for which the scientist is a collaborator or officer. Alternatively, an ethics researcher will be able to locate all firms reporting research programs on products for which there are

---

[2] Principal Investigators for the NSF-funded NanoBank project are Lynne Zucker, Michael Darby, Roy Doumani, Jonathan Furner, UCLA, and Evelyn Hu, UCSB (SES 03074727).

particular ethical concerns, while another might quantify university-to-firm knowledge flows or patents, articles, and products resulting from particular research funding programs.

NanoBank will also serve investors and firms seeking to allocate investment to promising new technologies, and policymakers attempting to assess the effects of alternative policy proposals. A NanoBank user might also seek all publications, patents, collaborations, alliances, and stock-price returns of firms working, for example, on a particular use of carbon nanotubes and trace all academic publications and research grants in nanoscience and technology tied to each firm involved in that use.

The key data elements that define the scope of content in NanoBank are outlined in Table 1.1. Related to each one is a series of specific elements (variables). We cannot review them in detail here, but the searchable fields found at www.webofscience.com and at www.uspto.com, plus the text found at www.edgar.gov, provide some feel for the underlying richness of variables. To data at those sites we add links on specific variables within and between sites, such as linking patents and research articles by the same person and venture capital received and products in development by the same company. In fact, a key aspect of NanoBank is that we will build links, supervised by Darby and Zucker,

Table 1.1
NanoBank: Examples of Defining Data Elements [1]

Name of Person
Patent inventor
Article author
Principal investigator (PI)
Dissertation author
Dissertation chair
Officer/founder of firm
Science advisory board chair
Science advisory board member
*Coinventor, author, etc.

Discipline of Person
Department, curent or former [2]
Department of dissertation [2]

Date or Time
Patent application & grant dates
Article publication date
Grant/contract begin & end dates
Dissertation filing date
Dissertation filing date
Directory/database dates
Firm founding date
Firm nanotech entry date
Financial reporting dates
Initial public offering (IPO) date
Merger or alliance dates
Venture capital round dates
*Interdisciplinary team start dates
*Dept., institute, center entry/change/ merger date
*New interdisc. journal areas/start date
*Existing journal new discipline/area entry date
*Fed. Instit., IGA program start date
*Date of move between disciplines [3]

Organization
Patent assignee
Affiliation on article
Grant/contract recipient
University lists--NRC, IPEDS
Firm directory listings
Public firm databases (filings)
Financial market databases
Mergers & alliances database
Venture capital firm database
Investment bank database
Federal laboratory listings
Research institute directories
Organization's parent org. (if any)
*Non-profit directories, tax filings

Industry of Organization
Firm/university/fed lab/res. Inst.
SIC or NAICS industry codes
Venture Economics industry codes
*Nonprofit tax codes [501(c)(3), etc.]

Science & Technology Area Codes
US & International patent classes
ISI journal area
PACS codes/text
Nano S&T subareas (VJNano *et al.*)
Z-D broad science/tech area codes
*NBIC product codes

Geo-location
Patent inventor's address
Patent assignee's address
Author address
Grantee address(es)
Organization address(es)

Inputs, Outputs, and Success Measures
These can be measured at person, organization or sub-organization level and aggregated (as appropriate) based on: organization; city, state, region or country; discipline, industry, science/technology area, time; or combinations (e.g., by firm, region, and year)
Patent: counts, citations, claims
Articles: counts, citations
*Employment (& membership for nonprofits)
*Interdisciplinary Collaborations: counts, classifications, citations for articles and patents
Products in development: counts, classifications
Products on the market: counts, classifications
Venture capital: round counts, round values
Offerings: IPO value, later offering values and types
Investment bank reputation rankings
Stock price history
Impact of risk assessment on stock price:
(1) Product failure, adverse event news
(2) NPO report, event news
Doctoral programs: ranking, graduates, faculty, funding
Awards: Nobels, NAS/NAE/IOM, Phi Beta Kappa, etc.
Grants/contracts: Federal, SBIR, ATP
*Interdisciplinarity
*Cross-discipline co-chair on dissertation: counts
*Cross-discipline co-authors, co-inventors: counts, citations, claims for patents
*Cross-discipline firm officers, firm science boards: counts
*Cross-discipline articles in old & new journals: counts, citations
*Cross-discipline membership: depts., instits., centers, IGAs: counts

***NBIC Interdisciplinarity Convergence**
*Analogies/images of cross-discipline concepts
*New cross-discipline analogies/tools
*Cross-discipline teaching, patenting, research

Notes: * indicates NBIC elements
[1] Identify and search on specific terms in all NBIC areas.
[2] Non-academics: use former department or dissertation department.
[3] E.g., if dissertation discipline is different from department of first job.

between data elements that theory identifies as especially crucial to knowledge transfer and to productivity in both science and industry.

We include some elements in NanoBank designed to track interdisciplinary convergence across nano-, bio-, info-, and cogno- areas of research and teaching (NBIC) and its outcomes in both nanoscience and nanotechnology. These elements are starred in Table 1.1, and they range from a variety of interdisciplinary measures to tracking changes in departments and schools that reflect and institutionalize interdisciplinary boundary changes. Two main NBIC themes are addressed:

1. Track amount/quality of interdisciplinary research and training and timing/ degree of new organizational structure that institutionalizes these changes, and the impact of this interdisciplinary convergence on products and success outcomes, with additional coding of products by NBIC subarea [18, p.17 Table 2 items B-F].
2. Use of analogies/images of cross-discipline concepts to: (a) Communicate clearly across discipline boundaries (in part, to decrease tacitness and hence natural excludability) and stimulate discovery of new knowledge; and (b) Facilitate borrowing of tools and other solutions across discipline boundaries, as in the new interdisciplinary area of computational biolinguistics.

NanoBank will also provide an important communication function for nanoscience and engineering generally, and for special initiatives such as the NSF National Nanotechnology Infrastructure Network (NNIN), through two archives on the site: (1) Vetted white papers dealing with nanoscience and engineering, business applications and issues, legal issues, and social and ethical impacts will provide a convenient source of reliable information for practitioners, other professionals, and an informed public; and (2) Preprints or links to preprints on an open basis subject to providing complete identification information on all affiliations and commercial interests will provide early access to nano-relevant research.

We are developing improved methods, including experimenting with alternative specifications for computer identification of nanoscale articles. In this paper, nanoscale articles are identified by the union of these two (overlapping) text searches: (1) for the string "nano"; and (2) for any of 475 nanoscale-specific terms. All measurement terms are excluded. Some initial results are displayed in Figure 1.1 using a dataset of high-impact (very highly cited) articles from ISI. The nanoscale articles are categorized by a broad science and technology classification scheme [19].

The number of articles rose initially most rapidly in semiconductors, but more recently the biology-medicine-chemistry and multidisciplinary categories have also seen dramatic growth. While the increase in information technology (IT) articles seems slight, other analyses not reported here show that, given the lower overall number of articles published in IT, the percentage increase is actually more dramatic than for the biology area.

As we develop new search strategies, we are benchmarking them against the *Virtual Journal of Nanoscale Science & Technology* (hereafter, VJNano, found online at www.vjnano.org). VJNano contains references to nanoscale articles published elsewhere as vetted by a distinguished scientific advisory panel of researchers actively working on the nanoscale. The search-based methodology used to produce Figure 1.1, discussed above, identifies about 65 percent of the articles in VJNano. This provides one test of the degree to which search terms and phrases are able to identify recent nanoscale articles. With Jonathan Furner, we are combining these and other methods with information studies techniques to develop computer algorithms that use probability-based methods of discriminating between nanotechnology and non-nano.

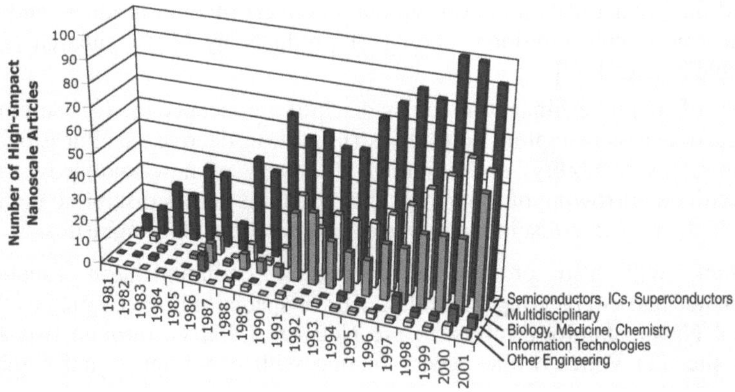

Figure 1.1. High-impact nanoscale articles by major S&T category and year.

Comparing Nanotechnology and Biotechnology

Fundamentally, nanotechnology and biotechnology are rooted in basic science, and thus we expect their development to follow roughly similar trajectories. While there are a number of different ways to measure this, to begin the process we look at the rate of development of the scientific knowledge base as indicated by scientific publishing and the rate of knowledge capture as indicated by patenting. While publishing alone is sufficient to build the science side of the process, establishment of intellectual property rights is necessary for much of the commercialization and its finance.

Figure 1.2 compares the remarkable increase in publishing and patenting that occurred during the first 20 years of the biotechnology revolution with what is occurring now in nanoscience and technology. The figure shows that the scientific and patenting growth of nanotechnology is of at least the same order of magnitude as biotechnology at a similar stage of development. We use 1973 as the base year for the start of biotech and 1986 for nanotechnology to compare them at similar points in their development (see the explanation of different years—and different methods of selection—below).

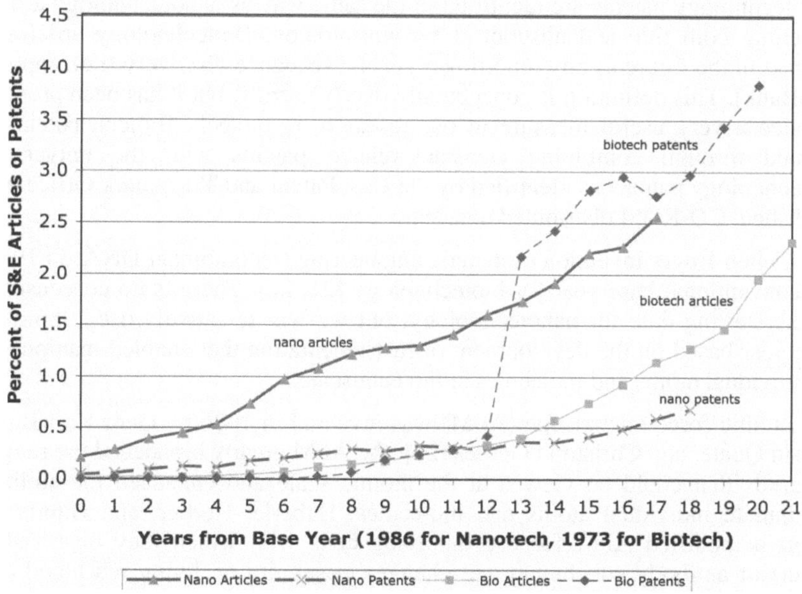

Figure 1.2. Comparing Nanotech (1986–2004) and Biotech (1973–1994) Publishing and Patenting Trajectories as a percentage of all patent issues.

For articles, nanoscience and nanotechnology is maintaining a growing lead over biotechnology articles. It is clear that nano nanoscience and nanotechnology has burst upon the science and engineering scene a bit less suddenly than one would judge by the current notices. In terms of publications, rapid growth began about 1990. Since 1990 the growth in nano nanoscience and nanotechnology articles has been remarkable, and now exceeds 2.5 percent of all science and engineering articles. Beginning in 1990, the percentage of nanotechnology articles was significantly greater than the 1981–1989 mean and increasing every year.

Figure 1.2 also shows steady growth in nanotechnology patents as a percentage of all patent issues. This growth is more dramatic considering that total patents also rise, increasing by about 150 percent over the same period. Actual counts of nanotechnology patents suggest a takeoff date for nanotechnology in the late 1980s. We observe that nanotechnology patents are ahead of biotech patents early in the process (through year 11) because very few patents were issued in biotech until the courts gave the go-ahead in 1980. Thirteen years into the biotech revolution (1986), biotech patenting took off as: (1) gene sequences were patented with little proof of their use, and (2) many variations on drug candidates were patented in an attempt to prevent quick competition from me-too drugs if one particular candidate were proved safe and effective.

For Figure 1.2, we identified nanotechnology articles using the text-search methodology described earlier, searching titles and abstracts for all articles in

Science Citation Index Expanded through 2003 [20]. Nanoscience and nanotechnology patents are identified in the same way as nanotechnology articles, searching both title and abstract at www.uspto.gov. Biotechnology articles are defined in the figure as any reporting a genetic sequence discovery (i.e., appear in GenBank). This definition is conceptually overly narrow, but it has been proven in practice a very useful measure in our research on biotech. Biotech patents are defined through combining GenBank-related patents with the universe of biotechnology patents as identified by the U.S. Patent and Trademark Office on its published, CD-ROM distributed data set.

The Cohen-Boyer invention of genetic engineering (recombinant DNA) in 1973 is the conventional base year for biotechnology [21, 22]. There is no consensus yet on the starting date for nanotechnology, but we will tentatively use 1986 as the base year based on the development of instrumentation that enabled manipulation of individual atoms and molecules at the nanoscale.

The atomic force microscope (AFM) was invented in 1986 by Gerd Karl Binnig, Calvin Quate, and Christoph Gerber [23]; the AFM greatly broadened the range of materials that could be viewed at the atomic scale and enhanced the ability to manipulate individual atoms and molecules. Haberle, Horber, and Binnig [24] report a modified AFM for use on living cells with which they observed the effects of antibody attachment and changes in salinity on living red blood cells. This built on earlier work developing the scanning tunneling microscope (STM) conducted at IBM's Zurich Research Laboratory in 1981 by Binnig and Heinrich Rohrer [25, 26]; they received the Nobel Prize in Physics in 1986 for their STM work. The STM works by moving a very fine pointer back and forth over a surface with each scan line displaced slightly from the next, called raster scanning in reference to the parallel lines that make up a television picture. A sensitive feedback mechanism maintains a constant distance relative to the surface so that a three-dimensional representation is obtained. The STM could be used only on conductive materials (metals) due to the electron tunneling method used to maintain the constant distance between pointer and surface. The STM was the first instrument to enable scientists to obtain atomic-scale images and ultimately to manipulate individual atoms on the surfaces of materials.

Darby and Zucker [27] argue that such inventions of procedures or instruments—not exclusively the paradigm shifts famous from Kuhn [28]—are the usual "inventions of a method of inventing" which set off major scientific and industrial transformations. Zvi Griliches [29, 30] was the first economist to study the class of breakthrough discoveries that he named an "invention of a method of inventing." His case was hybrid seed corn, a method of breeding superior corn for specific localities that effectively excluded farmers from reproducing the hybrid seed by saving part of their crop.

Instruments are particularly important because they effectively codify much of the "know-how" involved in a breakthrough discovery making it possible for others to access and apply the new knowledge without directly working with the discoverers and their students. For a parallel example, consider the gene splicing machines that made discovery of new genetic sequences so routine that by 1988 graduate students at major research universities could no longer get a Ph.D. by reporting the discovery of a new genetic sequence.

Geographic Concentration, Knowledge Transfer, and Firm Entry in Nanotech

There is concentration of knowledge in a few scientists and engineers who are pushing the frontiers of nanoscience and nanotechnology and in the laboratories in which they work, just as metamorphic technological progress is concentrated in relatively few firms in relatively few industries. This concentration is a notable characteristic of previous scientific breakthroughs, especially those that involve a significant degree of tacit knowledge—art learned by doing at the lab bench level. This tacit knowledge provides natural excludability that limits the diffusion of the new knowledge in cooperation with or even in the absence of explicit intellectual property rights of the discovering scientists and their organizations [12, 14, 15].

In Figure 1.3 as well as Figures 1.4, 1.5, and 1.6 that follow, we measure science base by the number of nanoscale-related publications in the ISI World of Science database. This database contains all the ISI indexed-articles from 1980 through 2003 and nanoscale publications are identified by searching for nanoscale-specific terms in the title and abstract (when available), as explained above.

Geographic Concentration

Figure 1.3 shows the geographic distribution of the nanoscience base in the United States with respect to years and functional economic areas identified by the U.S. Bureau of Economic Analysis (BEA). An address can be uniquely matched to a BEA area when the zip code is reported (which is the case in 95.56 percent of the observations). When the zip code is missing, the city and state information were used to infer the BEA area.

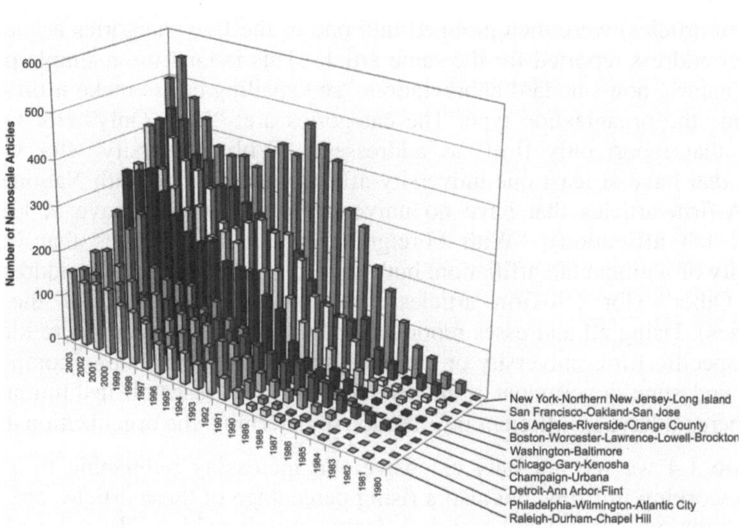

Figure 1.3. Nanoscience geographic concentration by region.

Ten regions with the most nanoscale-related papers (out of 172 BEA areas) account for 54 percent of the articles having at least one coauthor with a U.S. address. These 10 regions—New York-Northern New Jersey-Long Island, San Francisco-Oakland-San Jose, Los Angeles-Riverside-Orange County, Boston-Worcester-Lawrence-Lowell-Brockton, Washington-Baltimore, Chicago-Gary-Kenosha. Champagne-Urbana, Detroit-Ann Arbor-Flint, Raleigh-Durham-Chapel Hill, and Philadelphia-Wilmington-Atlantic City—are notable for the strength in nanoscience and nanotechnology of particular academic institutions and are not predictable by size, economy, or even overall strength of the science base. These regions can be compared with the relative importance of high-tech states in other studies [19, 31]. As a further illustration of the concentration, almost 28 percent of all nanotechnology articles are accounted for by the top-3 regions (New York-Northern New Jersey-Long Island, San Francisco-Oakland-San Jose, and Los Angeles-Riverside-Orange County).

Knowledge Flow from Universities to Firms

When commercialization is occurring close to the scientific frontier, it is more likely that natural excludability is a significant barrier to knowledge transfer from discovering scientists to those who are applying the knowledge to develop commercial products. Under these conditions, characteristic of both nanotechnology and biotechnology, participation at the lab bench level by top scientists who are making these discoveries is important to successful commercial application, necessary but not sufficient. Star scientist authorships of articles as or with employees of a firm were a potent predictor of the eventual success of biotech firms. Zucker, Darby, and Armstrong [15] showed that counts of articles authored by firm employees with authors at top-112 universities had a significant (although smaller) impact on firm success.

To identify knowledge flows to firms in nanotechnology, we selected out all of the articles that include a firm in California as one of its addresses. All such articles (CA-firm articles) were then grouped into one of the five categories according to the other address reported for the same article. This is far from a simple process: variant names, non-standard abbreviations, and spelling errors make it difficult to determine the organization type. The categories are: "Firm Only" (for CA-firm articles that report only firms as addresses); "With University" (for CA-firm articles that have at least one university-affiliated address); "With National Lab" (for CA-firm articles that have no university affiliation but have at least one national lab affiliation); "With Foreign" (for CA-firm articles that have no university or national lab affiliation, but reported at least one foreign address); and "With Other" (for CA-firm articles that don't fall into any of the above categories). Using all addresses reported in California, 95 percent were identified with a specific firm, university or national lab. The rest are mainly composed of federal and state government agencies and non-profit research institutions. Less than 5 percent have insufficient information to determine the organizational type.

In Figure 1.4 we see not only extensive and increasing publishing by scientist authors working in firms, but also a rising percentage of these articles are written in collaboration with scientists and engineers at universities. The university-firm

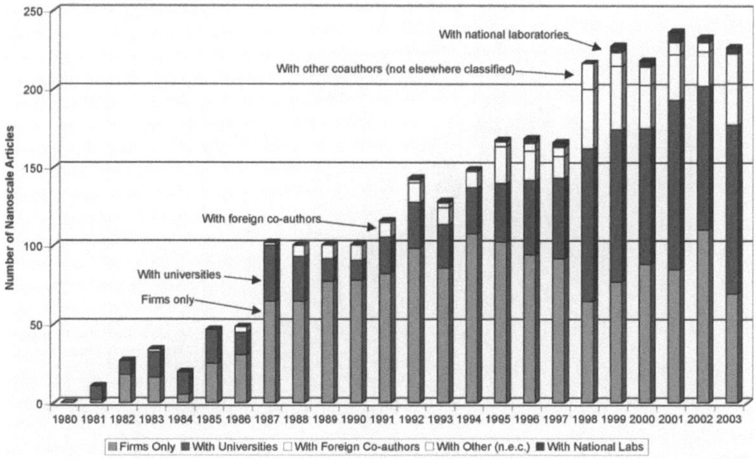

Figure 1.4. Tracing knowledge flow to commerce: California firm articles by year and co-authorship.

knowledge flows represented by these articles indicate not only the natural excludability that makes the costs of close collaboration across university-firm boundaries worth incurring, but also the expected commercial payoff of nanotechnology. While Figure 1.4 is based only on California data at this point, we expect the results to replicate more generally.

Birth of the Nanotechnology Industry

Figure 1.5 illustrates the number of firms first publishing a nanoscale-related article in the ISI database by region and publication year with the firm's region based upon the address given by the author at the firm. Less than 2.5 percent of all cases are unidentified. If a firm reports an address in s different areas in its first year in nanotech, each area is assigned 1/s firms. It can be argued that a parent company establishing branches in different areas is similar to establishing a new firm and should be counted as such; however, for the purposes of this paper, we count any branch of a corporation as part of the same firm. Hence, if an IBM Research Center in San Jose, California, enters nanotechnology earlier than a second IBM Research Center—located in Yorktown Heights, New York—only the California entry is reported in Figure 1.5.

The regions that have the most firms entering the field overlap with the regions where most nanoscale articles are being written, except that San Diego, Denver-Boulder-Greeley, and Minneapolis-St. Paul appear in the top 10 regions for firm entry.

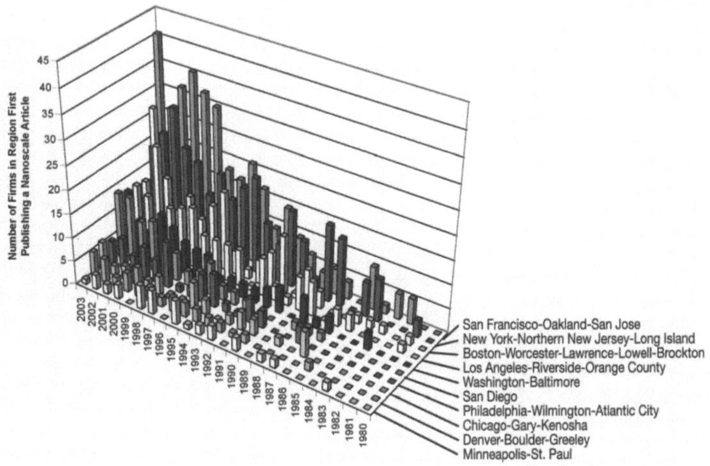

Figure 1.5. Birth of the nanotech industry: Comparing entry by region and year, 1980–2003.

Darby and Zucker [27] show that in a multiple-Poisson-regression context both the number of highly cited articles published in a region and its average wage level (a measure of labor-force quality) are significant determinants of where and when firms enter nanotechnology. The effects of federal research funding to and nanotechnology articles by authors from top-112 research universities, regional employment, and total venture-capital flows are not statistically significant when all of these variables are entered in the same Poisson regression, although these variables may be significant in regressions in which they are not competing with high-impact articles and/or average wages. It is difficult with small samples to measure separate effects of highly correlated variables such as high-impact articles (mostly authored by faculty with large federal research funding) and the amount of federal research funding. We expect some additional variables will be significant in future research when we can identify additional firms entering nanotechnology. The statistical insignificance of past venture capital flows is consistent with efficiency in that market.

There is in fact no census or widely accepted database to consult as to which firms are actively using nanotechnology in production or at least R&D activities. Over the next few years, we plan for NanoBank to fill that and other information gaps faced by both researchers in nanoscience and nanotechnology and those who study their impact. For now, the large number of articles in the ISI database provides a means of identifying firms with a sufficiently deep involvement to be either publishing highly cited research articles or articles coauthored with professors from universities or both. Based on the patterns observed in biotechnology, few other firms without such ties are likely to become significant players.

International Comparison

Figure 1.6 illustrates the international distribution of nanotechnology articles in the ISI database by year. Of articles written during 1980–2003, 72 percent have authors in one or more of the United States, Japan, and the European Union. The European Union articles are concentrated in Germany, France, and the United Kingdom. China (PRC) was also added as a separate group to illustrate its remarkable improvement in recent years. Considering the whole period, the United States alone accounts for 29.15 percent of the world's nanotechnology articles, establishing it as the most dominant player in nanotechnology.

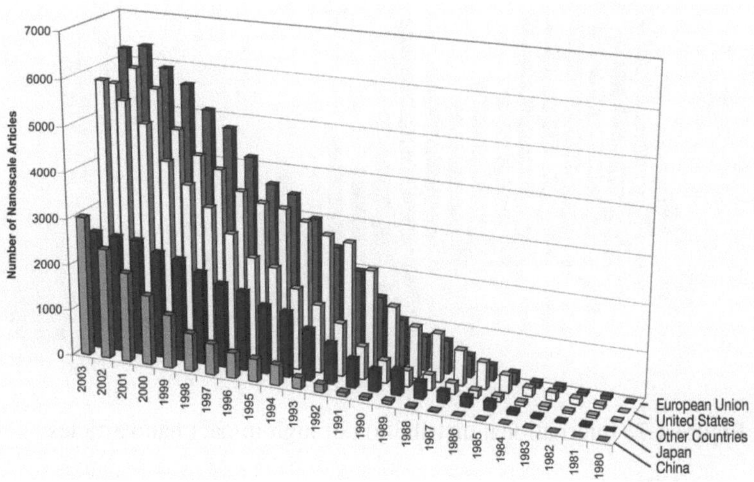

Figure 1.6. International distribution of ISI nano articles.

The data also suggest nano-related research became increasingly global throughout the last decade. Many countries that were not significant in the 1980s and early 1990s increased their production dramatically. Other than China, which eventually caught up with Japan, countries like South Korea and India can also be counted as examples.

There was also a great increase in the number of countries that engage in nano-related research. While nano-related articles were produced in 43 different countries in 1990, this number increased to 102 in 2003. Overall, almost 150 different countries were cited in the ISI articles in this time period. Both these factors cause a relative decline in the share of the U.S. nanotechnology articles, when compared to the initial stages of the nanotechnology improvement. Even so, more than 24 percent of the articles produced in the world in 2003 were in the United States, which is almost double the number by the next country, China.

Initial results adjusting for quality of research articles are shown in Figure 1.7. The distribution of high-impact (very highly cited) papers in the world further reinforces the picture of U.S. dominance, but also shows that scientists and engineers in other nations are increasingly publishing high-impact articles in the area of nanoscale research. China's great rise in nanoscience publications is evidence of a shift in effort, but its number of high-impact papers remains low relative to the overall increase in publishing rate. Taken as a whole, these data confirm that the strength and depth of the American science base points to the United States being the dominant player in nanotechnology for some time to come, while it also faces significant and increasing international competition.

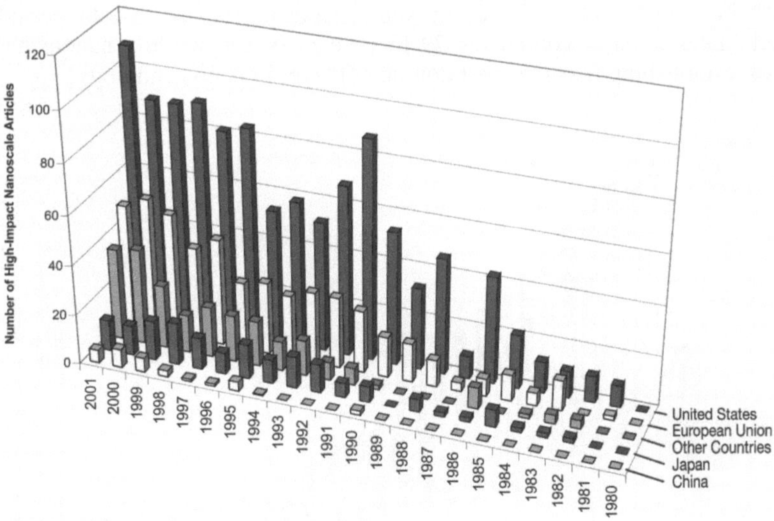

Figure 1.7. International distribution of high impact nano articles.

Summary and Conclusions

Nanoscale science and technology has all the earmarks of the kind of breakthrough metamorphic progress in which cascades of important scientific discoveries create the technological opportunities that transform existing industries and create new ones. We expect nanotechnology to account for a significant proportion of technological progress and economic growth over the next several decades. NanoBank will track these changes.

Nanotechnology is on a similar trajectory to biotechnology, stemming also from basic science breakthroughs. These breakthroughs include important instrument invention relatively early in its development to codify part of the most fundamental tacit knowledge: scanning probe and atomic force microscopy, similar to the gene sequencing machines. However, much of the knowledge remains tacit in nanoscience as in bioscience and is best transmitted by working at the lab bench by one of the discoverers or someone trained by him/her, yielding natural excludability. As in biotechnology, we find that nanotechnology companies are founded when and where top nanoscientists are publishing. And we have also presented early evidence that the knowledge flow via collaboration in the lab is increasing between university scientists and company scientists, as indicated by co-publishing.

Regional agglomeration is also evident, with the main clusters forming around major research universities publishing in nanoscience. While there is considerable overlap with the biotechnology pattern, i.e. the relative dominance of the New York region and both Northern and Southern California, there are also significant differences that we believe are due to different resource-allocation decisions made in the past. The same is true at the national level: the United States accounts for over 55 percent of the highly cited articles on the nanoscale identified as "High Impact Articles" by ISI, while the United States, European Union, Japan, and China account for over 88 percent. So the concentration of nanoscale work is quite high internationally, similar to that found within the United States.

It is too early to say where the most profitable commercial applications of nanotechnology lie. However, we can derive some early indicators from observing the pattern of areas in which firms enter nanotechnology, and over time decide to focus their efforts in product development, since both decisions are heavily conditioned by expectations held about eventual profits in different content areas of nanotechnology. Klevorick, Levin, Nelson, and Winter [32] have emphasized that profitability is based on the appropriability of returns by the pioneer(s) as well as upon technological opportunity. Griliches [29, 30] argued that the earliest applications of an invention or a method of inventing are to those areas with the greatest expected profitability—now known as the lowest-hanging fruit. This theory suggests focusing for analysis of early industrial formation and transformation on the regions with the strongest science bases in areas where profitability is expected to be highest.

The race to apply nanotechnology to new products and services will be a long one. The growth and changes in institutions necessary to support this revolution, from supporting new institutes to dealing with cross-pressures between disciplines in interdisciplinary research, will determine part of the outcome. Interest groups operating in the nanotechnology field will alter what is done, when it is done, and how it is done—and possibly even whether it is done. Policy issues on many fronts are already confronting nanotechnology, and must be successfully addressed for nanoscale research and commercialization to grow and prosper.

References

1. NSET, Committee on Technology, National Science and Technology Council. Nanotechnology definition (February 2000) (available at http://nano.gov/omb_nifty50.htm).

2. M. C. Roco, R. S. Williams, P. Alivisatos, eds., *Nanotechnology Research Directions: IWGN workshop report—Vision for nanotechnology in the next decade*, Boston : Kluwer (1999) (available at http://www.wtec.org/loyola/nano/IWGN.Research.Directions/).

3. R. W. Siegel, E. Hu, M. C. Roco, eds., *Nanostructure science and technology: R&D status and trends in nanoparticles, nanostructured materials and nanodevices*, Dordrecht, Netherlands: Kluwer (1999) (available at http://www.wtec.org/loyola/nano/).

4. M. C. Roco, From vision to implementation of the U.S. National Nanotechnology Initiative, *Journal of Nanoparticle Research* **3**(1), 5-11 (February 2001).

5. M. C. Roco, W. S. Bainbridge, eds., *Societal Implications of Nanoscience and Nanotechnology*, Dordrecht, Netherlands: Kluwer (2001).

6. L. G. Zucker, M. R. Darby, M. Torero, Labor mobility from academe to commerce, *Journal of Labor Economics* **20**(3), 629-660 (July 2002).

7. R. Jensen, M. Thursby, Proofs and prototypes for sale: The tale of university licensing, *American Economic Review* **91**(1), 240-259 (March 2001).

8. Agrawal, R. Henderson, Putting patents in context, *Management Science* **48**(1): 44-60.

9. J. Thursby, M. Thursby, Who is selling the ivory tower? Sources of growth in university licensing, *Management Science* **48**(1), 90-104 (January 2002).

10. L. G. Zucker, M. R. Darby, Virtuous circles of productivity: Star bioscientists and the institutional transformation of industry, National Bureau of Economic Research Working Paper No. 5342 (November 1995).

11. L. G. Zucker, M. R. Darby, Star scientists and institutional transformation: Patterns of invention and innovation in the formation of the biotechnology industry, *Proceedings of the National Academy of Sciences* **93**(23), 12,709-12,716 (November 12, 1996).

12. L. G. Zucker, M. R. Darby, M. B. Brewer, Intellectual human capital and the birth of U.S. biotechnology enterprises, *American Economic Review* **88**(1), 290-306 (March 1998).

13. M. R. Darby, L. G. Zucker, Growing by leaps and inches: Creative destructtion, real cost reduction, and inching up, *Economic Inquiry* **41**, 1-19 (2003).

14. L. G. Zucker, M. R. Darby, J. Armstrong, Geographically localized knowledge: Spillovers or markets?, *Economic Inquiry* **36**(1), 65-86 (January 1998).

15. L. G. Zucker, M. R. Darby, J. Armstrong, Commercializing knowledge: University science, knowledge capture, and firm performance in biotechnology, *Management Science* **48**(1), 138-153 (January 2002).

16. C. Harberger, A vision of the growth process, *American Economic Review* **88**(1), 1-32 (March 1998).

17. M. L. Tushman, P. Anderson, Technological discontinuities and organizational environments, *Administrative Science Quarterly* **31**(1), 439-465 (March 1986).

18. M. C. Roco, W. S. Bainbridge, *Converging Technologies for Improving Human Performance*, Dordrecht, Netherlands: Kluwer (2003).

19. M. R. Darby, L. G. Zucker, *California's Science Base: Size, quality and productivity, a report to the California Council on Science and Technology,*

February 3, 1999, Sacramento: California Council on Science and Technology (1999).

20. *Science citation index expanded*, electronic database, Institute for Scientific Information, Inc. (2003).
21. S. N. Cohen, A. Chang, H. Boyer, R. Helling, Construction of biologically functional bacterial plasmids *in vitro*, *Proceedings of the National Academy of Sciences* **70**(11), 3240-3244 (November 1973).
22. S. N. Cohen, H. W. Boyer, Process for producing biologically functional molecular chimeras, United States Patent 4,237,224 (2 December 1980).
23. G. K. Binnig, C. F. Quate, C. Gerber, Atomic force microscope, *Physical Review Letters* **56**(9), 930-933 (3 March 1986).
24. W. Haberle, J. K. H. Horber, G. K. Binnig, Force microscopy on living cells, *Journal of Vacuum Science & Technology B* **9**(2), 1210-1213 (March-April 1991).
25. G. K. Binnig, H. Rohrer, Scanning tunneling microscopy, *Helvetica Physica Acta* **55**(6), 726-735 (1982).
26. G. K. Binnig, H. Rohrer, Scanning tunneling microscopy, *Surface Science* **126**(1-3), 236-244 (1983).
27. M. R. Darby, L. G. Zucker, Grilichesian breakthroughs: Inventions of methods of inventing in nanotechnology and biotechnology, National Bureau of Economic Research Working Paper No. 9825 (July 2003).
28. T. S. Kuhn, *The Structure of Scientific Revolutions*, Chicago: University of Chicago Press (1962).
29. Z. Griliches, Hybrid corn: An exploration in the economics of technological change, Ph.D. Dissertation, Department of Economics, University of Chicago (1957).
30. Z. Griliches, Hybrid corn: An exploration in the economics of technological change, *Econometrica* **25**(4), 501-522 (October 1957).
31. L. G. Zucker, M. R. Darby, *California's Inventive Activity: Patent indicators of quantity, quality, and organizational origins, a report to the California Council on Science and Technology, October 11, 1999*, Sacramento: California Council on Science and Technology (1999).
32. K. Klevorick, R. C. Levin, R. R. Nelson, S. G. Winter, On the sources and significance of interindustry differences in technological opportunities, *Research Policy* **24**(2), 185-205 (March 1995).

MANAGING THE NANOTECHNOLOGY REVOLUTION: CONSIDER THE MALCOLM BALDRIGE NATIONAL QUALITY CRITERIA

William R. Boulton, Auburn University

With projected revenues of $1 trillion by 2015, nanotechnology-based industries will generate a need for approximately 2 million nanotechnology workers. Critical to meeting such demands will be an educational system and infrastructure that supports the development of future nanoscale development initiatives. U.S. National Nanotechnology Initiative (NNI) centers are being established to meet these needs and are expected to make future contributions to the development of nanoscale products and services to solve problems facing both business and society. The goals of nanoscale developments are to improve health care and ensure that our resources and an environment will support future generations.

In science and technology fields, we seldom discuss the basic needs for the effective management of multi-million dollar investments. The field of strategic management concerns itself with the effective use of resources and capabilities to achieve the goals and strategies that we set. In considering an approach to facilitate NNI contributions, the U.S. Baldrige National Quality Award, established in 1987, provides criteria for assessing organizational performance. Nearly 30 percent of its criteria assess the effectiveness of leadership, strategic planning, and the customer or market focus of the organization. Another 25 percent of the criteria address the human resources and organizational processes that create value for the targeted markets. But 45 percent of the Baldrige criteria rate actual performance outcomes [1]. Given the Baldrige framework for evaluating successful organizations, let's consider its potential application for managing today's nanotechnology initiatives.

Leadership, Strategic Planning, and Market Focus (30 percent)

Without vision, there is no way to chart a course or build the infrastructure and institutions needed for the future. The Baldrige criteria emphasize the need for strong leadership to set a direction and plan for how specific goals will be achieved. In the development of new science and technology (S&T), we seldom follow a preplanned course of action. Managing S&T innovations is often a complex and creative activity. The successful creation of useful products and services that are based on new concepts that were unknown or unthinkable just years earlier results in the growth of new markets and industries. A vision of today's integrated computer and communication world was first described in 1977 by NEC Corporation's chairman, Koji Kobayashi [2]. Miniaturization and integration of electronics technologies have now combined with telecommunication and Internet developments to support a mobile and virtual society for tomorrow. Today, nanoscale science and technologies give focus to a new vision and a vast range of new opportunities for innovation. According to Mihail C. Roco [3], first generation nanoscale innovations (~2001) were passive nanostructures like coatings, nanoparticles, and bulk materials. Second generation innovations (~2005) will include active nanostructures like transistors, amplifiers, targeted drugs and chemicals, actuators, and adaptive structures. Third generation innovations

(~2010) will include 3-D nanosystems with heterogeneous nanocomponents and various assembling techniques, bio-assembling, networking at the nanoscale level, and new architectures. By 2020, a fourth generation of molecular nanosystems will be developed with heterogeneous molecules, based on biomimetics and new architectural platforms. How we convert these new innovations into successful products and services will determine who wins the marketplace of the future. Management skills will certainly be an important contribution.

Henry Chesbrough argues that the open innovation model is the development paradigm of today. It incorporates "external as well as internal ideas as inputs to the innovation process, combined with employing internal and external paths to market for the results of innovative activities" [4, p.6]. Internal and external ideas flow into R&D processes, with outputs going to markets through both internal and external paths, as shown in Figure 1.8.

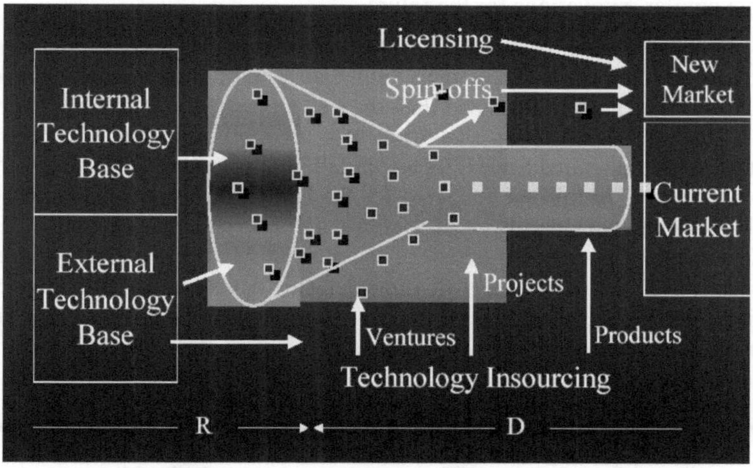

Figure 1.8. The Open Innovation Paradigm (Source: H. Chesbrough [4]).

Of particular importance is the flow of ideas and technologies into and out of the process throughout. Ideas can come into the process, for example, from internal research investigations, from external research, from licensing in another company's technology, or from an acquisition of a company's product. Similarly, ideas can flow out of the process to market in numerous ways. Many go to market through the company's own channels, while others may be licensed out, or spun out into a new venture, or into a new joint venture [4].

However, many successes are the result of serendipity, such as the well-known "Post-It" notes innovation, a serendipitous result of having developed glue that didn't stick well.

What the "open innovation" model does not explain is that technology inputs do not directly transform into products. Since science and technology are seldom stand-alone products or services, they must be integrated into devices or components as part of the product development process. Products, in fact, incorporate a wide range of technologies through their component technologies. A computer's hard drive, CD player, microprocessor, and monitor each consist of a range of technologies that have been packaged for computer applications. The charged-coupled device (CCD), for example, became a key component in allowing video cameras to be miniaturized. As shown in Figure 1.9, such component technologies often allow for the development of next-generation products, such as Sony's 8mm camcorder or today's cell phones with digital cameras. Large-scale integration of semiconductors has been critical for the miniaturization of consumer electronics products. The development of next-generation product platforms allows for a range of new features to address a wide range of market niches. For example, Sony offered shock-resistant sports versions of its 8mm video cameras in different colors and fitted for underwater applications. But we should remember that it took 16 years to develop the CCD [5]. NNI's support of nanoscale research in materials, electronics, health care, environment, energy, microcraft, manufacturing, security, and instruments should generate similar innovations for the future

Since the NNI's establishment in 2000, it has spurred support for nanoscale technology developments across the world. Japan and South Korea followed with initiatives established in 2001, and then the EC, Germany, and Taiwan entered the race in 2002. By 2003, over $3 billion was invested across a full spectrum of activities, with the United States, Japan, and the EU providing similar government support. Of the $770 million budgeted for the United States, the National Science Foundation (NSF), Department of Defense (DoD), and Department of Energy (DOE) accounted for three-quarters of the expenditures. In its first two years of existence, NNI's leadership has facilitated faster development than expected in materials, chemicals, pharmaceuticals, and electronics. Emerging areas include nanomedicine, energy conversion and storage, agriculture and food systems,

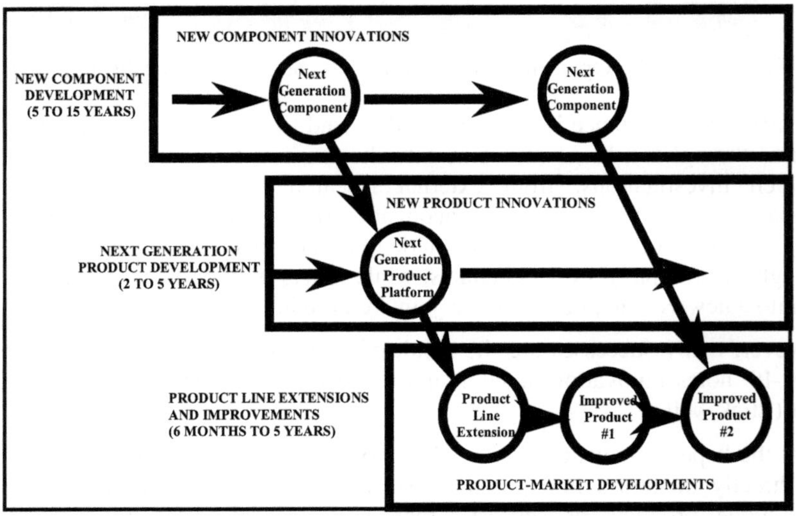

Figure 1.9. Technology and product development plan (Source: W. Boulton [5]).

molecular architectures, multi-scale simulations, environmental implications, and converging technologies from nanoscale developments.

Human Resources and Organizational Processes (25 percent)

The Baldrige criteria stress the importance of human resources and the processes needed to accomplish a given set of goals. With the need for some 2 million workers to be trained in nanotechnology over the next decade, NNI has provided a major impetus for developing educational programs through sponsored centers. NNI's 22 centers support nanoscale developments, including 10 centers funded through NSF; three, through DoD; five, through DoE; and four universities through NASA. All NSF centers require both education and outreach programs and include international educational opportunities. NNI supports curriculum development in K-12 and undergraduate programs. New teaching modules for nanoscale science and engineering are being developed to attract and develop students for tomorrow's workforce. While these centers involved over 7,000 students, technicians, teachers, and faculty by the end of 2003, we have hardly begun to address the numbers of skilled workers needed for the future.

The new Nanoscale Science and Engineering Education program was also initiated in 2004 to produce systemic changes in nanoscale science and engineering education. Centers for Learning and Teaching, Informal Science Education, Instructional Materials Development, and Nanotechnology Undergraduate Education are being established, and NSF plans to have 10 K-12 educational modules completed by the end of 2004. States are also leveraging these efforts by establishing local nanotechnology initiatives with over $200 million per year.

NNI is investing about 10 percent of its budget into future considerations of society and the environment. NNI is also addressing societal and educational implications of nanotechnology development, which will become increasingly critical as products begin moving into the marketplace. NSF's environmental centers and interdisciplinary groups are addressing a full range of concerns, including air pollution, water purification, environmentally responsible solvents and processes, environmental molecular science, nano-carbon particulates, and sensing devices for marine systems.

Performance Outcomes (45 percent)

The Baldrige criteria weigh actual performance outcomes as the most important measure of success. In terms of science and technology, Narin [6] specified the indicators of development as follows:

- **number of patents** indicates technology development activity
- **cites per patent** indicates the impact of an analytical unit's patent
- **current impact index** (*CII*) indicates patent portfolio quality
- **technology independence** *(TI)* indicates independence of an analytical unit's technology development (the number of self-citations divided by the total number of citations)

- **technology cycle time** (*TCT*) indicates speed of invention (the median age in years of the patent references cited for a unit's patents)
- **science linkage** (*SL*) indicates the relationship between an analytical unit's technologies and academic research results (the average number of scientific papers referenced in an analytical unit's patents)

While the United States has led in these indicators in the past, Shelton and Holdridge [7] have shown that U.S. leadership in science and technology has generally been on the decline. They showed that the European Union and Japan have surpassed the United States in overall scientific and engineering (S&E) publications, patent applications, production of S&E graduates, and S&E Ph.D.s, and have nearly matched the United States in share of triadic patents (patents in the United States, EU, and Japan). They further argued that, as a consequence of these trends, the U.S. market share in high tech trade has been continuing to decline. It is unlikely that this trend will be reversed as global economies become committed to S&T developments.

In nanoscale science and technology patent production, the United States has been the leader due to the efforts of U.S. corporations like IBM (2,092 patents), Xerox (1,039 patents), and 3M (809 patents). Huang et al. [8] explained the status of existing nanotechnology patent citations:

- The United States dominated most of the nanotechnology citations;
- Japan was the second largest patent citation center;
- Other patent citation centers included France, Great Britain, and Switzerland;
- Patents of Austria, Netherlands Antilles, Germany, Norway, and Singapore only interacted with the patents of the United States;
- Several local country citation networks include: (1) United Kingdom and England; (2) France, Sweden, Italy, and Netherlands; and (3) China (Taiwan) and Korea.

According to the Chinese Ministry of Science and Technology (MST), during the decade before 2001, China's total applications for nanotechnology patents numbered less than 1,000. By 2003, China had established more than 2,400 such patents, accounting for 12 percent of the world's total. Since the NNI is relatively new, most of its supported activities are still in their infancy.

Huang et al. [8] further analyzed the areas in which patents (the number shown in parentheses) were most prevalent:

- The fields of chemistry (530) and molecular biology and microbiology (435) dominated patent citation centers. These fields also formed an interconnected citation network with patents of "drug, bio-affecting and body treating compositions" (514), "drug, bioaffecting and body treating compositions" (424), "chemistry: analytical and immunological testing" (436), and "organic compounds—part of the class 532-570 series" (536). [8, p.357]
- Patents of "chemical apparatus and process disinfecting, deodorizing, preserving, or sterilizing" (422) had interacted intensively with the patents of

"chemistry: molecular biology and microbiology" (435) and "chemistry: analytical and immunological testing" (436). [8, p.357]

- Patents of "chemistry: electrical and wave energy" (204) had cited the patents "chemistry: molecular biology and microbiology" (435) and "chemical apparatus and process disinfecting, deodorizing, preserving, or sterilizing" (433) intensively. [8, p.357]
- Patents of "organic compounds—part of the class 532-570 series" (544 and 546) had cited the patents of "drug, bio-affecting and body treating compositions" (514) intensively. [8, p.357]
- Local technology field citation networks included (1) "active solid-state devices (e.g., transistors, solid-state diodes)" (257) and "semiconductor device manufacturing: process" (438); (2) "coating process" (427) and "stock material or miscellaneous articles" (428); and (3) "radiant energy" (250) and "optics: measuring and testing" (356). [8, p.359]

Product developments will require the continued development of nanotechnology tools and instruments (1995–2005), new materials and information capabilities (2002–2007), and molecular manipulations (2005–2015).

Paper citations will most likely provide early indications of the NNI's success in furthering nanoscale research and development. However, as the NNI infrastructure matures, we would expect to see growth in publications, patent applications, and the development of workers and Ph.D.s in the nanotechnology fields.

Challenges and Risks

Development of human resources will be no small challenge for the NNI. Math and science literacy scores of 15-year-olds ranked the United States below the average of OECD nations in 2000. In mathematics literacy, the United States ranked 18th. In science literacy, the United States ranked 14th [9]. In addition, Europe surpassed the United States by over 200,000 S&E graduates in 2000 [10]. Asian institutions also produce nearly 50 percent more Ph.D.s in S&E than the United States, and over half of those receiving U.S. Ph.D. degrees in S&E are foreign students. The U.S. challenge will be to develop the skilled knowledge base needed to maintain leadership in the nanotechnology arena [11].

A 2001 study from the Rand National Defense Research Institute described the nanotechnology revolution as multidisciplinary, global, and knowledge based. Rand's critical factors for success included the creation of multidisciplinary training and degree programs, and the development of new products based on local resource availability [12]. In developing growth scenarios, the report included the existing technologies that have already stimulated nanoscale product innovations. The potential barriers to future growth relate to costs, social and ethical acceptance, insertion of nanotechnologies into new products, and the ease of use and accessibility of new nanoscale technologies and their related devices (Figure 1.10).

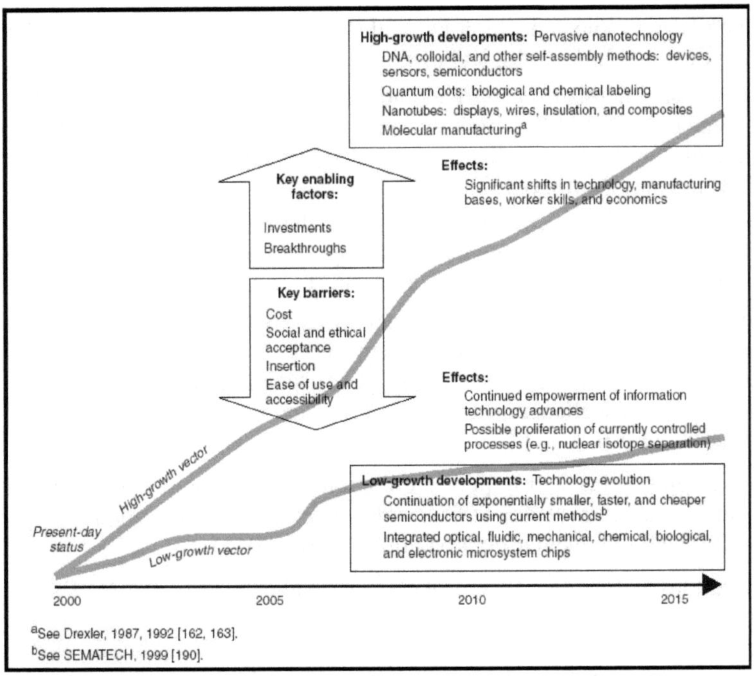

Figure 1.10. Potential Nanotechnology Growth Scenarios (Source: P. S. Anton, R. Silberglitt, J. Schneider [12]).

According to the Rand report, high-growth scenarios are dependent upon levels of investments and the generation of major technological breakthroughs. It has traditionally been the role of entrepreneurial corporations and angels to carry such breakthroughs into the marketplace with new product innovations. These two groups accounted for 34 and 25, respectively, of the funds used to develop and commercialize the products that incorporate such breakthroughs. In comparison, basic research and development has traditionally been the focus of federal and state investments. Federal and state governments have accounted for 29 and 5 of funding, respectively [10].

Most important is the transformation and integration of next-generation technologies into new products and services that will generate whole new industries. A full range of technology platforms will become nanotools, nanomaterials, nanostructures, and processes that can be developed and used in a wide variety of industry applications. In the field of nanotechnology coating applications, for example, Taylor [13, p.17] described the range of opportunities as enormous, as shown in Figure 1.11. The ability to use wet coating processes to apply nano particles creates an almost infinite range of opportunities for applications. A similar range of opportunities will exist for each nanotech platform and will provide unlimited potential for applications to be developed across industries. The barriers, however, will still come from investment limitations and the limited availability of resources needed to develop and commercialize the products that are conceived.

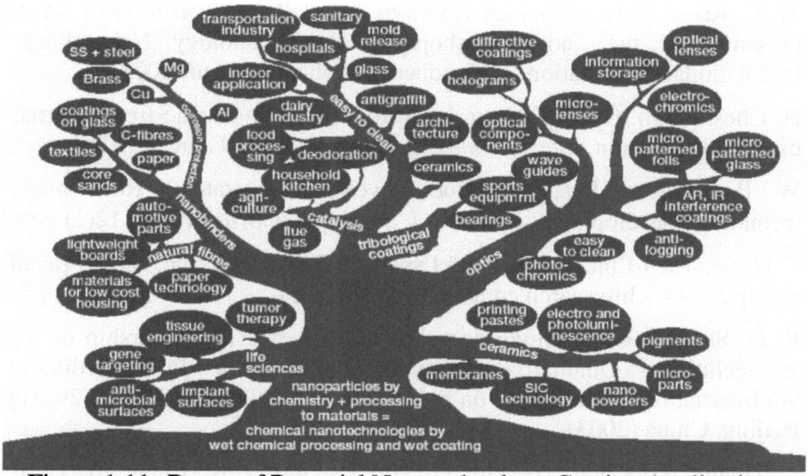

Figure 1.11. Range of Potential Nanotechnology Coating Applications (Source: J. M. Taylor [13]).

Conclusion

While there are unlimited opportunities for nanoscale visions, the development of basic capabilities and infrastructures to achieve these visions will determine our future economic growth. The Malcolm Baldrige National Quality Award provides insights and criteria for measuring successful organizations. These criteria can serve as a solid assessment tool and prototype for NNI activities. The key to leadership and planning is to develop applications for specific markets, thereby specifying the required human resources and processes or infrastructure required for success. These efforts are only successful if the outcomes meet the goals established for supporting activities. For NNI, much of the early funding is in support of infrastructural activities, including centers that target specific areas of investigation and development. Areas such as nanotools, nanomaterials, nano-structures, and related processes are needed for use across a wide variety of industry applications. Specific application developments will ultimately fall to organizations that seek to commercialize these applications, thereby generating jobs and revenues for the future. Good management will provide a competitive advantage to those organizations that enter the race. The Malcolm Baldrige National Quality Award provides the assessment criteria for such efforts. Let's incorporate them into our nanotechnology toolbox.

References

1. Malcolm Baldrige National Quality Award information. (available at http://www.quality.nist.gov).
2. K. Kobayashi, *Computers and communication: A vision of C&C*, Cambridge, MA: MIT Press (1985).

3. M. C. Roco, *Nanotechnology research and collaboration in the USA*. Paper presented at the 2nd Workshop on Nanotechnology Networking and International Cooperation, 12 October, Yokohama, Japan (2003).
4. H. Chesbrough, Open platform innovation: Creating value from internal and external innovation. *Intel Technology Journal*, 5-9 (19 August 2003).
5. W. R. Boulton, Japan's product development strategy: Road maps for technology development, *Business & the Contemporary World* **12**(2) (1995).
6. F. Narin, Tech-Line Background Paper, CHI Research, Inc. (2002) (available at http://www.chiresearch.com/techline/tlbp.pdf).
7. R. D. Shelton, G. M. Holdridge, The US-EU race for leadership of science and technology: Qualitative and quantitative indicators. In Proceeding of the 9th International Conference on Scientometrics & Infometrics, 25–29 August, Beijing, China (2003).
8. Z. Huang, H. Chen, A. Yip, G. Ng, F. Guo, Z-K. Chen, M. C. Roco, Longitudinal patent analysis for nanoscale science and engineering: Country, institution and technology field, *Journal of Nanoparticle Research* **5**, 3-4 (2003).
9. OECD Program for International Student Assessment (PISA) (2000).
10. U. Muldur, *Third European Report on science & technology indicators 2003; Towards a knowledge-based economy,* European Commission (March 2003).
11. R. E. Smalley, Our energy challenge. Presentations of the Houston Museum of Natural Science (April 2003).
12. P. S. Anton, R. Silberglitt, J. Schneider, *The global technology revolution: Bio/nano/materials trends and their synergies with information technology by 2015*, Santa Monica, CA: Rand National Defense Research Institute (2001).
13. J. M. Taylor, *New Dimensions for Manufacturing: A UK Strategy for nanotechnology*, London: Ministry for Science and Innovations (June 2000).

THE EMERGING NANOECONOMY: KEY DRIVERS, CHALLENGES, AND OPPORTUNITIES

James Canton, Institute for Global Futures

Can we anticipate and direct the impact of nanoscience on the economy? Will we be ready to develop the economic opportunities and navigate the challenges that nanoscience will bring? This paper examines a strategic view of the emerging NanoEconomy focusing on some of the change drivers, key challenges, and opportunities that will face the nation in the near future. Though nanoscience is in the early stages of invention, current breakthroughs indicate significant potential. The convergence of other key innovations in information technology and biotechnology, for example, will be accelerators moving nanoscience into mainstream commercial readiness. How might this occur? What other drivers of the nanoeconomy will be relevant? A possible nanoeconomic ecosystem will be reviewed. Attention to the key future-readiness factors that may influence the emerging NanoEconomy will be examined including: entrepreneurship, education, intellectual property, capital, jobs, talent, supply chains, competition, and nanoscience product or service offerings. An analysis of future-readiness factors that we need to adopt now, as a nation, to prepare for the coming NanoEconomy will be identified.

Towards the NanoEconomy

- A young child at risk of dying from a genetic defect is saved by a programmed nanoassembler in a telemedicine operation.
- A new nano-fuel cell enables hydrogen energy, bringing heat to a waiting community who live off the traditional power grid.
- A material is so adaptive that it keeps a firefighter safe from the intense heat of a warehouse fire so she can fight another day and return safely to her family the same night.

Each of these examples is human-enabling but also represents new innovations, new jobs, new industries, new companies, and a source of new economics—nanoeconomics: products born from nanoscience that may create entirely new industries.

Nanoscience, the design of matter at the atomic and molecular levels, represents a radical change in more then just materials science [1], but perhaps more fundamentally in the evolution of a new global economic system. It is too early to gauge this NanoEconomy scenario with accuracy, but it may be useful to speculate about it [2], given what we do know today about the rapid economic changes brought about by other technologies such as computing and the Internet.

There may be a new post-industrial economy emerging, shaped by nanoscience, and based on manipulating matter on-demand, thus fundamentally changing the products, services, markets, channels, jobs, and supply chains that we know today [3, 4]. This post-industrial shift, called the NanoEconomy here, may be just another stage, a marker in the evolution of economics as well as humanity.

We don't have economic theory sufficiently well developed to understand the impact of technology and science on the economy. This is a serious liability. Economists and scientists know too little about each other's worlds, but this must change if we are to better formulate social and scientific policy. Just as civilization and economics have evolved in parallel from hunter-gathers, to agriculture, to industrialism to information, nanoscience maybe the next stage in this evolution. We tend not to think about economics in evolutionary terms, like biology, but it is time we considered it.

Evolutionary Economics

An evolutionary economic model might provide a useful context for explaining the impact of core technologies and sciences that create economic change. Economic change may be described as the movement of capital, the invention of new products and services, the emergence of new markets, and the production of wealth. In addition the formulation of new economic opportunity such as job creation and investment potential would be characteristic of these economic changes as well. Competitive advantage may also be an outcome.

The basis of this approach to evolutionary economics is the recognition that core technologies such as computing represent economic shifts impacting on markets, customers, and industries. In fact, it is the convergence of these technologies that creates the most significant economic changes. It is becoming clear that as nanoscience becomes the next core scientific advancement, it also may represent the next evolutionary economic shift as well. Certainly the multi-billion dollar investment via the National Nanotechnology Initiative is a strong indication of how the U.S. scientific leadership views the potential economic value of nanoscience. Worldwide, the widespread and growing transglobal investment in nanoscience would indicate the emergence of a robust new economic force, not just the emergence of a new science. If this is the case, then we should rethink micro and macroeconomics and consider a new model of evolutionary economics that may provide insight into how core technologies may be key catalysts of economic change [5].

Towards a 21st Century Architecture

Examples of other core technologies that are creating large-scale socio-economic global change are information technology, Internet networking, and bioscience [6, 7]. Nanoscience is the next stage of core sciences that is creating an evolution in global market value: driving jobs, capital, investment, and product development, and provoking competition and opening new industries.

In a larger context here the sum is greater then the parts. These four power tools, collaboratively driving innovation, will vastly accelerate the transformation of society ushering in a new era of innovation. The next generation of nanoscience is perhaps more accurate to view in this larger systems model of other key technologies that together will make the most impact.

An emerging 21st century architecture, where the fusion of business and innovation, specifically the four power tools suggested here (Figure 1.12), will be the key to understanding the next global economic system. This NanoEconomy will be fueled not by the building blocks of the previous economy such as oil, steel or capital but rather on bits, atoms, neurons and genes. The NanoEconomy will be radically shaped by new innovations but will still operate within the rules of the global market economy.

Nanoscience may, much like computing, become embedded in every industry, every product, essential to every job and enterprise; even forging a higher quality of life for society [8]. If so, then the economic disruptions, changes, and opportunities will be largely based on how effective we are at forecasting this evolutionary shift as we prepare the nation.

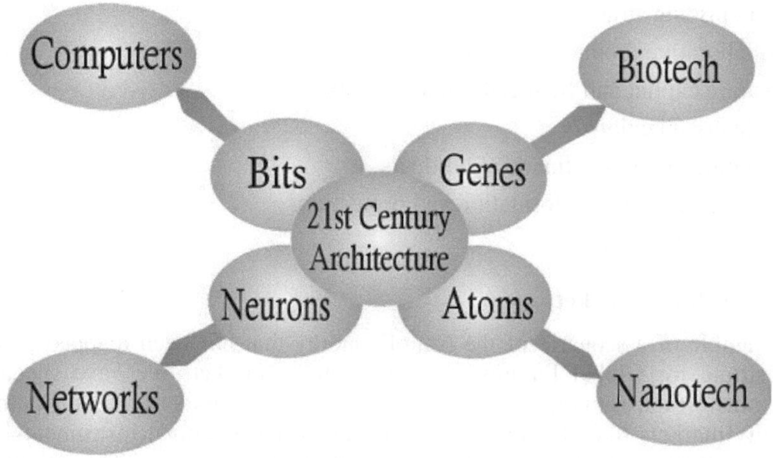

Figure 1.12. 21st century architecture schematic (Source: J. Canton [6]).

When so ubiquitous a science as nanoscience appears, cutting across many markets with so many varied applications in energy, medicine, materials, and others, it would be difficult to defend a forecast of low economic impact or change in the future.

Preparing for the Emerging NanoEconomy

The future-readiness of our institutions, especially education, government, and the private sector, will be the end product of our forecasts. Simply put, we must as effectively as possible prepare today for the future where evolutionary changes in science will bring evolutionary changes in economics and society. Fast, even radical change will most probably be the case where the timelines for adaptation will become smaller and institutions like education and government must be agile enough to be proactive.

Central to every economy, regardless of political ideology or model, is the idea of markets shaped by a number of common integrated features: buyers and sellers, pricing, capital, supply and demand, production and distribution. A NanoEconomy might also be characterized by not just the familiar products themselves being developed by nanoscience, but also the other aspects of what makes up an economy such as financial markets, intellectual property and the availability of qualified talent [9, 10].

A future NanoEconomy would be the totality of the commercial ecosystem that creates products and services that are offered through markets, enabling buyers and sellers to transact and conduct trade. The unique features of a NanoEconomy might be based upon such factors as:

1. real-time electronic banking transactions
2. on-demand production and manufacturing
3. globally interoperable distribution logistics

4. fluid pricing structures
5. collaborative product development
6. venture capital and financial liquidity markets
7. agile organizations
8. dynamically traded financial products
9. intellectual property formation and protection
10. talent availability

A Nano-Futures Market

One example of a scenario of the NanoEconomy that has been mapped by us at the Institute for Global Futures is a nano-futures market. We have speculated about a nano-futures market, available over electronic markets such as Bloomberg or equity markets like NASDAQ that enables virtual financial instruments to be traded, such as bonds or future contracts that are traded today. This might be a popular investment market based on the supply and demand of certain nano-commodities, branded nano-equities, or even forward contracts to deliver nanoscience products to certain markets.

Many of the traditional commodity markets today, from coffee to steel, to energy and even pollution credits, find an active financial services marketplace ready to broker buyers and sellers. The financial futures markets in many regards fuels the sales of real commodities and have become a vital financial aspect of the lifecycle and ecosystem of that industry.

Today there are many different—both traditional and fairly exotic—offerings of securities in the market. For example, specialty commodities like Bowie Bonds, based on the future projected revenues of the musician David Bowie, and Hollywood FX Futures, based on which movies will be successful box office hits based in Europe, represent highly speculative niche markets. It would not be difficult to construct a nano-futures exchange that exploits the intellectual properties, sales, products or return on investment for nanoscience. Where there are buyers and sellers, there will be markets and investment companies interested in brokering the speculation.

A nano-futures exchange is one end of the financial spectrum of fluid capital markets where attractive nanoscience offerings worthy of venture capital, debt financing and public offerings through stock markets would fuel nanoscience enterprises. We have seen the early stages of what is today even a billion dollar investment market for nanoscience. NSF predicts the market for nanoscience to be at $1 trillion by 2010. We think this is a modest forecast at best with nanoscience reaching over double this amount given the current forecasts at our firm.

Every technology wave in computing, biotech, and the Internet has set the stage for this evolutionary economic wave in nanoscience [6, 7, 9]. It is not difficult to see nanoscience become an engine of growth for the future global economy. The more complex speculation of financial instruments tied to nanoscience companies, products and research may provide an interesting backdrop to the future economy as nanoscience matures in time.

Robust liquidity and creativity in capital markets is emerging today in the venture capital marketplace. Early stage investments in nanoscience indicate the potential framework for next stage initial public offerings, mergers and acquisitions and other evolutionary economic indicators of a NanoEconomy: buyers and sellers of numerous financial and commercial products engaged in commerce.

An interesting way in the future to invest in nanoscience innovations might be to purchase the intellectual property futures of a certain company based on its receiving a patent for self-assembly as a debenture, a convertible debt issue, or security [5]. Then the secondary stage is to sell that instrument over an electronic supermarket, like Archipelago, an Electronic Communication Network. An entire new financial marketplace based on nanoscience might emerge with investment banks brokering global markets of buyers and sellers of nanoscience instruments.

Even the market for different types of self-assembly, for different industries such as medicine or energy, might spawn specialized electronic trading markets and investment schemes to create new investment markets for speculation. I could envision that nano-commodities as well, like buckyballs, could have their own financial markets where futures trading, hedges, wraps, and nano-bonds lie side by side with today's energy or pork bellies. Imagine that Rice U Bucky Ball Future Contracts sell for a premium on global stock exchanges such as the Hang Sen, Dax, or NY Stock Exchange.

NanoSupply Chains

Supply chains are the global distribution systems that facilitate how products get from the manufacturer, to the buyer and to the end-user. Most supply chains are becoming electronically linked, integrating finance, distribution, production, and transportation.

It is still early in the evolution of supply chains to be tech-enabled with global IT enablement. Supply chains would have to be changed to accommodate the nanofoundries that would be supplying the hard final products [11]. The framework for nanoscience is moving in this direction today, given the changes in supply chain logistics. The logistics supply-chain vendors will eventually become nanoscience-enabled. Today's logistics companies running global supply chains provide more then just the products they distribute for leading brands—they often assemble the products and even produce some parts, providing end-to-end solutions worldwide. This is certainly the trend.

There are also financial supply chains, logistics, payments, and transactions that would be needed to adjust to this change in the NanoEconomy. Many of these financial supply chains would be automated and Internet-enabled to capture orders on one end of the world, fulfilling orders on the other and fueling the nano-production somewhere else. There is much in place today to accommodate a future NanoEconomy.

Distribution would be in minutes or real time and might frustrate the tax collection because where the product was made would change based on where the most tax-advantaged sovereign domain would be. This would open up entirely new opportunities for the developing world [2]. Today, nations that are often capital-restricted cannot compete in the capital-intensive world of production but tomorrow they could host nanofoundries, either virtual or physical due to the tremendous cost reduction of a future nano-infrastructure [6]. Just as IT and biotech are driving prices down in the bio-IT infrastructure today, the nano-infrastructure of tomorrow will become a beneficiary of this value [12].

To companies like Intel, which spends an average of $5 billion per fabrication plant for making silicon chips, the cost reduction due to nanoscience would transform overnight the cost of computing, networking, and information technology. Whether these are medical devices or transportation systems, this one feature, reducing the cost of silicon by even one-third would have a radical impact on twenty plus markets, from autos to computers. This one feature would transform other commodities and finished products, causing drastic cost reductions and opening up new global markets for growth [5].

Additionally, the cost of one new drug is over $800 million today. The drug discovery process cost could be dramatically less, enabled by nanoscience supply chains. The implications for being able to produce new drugs, faster, more cost-effectively and with more accurate precision will transform both the medicine and economics of the bio-pharma industry [7].

Personalized nano-medicine will give us drugs we cannot create as yet today—such as a blood substitute or cancer treatment that can 'see' and 'react' at the nanoscale, preventing a mutation leading to disease [11]. We are headed in this direction now. Emerging drug delivery systems that might overnight change the way drugs are administered are being conceptualized given the nanobioscience platform strategy. An entirely new generation of nano-enabled cellular delivery systems may result. One of the obvious outcomes of this delivery system maybe the enhancement and repair of the human cell (or organ or system) via nanoscale biodevices, programmed for specific missions, with personalized DNA a "genomic awareness" (Figure 1.13).

Much of the transformation of supply chains is underway as the Internet and new inexpensive manufacturing capacity have opened up worldwide. It is fair to forecast that China and India will play an important role in the emerging nanoeconomy, as two of the largest manufacturing markets and software centers; they will become integral parts of this new future. Stiff competition will define the nanoeconomy. Innovation bars will be raised to new heights. The rush towards outsourcing to India and the manufacturing trade deficits with China point to this future. A NanoEconomy might be characterized as being born from the transformation of the existing high- tech economy [5]. Just as information technology and biotech have made significant inroads into the global economy, changing supply chains, contributing to the formation of new industries, and creating new product development as well as channeling capital and intellectual

Figure 1.13. Future cellular cybernetics as a biomedical commodity (Source: C. Ostman [7]).

property, so, too, shall nanoscience play a vital role, an evolutionary role in the future [6].

Nanofoundries in Asia receiving orders over the Internet from wireless broadband-based customers will be designed, programmed for self-assembly and distributed to waiting markets [11]. Some products, more exquisitely nano-engineered with more features, robustness, or better price will attract a higher price premium [13]. You will still get what you pay for; there may be more choices as companies will be co-designing collaboratively virtual in-silico products from DNA object libraries. Models may be available at lower costs, but with more features [6].

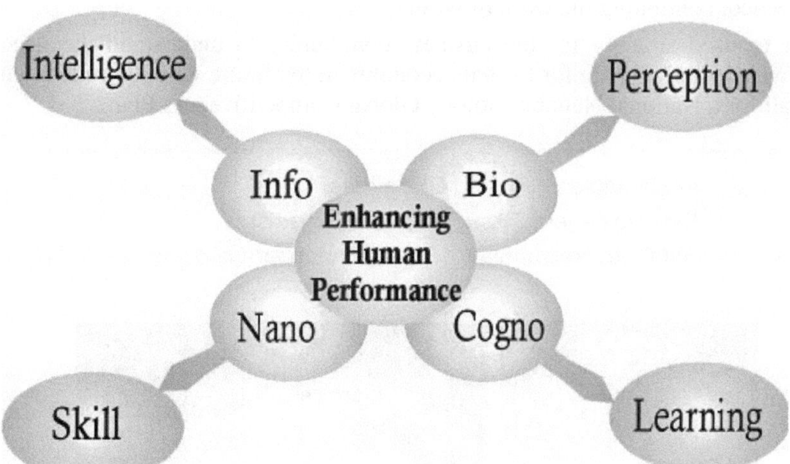

Figure 1.14. Human performance enhancement system dynamics (Source: J. Canton [6]).

Future-Readiness and the NanoEconomy

In preparation for the NanoEconomy there are key steps to take to become future-ready to take full advantage of the opportunities brought by nanoscience. Here are a few to consider:

- The need to change the educational system in radical ways to prepare nanoscience-ready students, starting from the primary school, will be paramount.
- Key disruptions in the readiness of U.S. business to tool-up, skill-up, and have the necessary vision and support for change will be essential.
- Investment capital will be needed in strategic areas such as manufacturing, nanomedicine, and energy that go beyond the paltry private capital we see today.
- The development of a post-graduate talent pool of nanobusiness-savvy executives will be needed.
- A competitive analysis of the nano-market landscape will be needed.
- Nanotechnology intellectual property development and protection will be vital.
- Supply chains need to be reconsidered, given the emerging nanoproduct pipeline.
- Nanotechnology needs to be conceptualized as a component driver, an accelerator of other convergent technologies such as IT, biotech, and neuroscience.
- A clear special-purpose effort to explain, promote, and market nanoscience to the public—emphasizing health enablement and quality of life—will be needed, continuing the work of NSF.
- A focused strategy for the business community to understand the need to invest in and prepare for the nanoeconomy in the future will be needed, with a call for a National Nanotechnology Global Competitiveness Plan.

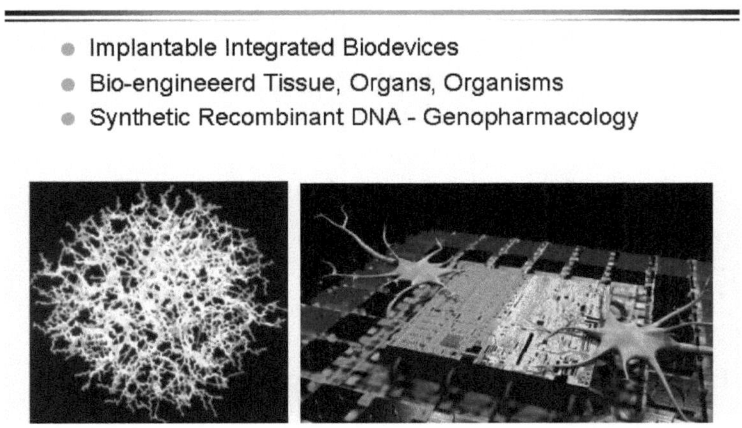

Figure 1.15. Future nanobio products (Source: C. Ostman [7]).

Nanoscience in a Democratic Society

Concerns about emerging technology and its impact on social and economic systems are not new [10]. There is a common misconception that scientists don't care about people and the risks of science and steam ahead with reckless abandon. Nothing could be further from the truth. With society embroiled in numerous debates today about the Internet, wireless broadband, cloning and stem cells, concerns are a vital part of the national debate in a free society. With every innovation, from electricity to automobiles, from biotech to atomic energy, there were debates questioning the value, ethics, safety, and impact on society. Even so, mistakes were made. Some harm was done. Scientists have learned to proceed with more caution today and will do so in the future.

The point is that in a free society and a democracy, dissent and debate are good, especially about new technologies that seem awesome in their potential for good as well as for harm. But since Big Brother's 1984 has come and gone, as nuclear proliferation is contained, and as privacy is still protected by law, we seem not to have fallen prey to technology's long-promised specter of destruction.

The sheer power of the innovations offered today, and those that may be invented tomorrow, may challenge our capacity to control nanoscience, some critics argue. Nanoscience—with its wide reaches, its sweeping potential, and with the possibility of radical change before us—justifiably requires special attention. Exploration, by definition, does not come without some danger, threat, or even risk. But that should not stop us from applying the Reasonable Person Rule of Law to science: Is it reasonable to continue the exploration? Is it reasonable to have controls? To both questions, I would answer yes.

Can we learn to control this technology? I would hypothesize yes, as we have as a society grappled successfully with other risks, such as nuclear proliferation. There are many unknowns about nanoscience that will challenge us, even maybe threaten society. Should we run and hide, placing a moratorium on nanoscience for fear about what has not even been proven possible to create, such as Bill Joy's nanobots on overdrive? This would be irresponsible and not consistent with our ethos as scientists or as a democratic nation to be free to explore the unknown. This does not mean we should throw out the controls, but quite the opposite.

There is too much at stake here to stop the research or drive it underground for fear of disaster. It is very early in our understanding of this science. Nevertheless, we owe it to future generations to proceed with caution, controls, and concern. I think, if properly guided, that nanoscience can be one of the single greatest contributions to enhancing the quality of life of humans. But, there is much more at stake.

If there is any real crisis coming, it is in not being ready for the socio-economic impact and changes that may come abruptly and disruptively to our economy. Jobs, businesses and people will be disrupted. We must be prepared for potential changes that will require change in education, jobs, markets, and talent [12, 16]. We must consider the radical innovations that nanoscience may bring sooner then anyone can imagine and prepare to meet this challenge, as a nation. We must become future-ready to deal with the economic impact of nanoscience to change our world.

If nanoscience is to provide an increase in the quality of life, as well as competitive opportunity for our nation, it will not come without some risk. We must learn to understand the risks associated with nanoscience and move rapidly to prepare our nation for the anticipated changes due ahead. I for one do not think there is a crisis coming, a nano-disaster waiting for us. I would prefer not to limit innovation so early in the evolution of this new science.

The Grand Challenge Revisited

If there is one overarching grand challenge that nanoscience presents to us, it is the possibility to do an immense amount of good for humanity. From an economic perspective, the emerging NanoEconomy promises to create more jobs, fuel more innovative products, perhaps heal and cure more patients, extending life and health. It is not about the concentration of power in the hands of the few, nor should it be viewed as a furthering of the gap between the haves and have-nots. I know it would be easy for some to think otherwise.

We embark on this journey to use science to better understand our world and to pass this benefit along to the generations to come. It will not come on the cheap or without risk. Nanoscience is a bold science; by its definition it will change many aspects of society. What may be speculation today may be the reality of science tomorrow.

There are many examples of sciences and technologies that have created competitive advantage and increased quality of life in global markets. At the very least this paper calls for an analysis of the socio-economics of nanoscience in preparation for a new future. The NanoEconomy is coming. Markets will be won and lost. Economies will rise and fall. The strategic question is, will we be ready and what constitutes readiness to meet the global challenge of the NanoEconomy? As the NanoEconomy emerges, this will create a number of unique challenges for individuals, businesses, and institutions.

References

1. R. Feynman, There is plenty of room at the bottom: An invitation to enter a new field of physics. Lecture at American Physical Society Meeting, 29 December (1959).

2. K. S. Taylor, Material life, markets and capitalism, in *Human society and the global economy* (available at http://distance.ed.bcc.ctc.edu/ econ100/ksttext/ chaplist.htm#rights)

3. B. G. Gower, Nanotech goes to work, *MIT Technology Review* (January 2001).

4. K. W. West, Nanotechnology heralds reinvention of global economy, *Charleston Regional Business Journal* (10 September 2001).

5. C. Ostman, Transactional process brokeraging of virtual commodity process assets: An emergent economic ecology. Paper presented at the 5th International Conference on Information Systems Analysis and Synthesis, 1–3 August, Baden-Baden, Germany (1999).
6. J. Canton, *Technofutures: How Leading Edge* T*echnologies* W*ill* T*transform* B*usiness in the 21st Century*, Carlsbad, CA: Hay House (1999).
7. C. Ostman, Bioconvergence: Progenitor of the nanotechnology age, *NanoIndustries newsletter* (July 2000).
8. M. Jacoby, New tools for tiny jobs, *Chemical and Engineering News* (16 October 2000).
9. D. Broderick, *The Spike: How Our Lives Are Being Transformed by Rapidly Advancing Technologies*, New York: Forge (2001).
10. V. Bayot, Nanotechnologies and the evolution of skills, UCL-CERMIN. Press briefing presented 14 June, Grenoble, France (2002) (available at http://europa.eu.int/comm/research/growth/pdf/nanotechnology-conference/bayot-presentation_en.pdf).
11. G. M. Whitesides, The once and future nanomachine, *Scientific American* (September 2001).
12. M. Pitman, Matching science and technology to future needs. Paper presented at the Ad Hoc Meeting of Experts on Government Foresight Exercises, 14 September, OECD, Paris (1994).
13. T. D. Schneider, Measuring molecular information, *Journal of Theoretical Biology* **201**, 87-92 (1999).
14. Personal fabrication on demand, *Wired* (9 April 2001).
15. A. Toffler, *Future shock*, New York: Bantam (1974).
16. M. Roco, W. S. Bainbridge, eds., *Societal Implications of Nanoscience and Nanotechnology*, Dordrecht, Netherlands: Kluwer (2001).

TRANSCENDING MOORE'S LAW WITH MOLECULAR ELECTRONICS AND NANOTECHNOLOGY

Steve Jurvetson, Draper Fisher Jurvetson

The history of technology is one of disruption and exponential growth, epitomized in Moore's Law, and generalized to many basic technological capabilities that are compounding independently from the economy. Thinking about Moore's Law in the abstract provides a framework for predicting the future of computation and the transition to a new substrate: molecular electronics. More than a niche subject of interest only to chip designers, the continued march of Moore's Law will affect all of the sciences, just as nanotechnology will affect all industries. An analysis of progress in molecular electronics also provides a detailed example of the commercialization challenges and opportunities common to many nanotechnologies.

Introduction to Technology Exponentials

Despite a natural human tendency to presume linearity, accelerating change from positive feedback is a common pattern in technology and evolution. We are now crossing a threshold where the pace of disruptive shifts is no longer inter-generational and begins to have a meaningful impact over the span of careers and eventually product cycles.

As early-stage venture capitalists, we at Draper Fisher Jurvetson look for disruptive businesses run by entrepreneurs that want to change the world. To be successful, we have to identify technology waves early and act upon those beliefs. We believe that nanotechnology is the next great technology wave, the nexus of scientific innovation that revolutionizes most industries and indirectly affects the fabric of society. Historians will look back on the upcoming epoch as no less important than the Industrial Revolution.

The aforementioned are some long-term trends. Today, from a seed-stage venture capitalist perspective (with a broad sampling of the entrepreneurial pool), we are seeing more innovation than ever before. And we are investing in more new companies than ever before.

In the medium term, disruptive technological progress is relatively decoupled from economic cycles. For example, for the past 40 years in the semiconductor industry, Moore's Law has not wavered in the face of dramatic economic cycles. Ray Kurzweil's abstraction of Moore's Law (from transistor-centricity to computational capability and storage capacity) shows an uninterrupted exponential curve for over 100 years, again without perturbation during the Great Depression or the World Wars. Similar exponentials can be seen in Internet connectivity, medical imaging resolution, genes mapped, and solved 3D protein structures. In each case, the level of analysis is not products or companies, but basic technological *capabilities*.

Moore's Law

Moore's Law is commonly reported as a doubling of transistor density every 18 months. But this is not something the co-founder of Intel, Gordon Moore, ever said. It is a nice blending of his two predictions; in 1965, he predicted an annual doubling of transistor counts in the most cost-effective chip and revised it in 1975 to every 24 months. The popular perception of Moore's Law is that computer chips are compounding in their complexity at near constant per unit cost. This is one of the many abstractions of Moore's Law, and it relates to the compounding of transistor density in two dimensions. Others relate to speed (the signals have less distance to travel) and computational power (speed x density).

Moore's Law drives chips, communications, and computers and has become the primary driver in drug discovery and bioinformatics, medical imaging, and diagnostics. Over time, the lab sciences become information sciences, modeled on a computer rather than through trial-and-error experimentation.

NASA Ames shut down its wind tunnels this year. As Moore's Law provided enough computational power to model turbulence and airflow, there was no longer a need to test iterative physical design variations of aircraft in the wind tunnels, and the pace of innovative design exploration dramatically accelerated.

Eli Lilly processed 100x *fewer* molecules this year than it did 15 years ago. But its annual productivity in drug discovery did not drop proportionately; it went up 100x over the same period. "Fewer atoms and more bits" is its coda.

Accurate simulation demands computational power, and once a sufficient threshold has been crossed, simulation acts as an innovation accelerant over physical experimentation. Many more questions can be answered per day. Recent accuracy thresholds have been crossed in diverse areas, such as modeling the weather (predicting a thunderstorm six hours in advance) and automobile collisions (a relief for the crash test dummies), and the thresholds have yet to be crossed for many areas, such as protein folding dynamics.

Unless you work for a chip company and focus on fab-yield optimization, you do not care about transistor counts. Integrated circuit customers do not buy transistors. Consumers of technology purchase computational speed and data-storage density. When recast in these terms, Moore's Law is no longer a transistor-centric metric, and this abstraction allows for longer-term analysis.

As Ray Kurzweil has pointed out, the exponential curve of Moore's Law extends smoothly back in time to 1890 (Figure 1.16), long before the invention of the semiconductor. Through five paradigm shifts—such as electro-mechanical calculators and vacuum tube computers—the computational power that $1,000 buys has doubled every two years. For the past 30 years, it has been doubling every year.

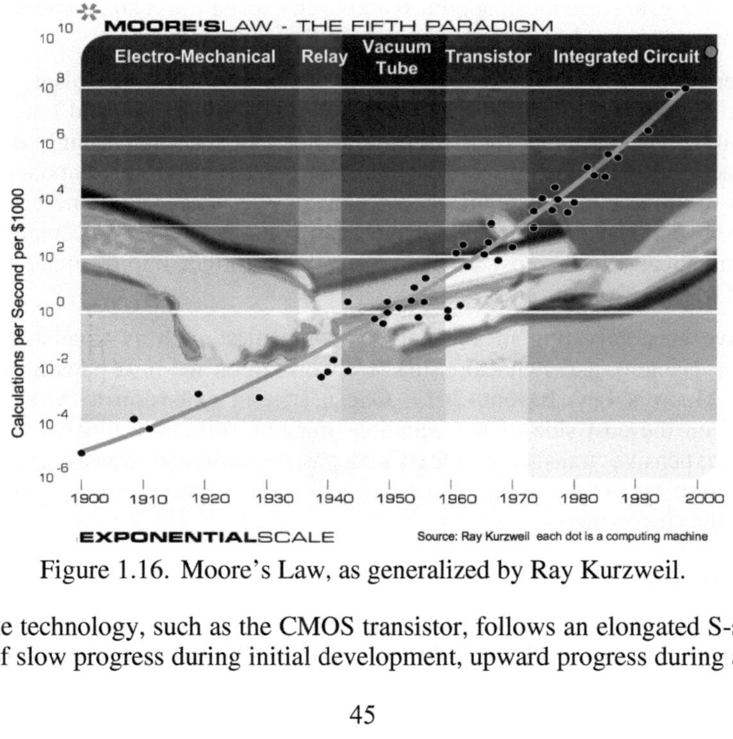

Figure 1.16. Moore's Law, as generalized by Ray Kurzweil.

Any one technology, such as the CMOS transistor, follows an elongated S-shaped curve of slow progress during initial development, upward progress during a rapid adoption phase, and then slower growth from market saturation over time. But a more generalized *capability*, such as computation, storage, or bandwidth, tends to follow a pure exponential—bridging across a variety of technologies and their cascade of S-curves. If history is any guide, Moore's Law will continue on and will jump to a different substrate than CMOS silicon. It has done so five times in the past and will need to again in the future.

Problems with the Current Paradigm

The traditional semiconductor chip is finally approaching some fundamental physical limits. Moore recently admitted that Moore's Law, in its current form, with CMOS silicon, will run out of gas in 2017. One of the problems is that the chips are getting very hot. This provides the impetus for chip-cooling companies, like Nanocoolers, to provide a breakthrough solution for removing 100 Watts per square centimeter. In the long term, the paradigm has to change.

Another physical limit is the atomic limit—the indivisibility of atoms. Intel's current gate oxide is 1.2 nm thick. Intel's 45 nm process is expected to have a gate oxide that is only 3 atoms thick. It is hard to imagine many more doublings from there, even with further innovation in insulating materials. Intel has recently announced a breakthrough in a nano-structured gate oxide (high k dielectric) and metal contact materials that should enable the 45 nm node to come on-line in 2007. None of the industry participants has a CMOS roadmap for the next 50 years.

A major issue with thin gate oxides, and one that will also come to the fore with high-k dielectrics, is quantum mechanical tunneling. As the oxide becomes thinner, the gate current can approach and even exceed the channel current so that the transistor cannot be controlled by the gate.

Another problem is the escalating cost of a semiconductor fab plant, which is doubling every three years, a phenomenon dubbed Moore's Second Law. Human ingenuity keeps shrinking the CMOS transistor, but with increasingly expensive manufacturing facilities—currently $3 billion per fab. A large component of fab cost is the lithography equipment that patterns the wafers with successive sub-micron layers. Nanoimprint lithography from companies like Molecular Imprints can dramatically lower cost and leave room for further improvement from the field of molecular electronics.

We have been investing in a variety of companies, such as Coatue, D-Wave, FlexICs, Nantero, and ZettaCore, that are working on the next paradigm shift to extend Moore's Law beyond 2017. One near-term extension to Moore's Law focuses on the cost side of the equation. Imagine rolls of wallpaper embedded with inexpensive transistors. FlexICs deposits traditional transistors at room temperature on plastic, a much cheaper bulk process than growing and cutting crystalline silicon ingots.

Molecular Electronics

The primary contender for the post-silicon computation paradigm is molecular electronics, a nano-scale alternative to the CMOS transistor. Eventually, molecular switches will revolutionize computation by scaling into the third dimension—overcoming the planar deposition limitations of CMOS. Initially, they will substitute for the transistor bottleneck on an otherwise standard silicon process with standard external I/O interfaces.

For example, Nantero employs carbon nanotubes suspended above metal electrodes on silicon to create high-density nonvolatile memory chips (the weak Van der Waals bond can hold a deflected tube in place indefinitely with no power drain). Carbon nanotubes are small (~10 atoms wide), 30x stronger than steel at one-sixth the weight, and perform the functions of wires, capacitors and transistors with better speed, power, density and cost. Cheap nonvolatile memory enables important advances, such as "instant-on" PCs. Other companies, such as Hewlett Packard and ZettaCore, are combining organic chemistry with a silicon substrate to create memory elements that self-assemble using chemical bonds that form along pre-patterned regions of exposed silicon.

There are several reasons why molecular electronics is the next paradigm for Moore's Law:

Size: Molecular electronics has the potential to dramatically extend the miniaturization that has driven the density and speed advantages of the integrated circuit (IC) phase of Moore's Law. In 2002, using an STM to manipulate individual carbon monoxide molecules, IBM built a 3-input sorter by arranging those molecules precisely on a copper surface. It is 260,000x smaller than the equivalent circuit built in the most modern chip plant.

- *Power:* One of the reasons that transistors are not stacked into 3D volumes today is that the silicon would melt. The inefficiency of the modern transistor is staggering. It is much less efficient at its task than the internal combustion engine. The brain provides an existence proof of what is possible; it is 100 million times more efficient in power/calculation than our best processors. Sure it is slow (under a kHz) but it is massively interconnected (with 100 trillion synapses between 60 billion neurons), and it is folded into a 3D volume. Power per calculation will dominate clock speed as the metric of merit for the future of computation.
- *Manufacturing Cost:* Many of the molecular electronics designs use simple spin coating or molecular self-assembly of organic compounds. The process complexity is embodied in the synthesized molecular structures, and so they can literally be splashed onto a prepared silicon wafer. The complexity is not in the deposition or the manufacturing process or the systems engineering. Much of the conceptual difference of nanotechnology products derives from a biological metaphor: complexity builds from the bottom up and pivots about conformational changes, weak bonds, and surfaces. It is not engineered from the top with precise manipulation and static placement.
- *Low Temperature Manufacturing:* Biology does not tend to assemble complexity at 1000 degrees in a high vacuum. It tends to be room temperature or body temperature. In a manufacturing domain, this opens the possibility of cheap plastic substrates instead of expensive silicon ingots.

- *Elegance:* In addition to these advantages, some of the molecular electronics approaches offer elegant solutions to non-volatile and inherently digital storage. We go through unnatural acts with CMOS silicon to get an inherently analog and leaky medium to approximate a digital and non-volatile abstraction that we depend on for our design methodology. Many of the molecular electronic approaches are inherently digital, and some are inherently non-volatile.

Other research projects, from quantum computing to using DNA as a structural material for directed assembly of carbon nanotubes, have one thing in common: they are all nanotechnology. Nanotechnology is often defined as the manipulation and control of matter at the nanometer scale (critical dimensions of 1–100 nm).

As venture capitalists, we start to get interested when there are unique properties of matter that emerge at the nanoscale, and that are not exploitable at the macroscale world of today's engineered products. We like to ask the startups that we are investing in: "Why now? Why couldn't you have started this business 10 years ago?" Our portfolio of nanotechnology startups have a common thread in their response to this question—recent developments in the capacity to understand and engineer nanoscale materials have enabled new products that could not have been developed at a larger scale.

There are various unique properties of matter that are expressed at the nanoscale and are quite foreign to our "bulk statistical" senses. (We do not see single photons or quanta of electric charge; we feel bulk phenomena, like friction, at the statistical or emergent macroscale.). At the nanoscale, the bulk approximations of Newtonian physics are revealed for their inaccuracy, and give way to quantum physics. Nanotechnology is more than a linear improvement with scale; everything changes. Quantum entanglement, tunneling, ballistic transport, frictionless rotation of superfluids, and several other phenomena have been regarded as "spooky" by many of the smartest scientists, even Einstein, upon first exposure.

For a simple example of nanotech's discontinuous divergence from the "bulk" sciences, consider the simple aluminum Coke can. If you take the inert aluminum metal in that can and grind it down into a powder of 20-30 nm particles, it will spontaneously explode in air. It becomes a rocket fuel catalyst. The energetic properties of matter change at that scale. The surface area to volume ratios become relevant, and even the inter-atomic distances in a metal lattice change from surface effects.

Innovation from the Edge

Disruptive innovation, the driver of growth and renewal, occurs at the edge. In startups, innovation occurs out of the mainstream, away from the warmth of the herd. In biological evolution, innovative mutations take hold at the physical edge of the population, at the edge of survival. In complexity theory, structure and complexity emerge at the edge of chaos—the dividing line between predictable regularity and chaotic indeterminacy. And in science, meaningful disruptive innovation occurs at the inter-disciplinary interstices between formal academic disciplines.

Herein lies much of the excitement about nanotechnology: in the richness of human communication about science. Nanotechnology exposes the core areas of overlap in the fundamental sciences, the place where quantum physics and quantum chemistry can cross-pollinate with ideas from the life sciences.

Over time, each of the academic disciplines develops its own proprietary systems vernacular that isolates it from neighboring disciplines. Nanoscale science requires scientists to cut across the scientific languages to unite the isolated islands of innovation. Nanotechnology is the nexus of the sciences (Figure 1.17).

In academic centers and government labs, nanotechnology is fostering new conversations. At Stanford, Duke, and many other schools, the new nanotechnology buildings are physically located at the symbolic hub of the schools of engineering, computer science, and medicine.

Nanotechnology is the nexus of the sciences, but outside of the science and research itself, the nanotechnology umbrella conveys no business synergy whatsoever. The marketing, distribution and sales of a nanotechnology solar cell, memory chip, or drug delivery capsule will be completely different from each other, and will present few opportunities for common learning or synergy.

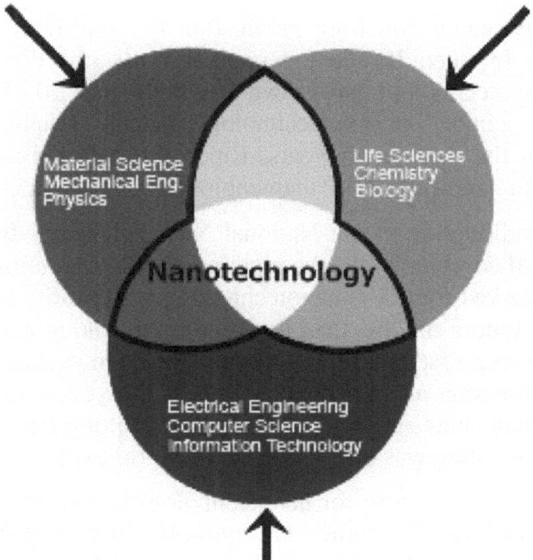

Figure 1.17. Nanotechnology is the nexus of the sciences.

Market Timing

As an umbrella term for myriad technologies spanning multiple industries, nanotechnology will eventually disrupt these industries over different time frames—but most are long-term opportunities. Electronics, energy, drug delivery, and materials are areas of active nanotechnology research today. Medicine and bulk manufacturing are future opportunities. Nanotechnology could have a trillion dollar impact on various industries inside of 15 years.

Of course, if one thinks far enough in the future, every industry will be eventually revolutionized by a fundamental capability for molecular manufacturing—from the inorganic structures to the organic and even the biological. Analog manufacturing becomes digital, engendering a profound restructuring of the substrate of the physical world.

Given that much of the abstract potential of nanotechnology is a question of "when" not "if," the challenge for the venture capitalist is one of market timing. When should we be investing, and in which sub-sectors? It is as if we need to pull the sea of possibilities through an intellectual chromatograph to tease apart the various segments into a timeline of probable progression. That is an ongoing process of data collection (e.g., the growing pool of business plan submissions), business and technology analysis, and intuition.

Two touchstone events for the scientific enthusiasm for the timing of nanotechnology were the decoding of the human genome and the dazzling visual images from the Scanning Tunneling Microscope (e.g., the arrangement of individual Xenon atoms into the IBM logo). They represent the digitization of biology and matter, symbolic milestones for accelerated learning and simulation-driven innovation.

And more recently, nanotechnology publication has proliferated, much like the early days of the Internet. Beside the popular press, the number of scientific publications on nanotechnology has grown 10x in the past 10 years. According to the USPTO, the number of nanotechnology patents granted each year has skyrocketed 3x in the past seven years. Ripe with symbolism, IBM has more lawyers working on nanotechnology than engineers.

With the recent codification of the National Nanotechnology Initiative into law, federal funding will continue to fill the pipeline of nanotechnology research. With $847 million earmarked for 2004, nanotechnology was a rarity in the tight budget process; it received more funding than was requested. And now nanotechnology is second only to the space race for federal funding of science. And the United States is not alone in funding nanotechnology and is not even in the lead. Japan outspends the United States each year on nanotechnology research. In 2003, the U.S. government spending was one-fourth of the world total.

Federal funding is the seed corn for nanotechnology entrepreneurship. All of our nanotechnology portfolio companies are spin-offs (with negotiated intellectual property transfers) from universities or government labs, and all got their start with federal funding. Often these companies need specialized equipment and expensive laboratories to do the early tinkering that will germinate a new breakthrough. These are typically lacking in the proverbial garage of the entrepreneur at home.

And corporate investors have discovered a keen interest in nanotechnology, with internal R&D, external investments in startups, and acquisitions of promising companies, such as AMD's recent acquisition of the molecular electronics company Coatue.

Despite all of this excitement, there are a fair number of investment dead-ends, and so we continue to refine the filters we use in selecting companies to back. All entrepreneurs want to present their business as fitting an appropriate timeline to commercialization. How can we guide our intuition on which of these entrepreneurs are right?

The Vertical Integration Question

Nanotechnology involves the reengineering of the lowest-level physical layer of a system, and so a natural business question arises: How far forward do you need to vertically integrate before you can sell a product on the open market? For example, in molecular electronics, if you can ship a DRAM-compatible chip, you have found a horizontal layer of standardization, and further vertical integration is not necessary. If you have an incompatible 3D memory block, you may have to vertically integrate to the storage subsystem level, or further, to bring product to market. That may require industry partnerships, and will, in general, take more time and money as change is introduced farther up the product stack. 3D logic with massive interconnectivity may require a new computer design and a new form of software; this would take the longest to commercialize. And most startups on this end of the spectrum would seek partnerships to bring their vision to market. The success and timeliness of that endeavor will depend on many factors, including IP protection, the magnitude of improvement, the vertical tier at which that value is recognized, the number of potential partners, and the degree of tooling and other industry accommodations.

Product development timelines are impacted by the cycle time of the R&D feedback loop. For example, outdoor lifetime testing for OLEDs will take longer than *in silico* simulation spins of digital products. If the product requires partners in the R&D loop or multiple nested tiers of testing, it will take longer to commercialize.

The "Interface Problem"

As we think about the startup opportunities in nanotechnology, an uncertain financial environment underscores the importance of market timing and revenue opportunities over the next five years. Of the various paths to nanotech, which are 20-year quests in search of a government grant, and which are market-driven businesses that will attract venture capital? Are there co-factors of production that require a whole industry to be in place before a company ships product?

Today's business-driven paths to nanotechnology diverge into two strategies to cross the "interface" chasm—the biologically inspired bottom-up path, and the top-down approach of the semiconductor industry. The non-biological MEMS developers are addressing current markets in the micro-world while pursuing an ever-shrinking spiral of miniaturization that builds the relevant infrastructure tiers along the way. Not surprisingly, this is very similar to the path that has been followed in the semiconductor industry, and many of its adherents see nanotechnology as inevitable, but in the distant future.

On the other hand, biological manipulation presents myriad opportunities to effect great change in the near-term. Drug development, tissue engineering, and genetic engineering are all powerfully impacted by the molecular manipulation capabilities available to us today. And genetically modified microbes, whether by artificial evolution or directed gene splicing, give researchers the ability to build structures from the bottom up.

The Top-Down "Chip Path"

This path is consonant with the original vision of physicist Richard Feynman (in his 1959 lecture at Caltech) of the iterative miniaturization of our tools down to the nanoscale. Some companies, like Zyvex, are pursuing the gradual shrinking of semiconductor manufacturing technology from the micro-electro-mechanical systems (MEMS) of today into the nanometer domain of NEMS. SiWave engineers and manufactures MEMS structures with applications in the consumer electronics, biomedical, and communications markets. These precision mechanical devices are built utilizing a customized semiconductor fab.

MEMS technologies have already revolutionized the automotive industry with airbag sensors and the printing sector with ink jet nozzles, and are on track to do the same in medical devices, photonic switches for communications, and mobile phones. In-Stat/MDR forecasts that the $4.7 billion of MEMS revenue in 2003 will grow to $8.3 billion by 2007. But progress is constrained by the pace (and cost) of the semiconductor equipment industry, and by the long turnaround time for fab runs. Microfabrica in Torrance, CA, is seeking to overcome these limitations to expand the market for MEMS to 3D structures in more materials than just silicon and with rapid turnaround times.

Many of the nanotechnology advances in storage, semiconductors, and molecular electronics can be improved, or in some cases enabled, by tools that allow for the manipulation of matter at the nanoscale. Here are three examples:

- *Nanolithography:* Molecular Imprints is commercializing a unique imprint lithographic technology developed at the University of Texas at Austin. The technology uses photo-curable liquids and etched quartz plates to dramatically reduce the cost of nanoscale lithography. This lithography approach, recently added to the ITRS Roadmap, has special advantages for applications in the areas of nano-devices, MEMS, microfluidics, optical components and devices, as well as molecular electronics.
- *Optical Traps:* Arryx has developed a breakthrough in nano-material manipulation. They generate hundreds of independently controllable laser tweezers that can manipulate molecular objects in 3D (move, rotate, cut, place), all from one laser source passing through an adaptive hologram. The applications span from cell sorting, to carbon nanotube placement, to continuous material handling. They can even manipulate the organelles inside an unruptured living cell (and weigh the DNA in the nucleus).

- *Metrology:* Imago's LEAP atom probe microscope is being used by the chip and disk drive industries to produce 3D pictures that depict both chemistry and structure of items on an atom-by-atom basis. Unlike traditional microscopes, which zoom in to see an item on a microscopic level, Imago's nanoscope analyzes structures, one atom at a time, and "zooms out" as it digitally reconstructs the item of interest at a rate of millions of atoms per minute. This creates an unprecedented level of visibility and information at the atomic level. Advances in nanoscale tools help us control and analyze matter more precisely, which in turn, allows us to produce better tools.

The Biological Bottom-Up Path

In contrast to the top-down path, the biological bottom up archetype is grown via replication, evolution, and self-assembly in a 3D, fluid medium. It is constrained at interfaces to the inorganic world and limited by learning and theory gaps (in systems biology, complexity theory and the pruning rules of emergence). Progress is bootstrapped by a powerful pre-existing hierarchy of interpreters of digital molecular code.

To elaborate on this last point, the ribosome takes digital instructions in the form of mRNA and manufactures almost everything we care about in our bodies from a sequential concatenation of amino acids into proteins. The ribosome is a wonderful existence proof of the power and robustness of a molecular machine. It is roughly 20 nm on a side and consists of only 99 thousand atoms. Biological systems are replicating machines that parse molecular code (DNA) and a variety of feedback to grow macro-scale beings. These highly evolved systems can be hijacked and reprogrammed to great effect.

So how does this help with the development of molecular electronics or nanotechnology manufacturing? The biological bootstrap provides a more immediate path to nanotechnology futures. Biology provides us with a library of pre-built components and subsystems that can be repurposed and reused, and scientists in various labs are well underway in re-engineering the information systems of biology.

For example, researchers at NASA Ames are taking self-assembling heat shock proteins from thermophiles and genetically modifying them so that they will deposit a regular array of electrodes with a 17 nm spacing. This could be useful for patterned magnetic media in the disk drive industry or electrodes in a polymer solar cell.

At MIT, researchers are using accelerated artificial evolution to rapidly breed T4 bacteriophage to infect bacteria in such a way that they deposit materials on semiconductors with molecular precision.

At IBEA, Craig Venter and Hammy Smith are leading the Minimal Genome Project. They take the *Mycoplasma genitalium* from the gut, and strip out 200 unnecessary genes, thereby creating the simplest organism that can self-replicate. Then they plan to layer new functionality on to this artificial genome, such as the ability to generate hydrogen from water using the sun's energy for photonic hydrolysis.

The limiting factor is our understanding of these complex systems, but our pace of learning has been compounding exponentially. We will learn more about genetics and the origins of disease in the next 10 years than we have in all of human history. And for the minimal genome microbes, the possibility of understanding the entire proteome and metabolic pathways seems tantalizingly close to achievable. These simpler organisms have a simple "one gene: one protein" mapping, and lack the nested loops of feedback that make the human genetic code so rich.

Hybrid Molecular Electronics Example

In the near term, there are myriad companies that are leveraging the power of organic self-assembly (bottom up) and the market interface advantages of top-down design. The top-down substrate constrains the domain of self-assembly.

Based in Denver, ZettaCore builds molecular memories from energetically elegant molecules that are similar to chlorophyll. ZettaCore's synthetic organic porphyrin molecule self-assembles on exposed silicon. These molecules, called multiporphyrin nanostructures, can be oxidized and reduced (electrons removed or replaced) in a way that is stable, reproducible, and reversible. In this way, the molecules can be used as a reliable storage medium for electronic devices. Furthermore, the molecules can be engineered to store multiple bits of information and to maintain that information for relatively long periods of time before needing to be refreshed.

The technology has future potential to scale to 3D circuits with minimal power dissipation, but initially it will enhance the weakest element of an otherwise standard 2D memory chip. The ZettaCore memory chip looks like a standard memory chip to the end customer; nobody needs to know that it has "nano inside." The I/O pads, sense amps, row decoders and wiring interconnect are produced with a standard semiconductor process. As a final manufacturing step, the molecules are splashed on the wafer where they self-assemble in the pre-defined regions of exposed metal.

From a business perspective, the hybrid product design allows an immediate market entry because the memory chip defines a standard product feature set, and the molecular electronics manufacturing process need not change any of the prior manufacturing steps. The interdependencies with the standard silicon manufacturing steps are also avoided given this late coupling; the fab can process wafers as they do now before spin coating the molecules. In contrast, new materials for gate oxides or metal interconnects can have a number of effects on other processing steps that need to be tested, which introduces delay (as was seen with copper interconnect).

For these reasons, ZettaCore is currently in the lead in the commercialization of molecular electronics, with a working megabit chip, technology tested to a trillion read/write cycles, and manufacturing partners.

Generalizing from the ZettaCore experience, the early revenue in molecular electronics will likely come from simple 1D structures such as chemical sensors and self-assembled 2D arrays on standard substrates, such as memory chips, sensor arrays, displays, CCDs for cameras, and solar cells.

Intellectual Property and Business Model

Beyond product development timelines, the path to commercialization is dramatically impacted by the cost and scale of the manufacturing ramp. Partnerships with industry incumbents can be the accelerant or albatross for market entry.

The strength of the IP protection for nanotechnology relates to the business models that can be safely pursued. For example, if the composition of matter patents afford the nanotechnology startup the same degree of protection as a biotech startup, then a "biotech licensing model" may be possible in nanotech. For example, a molecular electronics company could partner with a large semiconductor company for manufacturing, sales, and marketing, just as a biotech company partners with a big pharma partner for clinical trials, marketing, sales, and distribution. In both cases, the cost to the big partner is $100–300 million, and the startup earns a royalty on future product sales.

Notice how the transaction costs and viability of this business model option pivot around the strength of IP protection. A software business, on the other end of the IP spectrum, would be very cautious about sharing its source code with Microsoft in the hopes of forming a partnership based on royalties.

Manufacturing partnerships are common in the semiconductor industry, with the "fabless" business model. This layering of the value chain separates the formerly integrated functions of product conceptualization, design, manufacturing, testing, and packaging. This has happened in the semiconductor industry because the capital cost of manufacturing is so large. The fabless model is a useful way for a small company with a good idea to bring its own product to market, but the company then has to face the issue of gaining access to its market and funding the development of marketing, distribution, and sales.

Conclusion

While the future is becoming more difficult to predict with each passing year, we should expect an accelerating pace of technological change. We conclude that nanotechnology is the next great technology wave and the next phase of Moore's Law. Nanotechnology innovations enable myriad disruptive businesses that were not possible before, driven by entrepreneurship.

Much of our future context will be defined by the accelerating proliferation of information technology as it innervates society and begins to subsume matter into code. It is a period of exponential growth in the impact of the learning-doing cycle where the power of biology, IT, and nanotechnology compounds the advances in each formerly discrete domain.

So, at DFJ, we conclude that it is a great time to invest in startups. As in evolution and the Cambrian explosion, many will become extinct. But some will change the world. So we pursue the strategy of a diversified portfolio, or in other words, we try to make a broad bet on mammals.

SEMICONDUCTOR SCALING AS A MODEL FOR NANOTECHNOLOGY COMMERCIALIZATION

George Thompson, Intel

The semiconductor industry has successfully made the transition from Microelectronics to Nanoelectronics, and it makes an interesting case study for what may occur as other industries also make the transition to nanoscale technology. The transition from microelectronics to nanoelectronics is a direct result of the synergistic cost and performance improvements that result from the reduction in size of the basic building blocks of modern semiconductor devices. This dramatic improvement in cost and performance has made powerful computers widely available and was the direct cause of the "Information Revolution." This revolution caused major changes in our business systems, culture, and economic structure with complex impacts on society. The widespread availability of databases with high-speed access at low cost has opened up new opportunities for business and research. The widespread availability of computers for most of the population has changed the way we work, play and communicate; it has also created the "Digital Divide" where those without access to the technology are at a disadvantage [1, 2]. The impact of the Information Technology revolution and the implications of the pervasiveness of modern computers are so widespread that they have been the subject of detailed treatment elsewhere [2].

The potential for future scaling of semiconductor nanoelectronics is critical to both the semiconductor industry and the wider economy. The dramatic historical trends in both cost reduction and increased performance as a result of making the semiconductor devices smaller appear to be sustainable well into the foreseeable future. Cost reductions are primarily the result of more cost-effective wafer-processing equipment, materials, and the ever-increasing economies of scale realized through larger factories. These dramatic increases in manufacturing efficiencies more than offset those processes, such as lithography or substrate growth, which tend to have increasing manufacturing costs. Scaling could also in principle be interrupted by technological limits to the scalability of the current technology [3]. Technical limits have indeed been predicted several times in the past and changes in the technology were made in order to maintain performance and size scaling. Future technology enhancements and changes made in the manufacturing methodologies to support continued scaling in the future may be even more dramatic than those that have occurred in the past. It is possible that in the future technologies in other industries, such as pharmaceuticals, catalysis, and materials, may follow electronics and undergo a similar revolution in cost and performance, and trigger major changes analogous to those caused by the revolution in the electronics industry.

The Transition from Microelectronics to Nanoelectronics

The transition from microelectronics to nanoelectronics happened at Intel in the summer of the year 2000, as Intel's logic products, with transistor lengths of less than 100 nm, began to be produced in high volume. Modern integrated circuits can be economically produced in high volume today only as a result of the careful control and manipulation of atomic level properties to obtain unique performance characteristics that critically depend on size. Controlling the electrical performance of transistors by controlling the physical and chemical structures on a nanometer scale has long been a fact in the semiconductor industry. Well-known processes such as epitaxial layer growth and sidewall spacer formation are examples of nanoscale processing. The nanotechnology tradition continues with the recent development of strained silicon and silicon-on-insulator technology. The physical structures of current developmental transistors are also decreasing in size, as shown by the "Tri-Gate" transistor in Figure 1.18 and the nominally 60 nm transistor shown in Figure 1.19. Experimental transistors have been made with 10 nm gate lengths.

Various self-assembly schemes are also being investigated in several universities for future applications in lithography, thin film dielectrics, and transistor gate formation. While many industries are now charting their course as they enter the nanotechnology era, the semiconductor industry is already in the nanotechnology era, and never turning back.

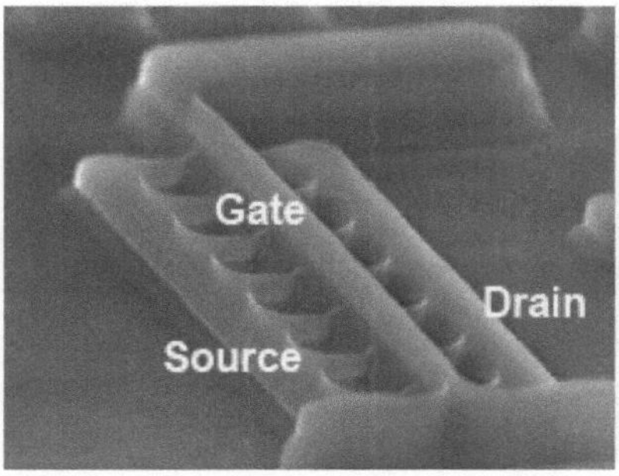

Figure 1.18. Tri-gate Architecture (courtesy of Intel).

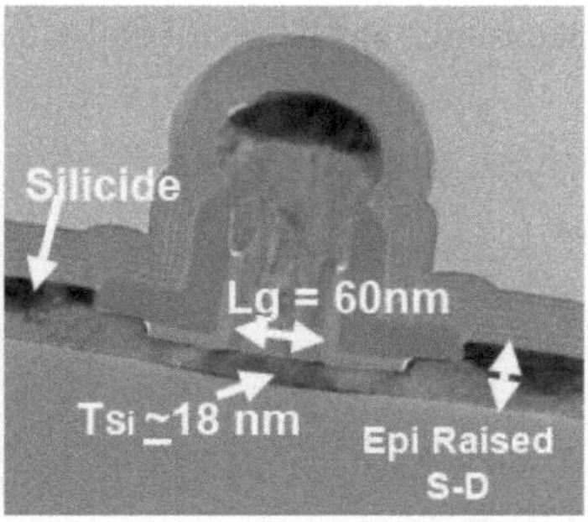

Figure 1.19. Experimental 60 nm device (courtesy of Intel).

Moore's Law as the Driver to Nanotechnology

This synergy justified massive investments in research, development and new factories filled with the latest equipment. The dramatic trend in the reduction in transistor size was first predicted by Gordon Moore, and has since been referred to as Moore's Law. Figure 1.20 shows the exponential increase in the number of transistors in time, which is a direct result of the decrease in the size of the individual transistors. (There is little doubt in the industry that the 1 billion transistor milestone will be reached by 2007, if not sooner.) Figure 1.21 shows the corresponding reduction in cost per transistor. This dramatic increase in

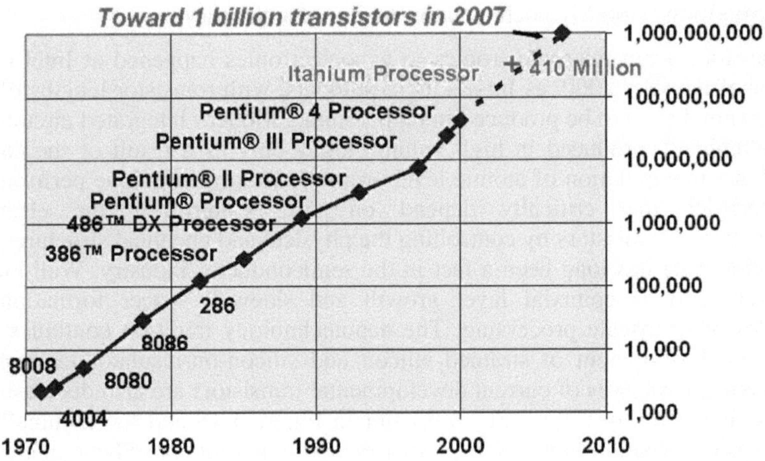

Figure 1.20. Moore's Law: The number of transistors per chip of Intel lead products by year (courtesy of Intel).

performance and decrease in cost has been passed on to consumers, as evidenced by the remarkably low prices of high-power home PCs that are available today from any major computer supplier.

The transition of the semiconductor industry from microelectronics to nanoelectronics has been driven by the industry mantra "cheaper, smaller, better."

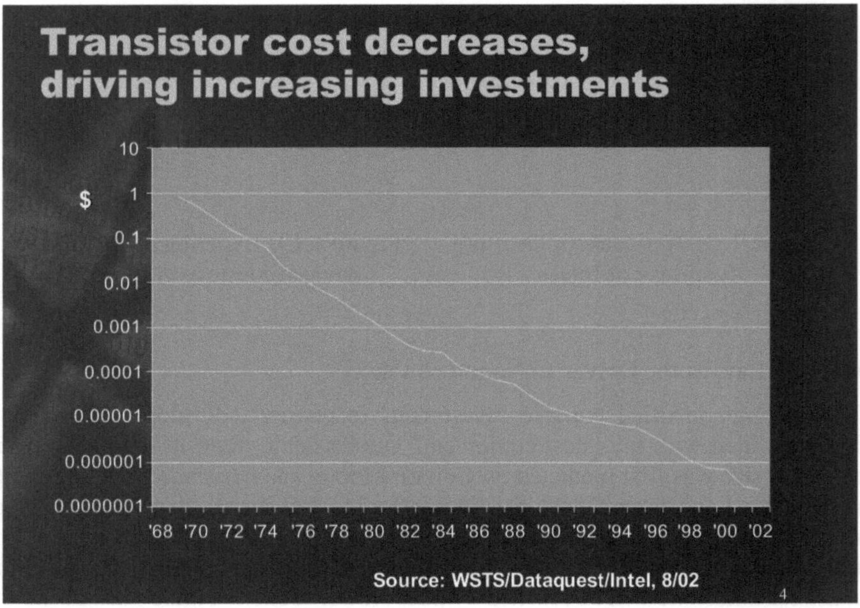

Figure 1.21. Transistor cost trend, a long-term 25 to 30 percent cost reduction per transistor per year.

In 1990 transistors had gate lengths on the order of 1000 nm, and by the year 2000 this had been reduced to less than 100 nm. This 10-fold reduction in size in a decade is a direct result of the synergy that exists between decreasing size and cost with increased performance. This synergy justified massive investments in research, development and new factories filled with the latest equipment. The dramatic trend in the reduction in transistor size was first predicted by Gordon Moore, and has since been referred to as Moore's Law. Figure 1.20 shows the exponential increase in the number of transistors in time, which is a direct result of the decrease in the size of the individual transistors. (There is little doubt in the industry that the 1 billion transistor milestone will be reached by 2007, if not sooner.) Figure 1.21 shows the corresponding reduction in cost per transistor. This dramatic increase in performance and decrease in cost has been passed on to consumers, as evidenced by the remarkably low prices of high-power home PCs that are available today from any major computer supplier.

It has been argued that Moore's Law is not sustainable as a result of cost factors. This has been coined by some as "Moore's Second Law." Even if such an economic argument for the limits of scaling had some validity, it was never advocated by Dr. Moore and should not bear his name. The arguments for an end to the cost scaling of semiconductors have generally been based on either the potential increase in the cost of particular process modules, such as lithography or wafer fabrication, the cost of ever increasingly complex designs, or meeting reliability requirements. It has also been argued on the basis of the increasing capital cost of new factories. The semiconductor industry has been aware of these cost risks, and managed them aggressively by increasing the capacity of the highest-cost process tools in order to maintain the critical cost-per-device, in part by the transition to larger silicon wafers. There has also been a trend to build larger factories to achieve greater economy of scale. These developments, although technologically challenging, maintained the historical trends in cost reduction. Intel's current state of the logic process has a production cost, per transistor, of less than one-third that of the previous generation processor.

The trend in wafer cost is supported by Intel internal data. Factories, or "Fabs" constructed in 1993 cost approximately $0.9 billion and could produce typically 20,000 200 mm wafers per month. Fabs constructed 10 years later, in 2003, cost approximately $3 billion. This large increase in Fab construction cost has alarmed some, but the increase in cost to build a new Fab must be taken in context. New Fabs typically produce 40,000 wafers per month, and the wafers are 300 mm in diameter. The cost of each Fab has increased by a factor of 3, but the cost of the Fab, when adjusted for its production volume of total square centimeters of silicon per month, has actually decreased. This strategy has been successful so far, although its impact on low-volume technology manufacturing has not yet been fully comprehended. There is no clear evidence that this strategy for managing the total cost will fail in the foreseeable future.

Technical factors are often raised as potential limiters to Moore's Law. Several potential limits to scaling have been proposed, lithographic resolution and device stability in the presence of background radiation were often suggested as fundamental limits to the construction of devices of less than 100 nm. Solutions to these challenges were found without any impact on the scaling trend. Eventually fundamental limits to the scaling of charge-based devices will be reached as a result of fundamental quantum mechanical effects associated with the Heisenberg uncertainty principle [4]. These limits are not expected to become significant within the next 15 years. These limits may require a change in the technology of the computing devices in order to maintain the general scaling trend in computational power described by Moore's law.

The scaling of nanoelectronics is a result of the synergy that exists between the performance and the cost of the product. Smaller devices are both cheaper and have higher performance. The increased performance enables the integration of additional features into the product, which in turn enables new functionality. The increase in functionality creates new applications and markets. This in turn justifies the massive industry investment in research and development, and new, more efficient factories.

Information Technology and the Creation of Wealth

The Information Technology (IT) industry has been hailed as a major creator of wealth, innovation, and high-quality American jobs [5]. While the positive aspects of the creation of wealth are clear, the risk of the creation of a "Digital Divide" has become apparent as businesses or individuals who cannot use modern information technology in their daily routines run the risk of professional and perhaps even social marginalization, as more aspects of commerce, industry, and communication become IT-based. The economic impact of the IT revolution has been described in detail in an excellent Department of Commerce report authored by David Henry, et. al., in 1999 [6]. In the period from 1987 to 1996, the value added of the semiconductor industry to the U.S. economy moved from 17th place to first place. In the five years ending in 1998, the contribution of IT and the communications-based industries to the U.S. GDP increased from 16 to 24. This was done while maintaining a highly competitive employee compensation strategy. The average salary increase for all industries from 1989 to 1997 was approximately 35 percent, while for the IT-producing industries the increase was over 50 [6].

In addition to the general increase in the efficiency of industries that adopted IT-intensive strategies, another major impact of the IT industry has been its contribution to the low level of inflation during the 1990s in the United States [6]. The low rate of inflation in the United States, and the subsequent lowering of interest rates, was one of the factors in the major economic expansion that occurred in the 1990s.

Conclusion

The semiconductor industry is unambiguously a leader in nanoscale technology. This position of leadership has been achieved by a synergistic business model that is often described as "smaller, faster, cheaper." This technological revolution provided the physical infrastructure that enabled the Information Technology revolution, which has resulted in the creation of wealth, a reduction in inflation, and high-quality employment for many. The impacts from the transition to nanotechnology in the semiconductor industry are a result of the overall Information Revolution that has occurred, and the attached references are just an introduction to a wide body of literature on the subject.

This dramatic trend in size and cost reduction has been the result of numerous technological innovations and the ability of the increased functionality of the devices to create new valued-added applications for the technology. This has generated the cash flow needed to maintain the investments needed for additional innovation and size reduction. There is no evidence found to support the notion of an inherent cost constraint to scaling, often described by the misnomer Moore's Second Law. Technological changes may be required in the future in order to insure the continuation of scaling, and it is a possibility that in the next 15 to 20 years, a transition from charge-based devices to computing-based on another physical parameter may occur.

In addition to the direct implications from the development of nanoelectronics technology, there is also the possibility that other technologies, such as pharmaceuticals, catalysis and materials, may make a similar transition to nanotechnology with the concurrent synergistic improvements in performance and cost. This would drive the technological transition forward quickly as massive investments are made to exploit the "smaller, faster, cheaper" business model, as happened in the semiconductor industry. In such a case these industries may also create similar increases in wealth, standards of living and quality of life for those societies that embrace the new technology. If, on the other hand, the new nanoscale industries do not arrive at a "smaller is better and cheaper" strategy, then those nanoscale technologies, while possibly creating significant technical advancements, will be less likely to revolutionize major aspects of our economy and society.

References

1. R. S. Rosenberg, *The Social Impact of Computers*, Burlington, MA: Academic Press (1997).

2. J. McConnaughey et al., *Falling through the net:* D*efining the digital divide, a report on the telecommunications and information technology gap in America*, Washington, DC: U.S. Department of Commerce (July 1999).

3. V.V. Zhirnov, R. K. Cavin, III, J. A. Hutchby, G. I. Bourianoff, Limits to binary logic switch scaling—A gedanken model, Proceedings of the IEE*E* **91**(11), (November 2003).

4. T. Fukushima, ed., International technology roadmap for semiconductors, (2003) (available at http://public.itrs.net/).

5. R. J. Damuth, *America's Semiconductor Industry: Turbocharging the U.S. Economy*, San Jose, CA: Semiconductor Industry Association (1998).

6. D. Henry et al., *The Emerging Digital Economy*, Washington, DC: U.S. Department of Commerce (June 1999).

NANOTECHNOLOGY AND ZETTABITS: IMPLICATIONS OF INFORMATION ANYTIME AND EVERYWHERE

Ray Tsui, Motorola

To set the stage for discussions, let us define nanoscale science and technology (Nano S&T) as the knowledge and engineering of working at the sub-100-nm length scale. Within this broad context, Nano S&T encompasses many diverse subjects and undoubtedly will have significant impact on numerous aspects of society, both in the near term and over the coming decades.

One area where there will be tremendous impact is on "information" in the broad sense of the word, perhaps forever changing how information will be collected, stored, processed, distributed, and handled. The Internet as we know it today is only the very beginning. Imagine a world where information on anything can be accessed anywhere, utilized and/or modified in a variety of ways, and then stored and/or disseminated with no reasonable limits on processing speed and storage capacity. Smart sensors will collect and transmit the information, which will be viewed in highly portable displays, and stored and handled in ultra-dense and super-fast data processing devices that are equally portable, all made possible by Nano S&T. And the information available will be wide-ranging. It can encompass the world of business, the environment, arts and sciences, governments, as well as personal data on finance and health.

One may argue that it will require many years of research, development, and commercialization before society will be faced with such a scenario. However, we should bear in mind that the information proliferation need not be via electronic means only. Nano S&T will give us light-weight materials to build more efficient and less expensive vehicles as well as alternate energy sources that are cost-effective, making it possible for more people to travel to more places than ever before. Information will be carried back and forth by these travelers, and one can imagine the impact in particular for people from less developed regions or from countries with more restrictive governments.

It is obvious that access to such a wealth of information raises questions related to privacy, copyrights, the well being of individuals, organizations, and communities, and national security, as well as productivity and equity. But it will introduce other questions also. What if we have certain information and do not know how to act on it? An example could be detecting a disease at a very early stage but having no cure for the individual. Will it change the way humans communicate? And last but not least—can the human mind adapt to handle and sort through such a wealth of information?

Impact on Productivity and Equity

Given the earlier definition of Nano S&T (i.e., sub-100-nm length scale), the semiconductor integrated circuits (IC) industry is already part of a nano-based economy. Recent advances in scaling have resulted in more powerful data processing and storage capabilities, with a lower cost per bit of information processed or stored than before [1]. One would surmise that this contributed towards an increase in the productivity of skilled individuals and the organizations that employ them. However, it has not prevented the occurrence of a major slump in the high-tech industry in the last three years. Other economic forces and world events would appear to have a dominating effect still.

The case can be made that an overriding positive impact of Nano S&T on the economy would not occur until a truly disruptive scenario is commercialized through advances in R&D. This can come in the form of a new manufacturing process that is of *significantly* lower cost for one or more existing products, or the introduction of a new capability/market not previously available. In their recent book, *The Innovator's Solution*, Christensen and Raynor have advocated for the former case of "disruptive innovation" in particular [2]. As for the latter case, the emergent convergence of nanotechnology and biotechnology may provide the type of breakthroughs in health care and life sciences that can generate new capabilities. Examples include early disease detection (at the molecular level), microscopic implantable augmenters, and eventually in-vivo therapeutic agents.

Whether in an incremental or revolutionary manner, advances in Nano S&T will undoubtedly give us more powerful computers as well as highly sophisticated sensors and displays. This will drive an escalating increase in the generation and availability of information. The implication on productivity could be quite significant because having the appropriate information immediately will increase the efficiency of carrying out the task at hand. The key, of course, is getting the *right* information. In fact, being flooded with too much information of which the majority is irrelevant and having to sort through it all can potentially cut down on productivity. This suggests that the increase in information accessibility will stimulate and generate new businesses geared towards the handling, processing, and distribution of the generated information (e.g., data mining, real-time analysis, ultra-wide-band communication, etc.). In turn, there could be corresponding shifts in wage/profit equity from manufacturing to these other areas, since the former is becoming more efficient and less dependent on skilled workers.

At a broader level, some other points to consider that are related to the increasing quantity of information include the following:

- Will information become accessible to all equally, or will the "Information/Knowledge Divide" become even wider? This is more than an access issue for underserved populations. The role of education is also critical since the information is only useful if it can be properly understood and appropriately utilized to improve productivity or the quality of life.
- Will readily available information on Nano S&T-based manufacturing allow competing or new industries to start up and prosper in less-developed regions and countries that hitherto lacked the resources and know-how to do so?
- How will we balance individual versus public versus government control and processing of the ever-increasing amount of information?

Elsewhere in this volume, other authors have ably addressed these issues. Instead of duplicating the information in this narrative, the interested reader can find many insightful discussions there.

Finally, it is necessary to mention the importance of developing better metrics to measure productivity. Even for the more mature technology of computerization, the conclusion on its effect on productivity is not certain. It is generally accepted that Nano S&T is crosscutting in nature, and will impact many areas and products without being obvious to the end user in many instances [3]. This will make meaningful measurements even more challenging. Research funding to develop more insightful approaches is highly recommended.

References

1. G. Thompson, Semiconductor scaling as a model for nanotech commercialization. Presentation made at NNI Workshop on Societal Implications of Nanoscience and Nanotechnology, 3–5 December, Arlington, VA (2003).
2. C. M. Christensen, M. E. Raynor. *The Innovator's Solution: Creating and Sustaining Successful Growth*, Cambridge, MA: Harvard Business School Press (2003).
3. M. Roco, Nanotechnology. Presentation made at NNI Workshop on Societal Implications of Nanoscience and Nanotechnology, 3–5 December, Arlington, VA (2003).

SUSTAINING THE IMPACT OF NANOTECHNOLOGY ON PRODUCTIVITY, SUSTAINABILITY, AND EQUITY

Louis Hornyak, University of Denver

Sustainability, in the context of this paper, implies the traditional environmental meaning of the word, but is also intended in the economic sense. For example, how long can a company remain within the community from which it was germinated, gestated, and born and provide productivity and socio-economic equity that exceeds short-term profitability? How does one predict the effect of nanotechnology, a certainly disruptive and influential technology, on future productivity and equity in an economy that is controlled by powerful forces? Will nanotechnology simply map onto pre-existing infrastructure, one that defines current American economic practice? Or is nanotechnology intrinsically capable of altering paradigms in such a way as to encourage responsible change and longevity? I propose a three-pronged agenda to ensure that nanotechnology "hangs around" beyond the initial luster and polish phases by: (1) practicing sustainable business development, (2) forming strategic economic clusters and partnerships and (3) educating and creating a highly trained workforce.

According to a Battelle Press publication entitled *Sustainable Development: A Business Perspective* [1], author Joseph Fiksel states that "following a decade of globalization and increasing concerns over security, human rights, and environmental stability, the need for foresight and equity in the satisfaction of human needs is more important than ever." [1, p.4] He states further that a policy invoking a "Triple Bottom Line" is required:

- Economic prosperity and continuity for the business and its stakeholders
- Social well being and equity for both employees and affected communities
- Environmental protection and resource conservation, both local and global

Let's begin by addressing the first item in the above agenda. Whether perceived as good or bad, environmental protection and resource conservation concerns often are regulated, legislated-out in favor of business development, or simply fail to acquire feasible market drivers that make environmental concerns practical in the short-term. Fiksel proposes that "nanotechnology, in particular, may enable what once seemed an impossible dream—products that deliver the same functionality with a resource footprint two or three orders of magnitude smaller," [1, p.9] and that "companies that ignore sustainability may find themselves displaced by the entry of new, more agile competitors. Some have delayed adoption of sustainability because they perceive it as a threat similar to more stringent environmental regulations, and are adopting a 'wait and see' attitude. Ironically, the real threat may come not from regulatory agencies but from competitive challenges in the marketplace" [1, p.9]. Nanotechnology, in addition to its vast economic potential, also has the potential to make sustainable business practical.

If nanotechnology, a market-driven enterprise, can play a role in promoting environmental protection and resource conservation while creating long-term equity towards the "satisfaction of human needs," then it will be the single-most important technological breakthrough ever, making one think of the promises of nuclear energy. Nobel prize winning scientist Richard Smalley states that one of the most important challenges of the 21st century is creating enough energy to supply the needs of a burgeoning population [2]. According to Smalley, nanotechnology will be key in mitigating impending energy demands. A business that incorporates sustainable affordable energy should be able to contribute significantly to productivity and equity. By creating businesses that are built on profitable but sustainable models (in the environmental sense), the probability of long-term sustainability (in the economic sense) of nanotechnology productivity and equity increases. Many indicators concerning nanotechnology certainly support this scenario [3]. Will a nanotechnology company consider relocation if sustainable business practices that serve community well being are strongly entrenched in the fabric of the businesses and the community? Is sustainable practice alone enough to keep businesses from relocating elsewhere, especially overseas?

Moving on to the second agenda item, if one were to evaluate the three dimensions listed above as they apply in today's world, issues associated with economic prosperity always draw the most attention. On the other hand, social well being and equity, with jobs going overseas and pensions in danger of being minimized, have diminished significantly in the past decade. The two dimensions are inextricably entwined. On the economic front, will nanotechnology just simply become "business and government as usual" or will nanotechnology be able to, due to its unique intrinsic qualities and accelerated fanfare, influence the way business and government are accomplished? The answer to the second component involving governmental aspects has already been addressed at the federal level—

in the affirmative. Nanotechnology has already influenced federal perspectives on how to launch a new technology by stressing the importance of the societal impacts of a new technology (like never before) [4].

On the other hand, the "business as usual" paradigm is paradoxical as business is always subject to innovation. Changes in the way business is accomplished have been demonstrated over and over during the past century with the advent of, for example, mass production and computer management. As a result of these innovations, American economic productivity and equity proceeded unparalleled. Nanotechnology is capable of affecting the way business will be accomplished, not just by the effects of new products, but by the way business itself will be done. However, since globalization is not expected to abate anytime soon, what can nanotechnology do to boost productivity and equity in the United States and keep it here? Ironically, just by doing "business as usual" will certainly accomplish that goal for the short-term. Any kind of economic development is, after all, economic development and will bolster all the proper short-term indicators.

The promise of nanotechnology lies within its capability to stimulate a solid manufacturing base, a base that has been eroded over the past decade, and its reliance on forming academic, business and government partnerships (like never before). How does one go about bringing nanotechnology clusters to the marketplace? Larry Bock, CEO of NanoSys, inc., stated that nanotechnology will be driven in large part by academia [5]. Does this statement and others like it imply that productivity will, as a result, be increased if these components are in place, and that socio-economic equity, hand in hand, will be spread across a broader landscape due to the partnerships? The answer is yes, depending on how long we can hold on to it: i.e., how sustainable (in the economic sense) is our nanotechnology-derived productivity and equity? Building a solid manufacturing-based nanotechnology cluster characterized by strong inter-institutional relationships with facilitated intellectual property/ technology transfer/ patenting mechanisms in place, would make relocation or buy-outs more difficult.

The roles of local economic development councils (EDC) and local government are vital ones. Since EDCs understand the economic makeup of their district, they must play a major role in putting the cluster together and maintaining viability. But, what if a mega-multinational corporation tried to buy-out the key company that serves the cluster with the intent to move it overseas? Would they have to buy out the entire cluster? Do the citizens of the EDC district have a stake in the cluster? Part ownership? Own their own equity? Once again, economic sustainability has to be more than short-term profitability. Would this kind of hijacking be more difficult to accomplish if a strong viable cluster is in place? But money does talk and for a cluster to retain its status, it must offer better economic incentives, plain and simple.

The third item on the proposed agenda concerning the high-tech workforce is a critical one. Specialized economic clusters and infrastructure built from the ground up, required for nanotechnology to happen, including development of a highly trained workforce akin to Dr. Mary Ann Roe's "Gold Collar Worker" [6], should make nanotechnology stick for the long term. Part of any cluster would require a qualified workforce that can innovate and problem-solve [6]. According to experts [2], there is a brain drain in the United States that is likely to threaten the future of nanotechnology. Academic institutions, whether research or educational, that are involved intricately with an economic cluster will help maintain and retain the economic viability of the community by research dollars, providing expertise and the workforce. The importance of EDCs, once again cannot be understated. Academic entities are often members of governing councils of EDCs.

Economic prosperity and continuity for the business and its stakeholders *should* provide social well being and equity for both employees and affected communities. Environmental protection and resource conservation based on nanotechnology *should* provide for economic prosperity and continuity. The promise of nanotechnology *should* be able to influence positive outcomes by developing cluster-based manufacturing partnerships that will be difficult to relocate by possessing intrinsic sustainable characteristics (both senses). Dr. Daniel Chiras, President of Sustainable Systems Design, Inc., states that intra-generation equity is a good thing, but we must also consider inter-generation sustainability both in the environmental sense as well as in the economic sense [7]. Chiras also promotes a strategy that rewards and enriches preexisiting companies (or economic clusters) by providing incentives to them rather than to big companies that are considering relocating into an EDC's district. Aside from stressing the infrastructure by increasing the population, reducing tax base and living with the risk that the "big company" will relocate, perhaps overseas, what happens to the cluster? Hopefully, the cluster will have a well-thought out strategy in anticipation of such an event and that 1, 2, and 3 are all strongly in place.

References

1. J. Fiksel, *Sustainable Development: A Business Perspective*, Columbus, OH: Battelle Press (2002).
2. R. Smalley, Our energy challenge. Public Lecture presented at Low Library, Columbia University, 23 September, New York (2003).
3. D. Mulhall, *Our Molecular Future*, Amherst, NY: Prometheus Books (2002).
4. M. Roco, W. S. Bainbridge, eds., *Societal Implications of Nanoscience and Nanotechnology*, Dordrecht, Netherlands: Kluwer (2001).
5. L. Bock, Thinking Small, *Forbes/Wolfe Nanotech Report,* (March 2003).
6. M.A. Roe, Cultivating the gold collar worker, *Harvard Business Review* **79**(5), 32-33 (May 2003).
7. D.D. Chiras (President, Sustainable Systems Design, Inc.), Private Communication (November 2003).

NON-NANO EFFECTS OF NANOTECHNOLOGY ON THE ECONOMY

Richard Freeman, Harvard University

Economic analysis has three messages about the effects of new technologies on market outcomes, and thus on the likely impact of nanotechnology on U.S. economic performance and the American job market.

Effect on Labor

The first message is that labor, rather than capital or natural resources, has been and is likely to remain the main beneficiary of technological change. By technological change, I mean improvements in the mode of production that reduce costs of production by using less inputs to produce a given output or by allowing the same inputs to create a higher-quality product, and inventions of new products that meet consumer or business demands better than previous products. A good example of cost-reducing technology improvements in the mode of production is Henry Ford's assembly line, which reduced the cost of automobiles. A good example of the introduction of a new product would be the invention of the daguerreotype in the mid-1800s. In practice, the distinction between whether a given technological change should be classified as a new product or service or as a cost-reducing improvement in an existing product is somewhat arbitrary. From one perspective, digital photography represents a new product; from another it is simply a higher quality version of the camera that the photography industry has produced for years.

In any case, virtually all models of technological change predict that the main beneficiary of technological improvements is the most inelastic factor of production. In the modern world, where the supply of labor is inelastic relative to the supply of capital and where the supply of economically useful natural resources has increased as a result of the discovery of new sources of natural resources and/or development of substitute resources, this makes labor the main beneficiary of technological advance.

The easiest way to see that inelastic factors gain from technological advance is to assume that goods and services, including new products, are produced by two inputs, capital and labor, and that the economy is sufficiently competitive that any technological advance reduces the prices of those goods and services in the market. Evidence on the relation between productivity growth and price changes across industries supports the assertion that technological advance does indeed show up largely in declines in the prices of industries with productivity growth relative to others [1, 2].

The following price-cost equation presents the argument in simplest form. It relates percent changes in the price of output p', (where ' after a variable denotes percent changes) to percent changes in wages (w'), percent changes in the price of capital (r'), and the rate of technological improvement (t'):

(1) $p' = -t' + a\,w' + (1-a)\,r'$

The rate of technological improvement t' is defined as the percent decline in the average amount of labor and capital used to produce a given amount of goods or services. A technological change that saves on labor and capital proportionately is called a neutral technological change. While the analysis fits the case of neutral technological changes most readily, it applies as well to changes in technology that save more on one input than on the other. In that case, the measure of technological change is the sum of the percentage change in each input per unit of output due to the technology weighted by the factors' share in the cost of production. If labor is a larger share of the cost, a labor saving technological change that reduces labor usage by 10 percent with no effect on capital will have a greater impact on t' than a comparable technological change that reduces capital usage by 10 percent with no effect on labor usage.

The parameter a is the share of labor in the cost of production. When the share of labor in cost is high, percentage changes in wages have a larger impact on the percentage changes in prices than when the share of labor in costs is low. The parameter 1-a is the share of capital in the cost of production.

Finally, since factor supplies depend on the real returns to factors, note that the real wage is the nominal wage divided by the price, w/p, and that the real return to capital is the cost of capital divided by the price, r/p.

The key assumption of the analysis is that capital is more elastic with respect to its real return than labor is responsive to the real wage. Indeed, over the long run the supply of capital appears to be nearly infinitely elastic at a given real rate of return, (r/p)*. With the real return to capital fixed at (r/p)*, the price of capital changes at the same percentage rate as the price of goods and services: r'= p'. In some periods of time, the real return to capital may rise; in other times it may fall, but in the long run the real return to capital hovers around (r/p)*. Assume for simplicity that the supply of labor is completely inelastic with respect to the real wage. Then it is easy to demonstrate that labor is the beneficiary of technological change. Substitute p' for r' in (1) and rearrange terms to obtain:

(2) $w'-p' = t'/a$

Technological change that improves productivity and lowers cost by t percent raises the real wage by a multiplier effect dependent on labor's share in cost. This does not mean that real wages increase more than labor productivity, measured by output per unit of labor. The rate of technological change t' is defined as the savings in labor per unit of output and in capital per unit output. If technological change is solely labor saving and measured by output per unit of labor, then t' = a multiplied by the change in labor per unit of output due to the labor-saving technology. This yields the familiar result that changes in the real wage equal changes in labor productivity as long as the share of labor in national output is fixed.

Assuming a fixed coefficient technology, Simon [3] applied this model to both labor-saving technologies and capital-saving technologies and showed that it does not matter whether technological change is biased toward using less labor or less capital. Going back further, classical economists used the same mode of analysis to develop a very different picture of the effects of economic growth on the economy. They believed that labor was the elastic factor due to Malthusian population responses to incomes while land was the inelastic factor. In that case, the model implies that land gains from technological changes while labor does not [4]. Club of Rome concerns over the economic effects of limited natural resources in the 1980s rest on a similar belief: that those resources are the inelastic factor while labor is elastic [5, 6].

The facts, however, support the generalization that labor is the most inelastic factor input and thus the main beneficiary of technological progress. Real interest rates vary from decade to decade, but show no long-term historical trend [7]. Government bonds paid the same in the latter half of the 1900s as in the 1800s as in the 1700s. Interest rates were not markedly different in biblical days than today. These observations imply that the supply of capital is highly elastic over the long run, so that we can increase capital/labor ratios without paying capital an increased rate of return. By contrast, real wages have trended upward over time. The demographic transition has proven that Malthusian fears that higher wages lead people to produce more children and reduce wages to subsistence were misplaced. Similarly, the real price of natural resources has not risen over time, contrary to the fears that natural resources are the inelastic factor whose scarcity might reduce economic growth and gain a larger share of GDP [8, 9].

How does nanotechnology fit into this analysis? As a new driver of technological progress, nanotechnology should have similar impacts on the well being of citizens over the long run as previous technological advances. It should raise real wages and living standards. To the extent that nanotechnology advances improve or create better materials, moreover, nanotechnology should reduce the danger that natural resources will be an inelastic factor that limits economic growth. It is this aspect that makes nanotechnological advances potentially more advantageous than advances that might use more natural resources and more friendly to the environment than many other possible technological changes.

What about concerns about the effect of nano-induced technological progress on employment? Will nanotechnology create jobs or destroy them?

Technological revolutions are invariably associated with changes in the nature of work and thus in the composition of employment by industry and occupation. New technologies usually give rise to fears that they will displace labor and create unemployment. During the great "automation scare" of the 1960s, the government was sufficiently worried that automated factories would eliminate jobs that it established the National Commission on Technology, Automation, and Economic Progress to investigate the effects of automation [10]. The fears proved false, as the economy adjusted to automation. Advocates of nanotechnology currently argue that it will create jobs. On the basis of what has happened in the past, however, a better prediction would be that nanotechnology will have no substantive positive or negative impact on the overall level of jobs or on the unemployment rate. Rather, nanotechnology will impact rates of pay and the quality of jobs. Claims that advanced technologies produce or destroy jobs miss the point about how the economy adjusts to technological change. In the long run *technology affects living standards and quality of work, not quantity of work.* If higher living standards induce people to take more leisure, the amount they work will fall (as appears to be the case in most advanced European countries). If the higher real wage that technology creates induces people to want to work more, employment and hours worked may rise (as has occurred in the United States in recent decades). What differentiates successful from unsuccessful societies is not rates of employment but GDP per capita or real wages. Employment is quite high and unemployment rare in some developing countries that have very low living standards.

Effect on Employment

The second message is that technological change does not create proportionately more jobs in the technically advanced industries.

Technological change in a given sector affects the demand for labor in that sector in three ways. First, since a new technology almost always lowers the amount of labor needed to produce goods or services, it will reduce the demand for labor at given levels of output. If output did not change, this would lower employment in the affected sector. This *displacement effect* of technology is what generates fears that new technology will destroy jobs and create technological mass unemployment. Second, however, new technology also increases the demand for products. Productivity-induced reductions in the cost of goods and ultimately prices, as in equation (1), generate greater demand for the product and thus greater demand for labor to produce it. The massive decline in the price of computing power that results from technological improvements has spurred huge increases in the sales of computers. Whether the induced increase in output and demand for labor overpowers the displacement of labor due to the increased productivity depends on the elasticity of the demand for the product. When prices fall, consumers increase their purchases massively for products with a high elasticity of demand but increase purchases only modestly for products with a low elasticity of demand. On the assumption that elasticities of demand are greater for new products than for existing products, I expect that technological changes that create new products are more likely to increase demand for the labor in the relevant sector than technological changes that reduce the price of existing products.

For the past 50 years or so, the displacement effect of technological change has exceeded the employment-increasing effect of expansion of production. Employment declined in manufacturing and agriculture, where technological change is most rapid, and shifted toward services, where technological change is modest. Within manufacturing, employment has shifted to the more technologically advanced sectors, and from less technologically advanced to more technologically advanced goods, but this has not been sufficient to counterbalance declining employment elsewhere in manufacturing. Still, there has been no adverse effect on overall employment nor on the quality of jobs. In the United States, though not in Scandinavian countries where central union federations and employers' groups determine wages, the wages of workers in sectors that have rapid productivity advances tends to rise relative to wages of comparable workers in other sectors [2].

Since nanotechnology is likely to become part of the way many industries operate—a general purpose technology that affects everything, analogous to electricity or IT—rather than a technology that creates a new industry, it will take a long time to affect labor demand throughout the economy. Given the dominance of the displacement effect, successful nanotechnology-initiated improvements in production are likely to raise U.S. manufacturing output while reducing employment in manufacturing or at least without creating many new jobs in manufacturing. Nanotechnology should be seen as a productivity-enhancing technology that permeates the economy, not as a job-creating technology. Its effect on employment will be indirect, as productivity-induced higher real wages and living standards generate additional consumer demand for all sorts of products and services and thus contribute to the growth of employment.

The third way new technology affects demand for labor is in its use of particular factors of production. New technologies invariably use a different mix of factors than older technologies. During the development of a new technology, firms are likely to have high demands for the scientists, engineers, and technicians who have to build and integrate the new ideas into processes and products. In addition, there is need for supporting labor services, which creates job opportunities for other workers. Even the most high-tech industry hires many persons in sales, clerical and office work, and employs blue collar workers and service workers of different types. As the technology matures, demands are likely to shift toward workers with lesser skills. In the past 30 years or so new technologies have been skilled-labor-using, which has contributed to the increased earnings of educated workers relative to less educated workers, despite the increase in the relative supply of the educated.

Given that nanotechnology is a general technology, I am skeptical that it will have any large impact on the relative demand for skilled as opposed to unskilled workers beyond its initial impact on the demand for science and engineering workers skilled in nanoscience and engineering. Studies of the impact of computer technology on demand for skills indicate that only a modest part of the rise of inequality in the United States in the 1980s and 1990s can be attributed to computerization [11]. Nanotechnology is likely to have even less effect on labor market inequalities. This is because most of us need not become literate in nanotechnologies any more than we are literate about computer chip design.

Effect on Productivity

The third message from economics is that *the impact of technology on the economy depends on the share of cost of the final good that the technology influences*. The larger the share of cost of the inputs affected by the new technology, the greater the economic gains from it. Nanotechnology improvements in materials that make up 5 percent of the cost of a product will have a smaller economic benefit than improvements in materials that make up 10 percent of the cost. Since nanotechnology is likely to affect lots of sectors in small ways rather than affecting one sector massively, I expect nanotechnology to have relatively smooth impacts on the economy, rather than to cause substantial upheavals in production and employment. Nanotechnology improvements in one sector will lead to increased output, lower prices, and will alter labor demands in that sector. Different improvements in another sector will cause comparable changes. With enough sectors and improvements, nanotechnology will boost overall productivity without dislocating production or employment. In any case, adjustments will occur over relatively long periods. The economy took a long time to adjust to electricity and is still adjusting to computerization and the Internet. In 1987, Robert Solow said "we see computers everywhere but in the productivity statistics"[12, p.36]. Now we see computerization in the productivity statistics. I anticipate a similar long-time delay before nanotechnology affects the productivity statistics.

The fly in the ointment in these expectations is the potential that, shades of Eric Drexler [13], some nanotechnology-related advance could create new products that would shake up the structure of the economy and greatly change the structure of employment. Indeed, given that scientists tend to be conservative in their assessments of future changes and expect major breakthroughs to take more time to develop than in fact they have done historically, it is probably safe to predict that there will be some such breakthrough product in the next decade or so, associated with nanotechnology broadly defined. But I would still expect the economy to take two or so decades to adjust to any breakthrough product and for the effects on the allocation of the workforce to be modest rather than massive.

Conclusion

The bottom-line message from economics is to think of nanotechnology not as a job-creating technology in manufacturing but rather as a productivity-enhancing technology that will operate throughout the economy. Given worldwide competition in advancing nanotechnology, with the European Union, Japan, and China as well as the United States supporting large research investments in the area, the key policy issue is much more about how the United States can effectively develop and use the technology to improve well being than in worrying about its impact on the economy and employment. Research has shown that with the same technology and comparably skilled workers, U.S. firms have done a better job in adjusting to IT and using it productively than EU firms [14], which suggests that the U.S. edge is organizational, associated with startup businesses and business practice more than with science and engineering research per se. Consistent with this, in nanotechnology, while the United States has a sizeable proportion of scholarly papers, it has a larger proportion of patents. One way to maintain this edge in using new technologies would be to introduce nanotechnology courses into MBA programs and into shorter executive education programs. This will alert firms to new opportunities, as well as add some exciting science and engineering into business curriculum. In addition, since new products are likely to have a larger impact on the structure of the economy than new processes, policymakers should pay special attention to identifying those nanoscience advances that are most likely to engender brand new products and industries and to seek ways to give firms incentives to think about new products as well as processes. The agglomeration of economic activity, whereby firms tend to buy products from other firms more in the cities and states where they are located, remains significant despite globalization, and the country that succeeds in linking nanotechnological advances to traditional production and services, is likely to be the main winner in the nanotechnology revolution. But I reiterate, the payoff will be in higher living standards and wages and in the quality of employment, not in the number of jobs generated.

References

1. W. E. G. Salter, *Productivity and Technical Change*, 1966 edition, Cambridge, UK: Cambridge University Press (1966).
2. L. Bell, R. Freeman, The causes of increasing U.S. inter-industry wage dispersion, *Industrial and Labor Relations Review* (January 1991).
3. H. Simon, *The Shape of Automation for Men and Management*, New York: Harper & Row (1965).
4. T. R. Malthus, *An Essay on the Principle of Population*, London: J. Johnson (1817).
5. D. H. Meadows et al., *The Limits of Growth: A Report to the Club of Rome on the Predicament of Mankind*, New York: Universe Books (1972).
6. D. H. Meadows, D. L. Meadows, J. Randers, *Beyond the Limits*, Post Mills, VT: Chelsea Green Publishing (1992).
7. S. Homer, R. Sylla, *A History of Interest Rates,* New Brunswick, NJ: Rutgers University Press (1996).
8. P. Linder, Long-term trends in America's farmland values, *Agricultural History* **62**(3), 45-85 (1988).
9. W. Nordhaus, Lethal model 2: The limits of growth revisited, *Brookings Papers on Economic Activity* **2,** 1-59 (1992).
10. National Commission on Technology, Automation, and Economic Progress, *Technology and the American economy*, Washington, DC: U.S. Government Printing Office (1966).
11. J. Bernstein, L. Mishel, Six reasons for skepticism about the technology story of U.S. wage inequality, in I. Berg, A. Kalleberg, eds., *Sourcebook of labor markets: Evolving structures and processes*, Dordrecht, Netherlands: Kluwer (2001).
12. R. Solow, We'd Better Watch Out, *New York Times Book Review* (12 July 1987).
13. E. Drexler, *Engines of Creation*, New York: Anchor (1987).
14. S. Basu, J. Fernald, N. Oulton, S. Srinivasan, The case of the missing productivity growth: Or, does information technology explain why productivity accelerated in the US but not the UK? NBER Working Paper No. W10010 (October 2003).

2. SOCIAL SCENARIOS

NAVIGATING NANOTECHNOLOGY THROUGH SOCIETY

Davis Baird, University of South Carolina

Obstacles, Opportunities

Work on the societal and ethical implications of nanotechnology faces major obstacles. But, at the same time, these obstacles present significant opportunities to re-conceptualize how societal and ethical concerns can work with the many and various forces that drive technological change, in particular nanotechnological change.

There is no single definition of what nanotechnology is [1], and, whatever it is, nanotechnology is not singular. Among other things, nanotechnology embraces work on molecular electronics, novel materials, and biomedical technologies that run from new systems for drug delivery to work on the human/machine interface [2, 3]. Each individual area presents its own special set of problems and concerns to those thinking about the societal and ethical implications of nanotechnology [4, 5]. The definition itself—or the struggle over its specifications—poses societal and ethical issues. The definition affects the distribution of resources from federal grants and other sources. At a deeper level, nanotechnology challenges traditional disciplinary boundaries, and exactly how these boundaries are breached may have far-reaching consequences for the various institutions involved in the development of nanotechnologies, and indeed for the final form of the technologies developed.

The fact that nanotechnology runs roughshod over disciplinary boundaries provides powerful momentum toward transgressing boundaries between C.P. Snow's "two cultures" [6], and it is these boundaries that have kept societal and ethical deliberations about new technologies from constructively engaging the development of these new technologies.

Our time is characterized by a fundamental and deep paradox. On the one hand, developments in information technology, transportation technology and the rise of powerful multinational corporations, give genuine meaning to the oft-repeated slogan that we live in a "global village" [7, 8]. The environmental protection, progress on unraveling the interconnected complexities of global climate change, and even the images that space exploration has provided of the earth floating in the vast abyss of space, powerfully bring home the point that we all share a finite planet, "lifeboat earth" [9].

While we may all live in a global village, we do so in increasing disciplinary isolation. C. P. Snow lamented two disconnected cultures, a scientific culture and a humanities culture [6]. If anything Snow's concerns have ramified. We now have a dizzying multiplicity of sub-sub-sub cultures where we seem only to engage fellow travelers in our own special interest group. Identity politics has

M.C. Roco and W.S. Bainbridge (eds.),
Nanotechnology: Societal Implications — Individual Perspectives, 75–130.

replaced the politics of the melting pot [10, 11, 12, 13], and one can get the impression that there is little engagement between sub-cultures beyond each jockeying for a larger piece of the pie. In the sciences, the drive toward specialization and the multiplicity of technical jargons that has come with specialization have sharply limited the ability of researchers to engage across disciplines—let alone engage with those in the humanities and social sciences.

So we live in a time when we have some awareness that we are "in this together" and we need to work together to create a better world for all—as well as avoid some international political, military, or climate-related disaster. At the same time, we have sharply diminished skills for multi-disciplinary and multi-cultural engagements, and it is exactly these skills that we need to make progress with the multi-dimensional issues that socio-technological change produces.

The good news is that there is widespread and growing recognition of this situation. Consider the following from the National Science Foundation's Integrative Graduate Education Research and Traineeship (IGERT) program: "The program is intended to catalyze a cultural change in graduate education, for students, faculty, and institutions, by establishing innovative new models for graduate education and training in a fertile environment for collaborative research that transcends traditional disciplinary boundaries" [14, p.1]. Nanotechnology lies at the intersection of a variety of disciplines—chemistry, physics, biology and several engineering disciplines. There is a growing understanding that new patterns of interdisciplinary engagement will be necessary to realize the potential nanotechnology presents. Again, the calls for nano-related research proposals issued by the NSF demonstrate the agency's awareness of this situation.

In addition to numerous science and engineering disciplines, nanotechnology also fundamentally involves the humanities and social sciences. These disciplines, in their different ways, work to provide a better, more socially contextualized understanding of socio-technological change. Armed with a better understanding, we are better able to manage socio-technological change, and in particular those transitions that nanotechnologies will prompt. Thus, in addition to making demands on the traditional science and engineering disciplines to engage and work with each other, nanotechnological progress also makes demands on those who have concerned themselves with the societal and ethical implications of technology to work with each other, and to work with those in nano-science and nano-engineering. Nanotechnology presents an opportunity to invent a new and better way for ethical and societal concerns to inform and be informed by scientific, engineering, commercial, political, and geo-political concerns.

Post-hoc Versus Therapeutic Ethics

Consider the motto of the 1933 Chicago World's Fair, "Science Discovers, Technology Applies, Man Conforms" [15]. How strange and jarring this sounds today—and in so many different ways. The motto concisely captures the seductive, but historically unsupportable notion of change uni-directionally flowing out from the Olympian brains of scientists to the forge of technology and thence to a compliant society. But the piece that jars is this last bit, society—or

"Man"—conforms. As the 1933 World's Fair had it, society was simply a receptacle for the fruits of technology. Society was not even a choosy consumer, let alone an active contributor to the process of technological change.

This was an era of technological enthusiasm [16]. The World's Fair was devoted to "a century of progress" [17]. Society did not grudgingly conform, society gobbled up the amazing fruits of technological change. When I was 10 years old, at my grandmother's funeral in 1964, I was struck by the minister's relating my grandmother's sense of awe in having lived from 1882 and the age of horse-and-buggy to space exploration. This was indeed a century of progress.

Times change. Rachel Carson's Silent Spring [18], Three Mile Island, Chernobyl, the Challenger and Columbia accidents all have done their part to promote a more reflective attitude toward technology. Now we have environmental impact statements; we have careful investigations into what happens when O-rings freeze. When bad things happen we want experts to tell us why and we want other experts to develop analytical tools that we can apply to proposed technologies to see if they will be bad, too. Disaster has been the springboard for the desire for societal and ethical assessment.

The waning of society's naïve technological enthusiasm, and the consequent rise in the demand for societal and ethical assessment, have occurred at the same time as the waxing of disciplinary isolation that Snow was vexed by. I doubt this simultaneous waning and waxing is anything but coincidence, but it is a coincidence with consequences. The ethical assessment of technology has been conceived something like this. The ethicist, like an expert consultant in any other area, uses a conceptual probe—an "instrument"—to determine whether some social structure tied to a technology is good or bad. To do this, the social structure and its allied technology need to be fairly well developed. Instantiating Snow's "two cultures," the ethicist from one culture examines and, with the help of the conceptual probe, passes judgment on the efforts or products of scientists and engineers working in their other culture. I call this *post hoc* analysis.

Post hoc analysis is unsatisfactory. This picture of "expert ethical analysis" assumes that ethicists are inherently moral in a way that scientists and engineers are not—an assumption both false and insulting. The insult further entrenches a cross-disciplinary antagonism and entrenches the mutual dismissal Snow was bothered by. Under such conditions ethicists cannot work *with* scientists and engineers, but at best examine scientific and engineering work on completion and *in situ*. But analyzing established technologies or technologies at a point where much research and development has already been invested in them is too late. Change—"for the better"—is much more costly late in the game. "Technological momentum" [19, 20] is a real phenomenon that can only be ignored by those doing societal and ethical assessment at risk of being irrelevant.

In light of these problems with *post hoc* analysis, the efforts to frontload work on the societal and ethical implications of nanotechnology are important. The effects of the National Nanotechnology Initiative to promote and support serious efforts to think about these issues, in co-evolution with the technology itself, demonstrates an awareness of the need to frontload work on the societal and ethical implications of nanotechnology [21, see also http://www.nano.gov].

Such co-evolution makes two demands on this work. First it needs to start now, while the space of possible nanotechnological development is open and informed choices can be made unencumbered by a large load of technological momentum. Second, this work needs to be pursued in a multi-disciplinary manner where the genuine prospects of nanoscience and nano-engineering inform societal and ethical reflection, and vice versa, serious societal and ethical scholarship informs the science and engineering.

Multiple Points of View

Multiple points of view will be necessary to fully consider the societal and ethical implications of nanotechnology. Consider the following—no doubt, incomplete—list:

Technical drive: J. Robert Oppenheimer spoke of an "organic necessity" to tracking the technological promise of solutions that were "technically sweet" [22, p.317; 23, p.1]. He used these terms when speaking of Teller's solution to building a hydrogen bomb. Fully aware of the significant social consequences of this work, he acknowledged a kind of drive from our human curiosity with the technology itself.

- *Economic drive*: Technologies develop in economically competitive environments. If one company—or one country—hesitates because of, for example, concerns over potential toxicity, a competitive advantage is lost. This drive works within the United States, and between U. S. firms and those of Japan, the European Union, and other countries, any one of which would be very happy to get a competitive advantage in nanotechnology. Jobs, access to new and/or less expensive consumer goods, and broad indices of health and wealth are tied to economic competitiveness.
- *Ethical norms*: There are individual, social and political norms concerning what we should and should not do. The "prudential principle," which has gained some currency in bioethics [24, 25; see also http://www.biotech-info.net/precautionary], urges us not to take unnecessary risks. The demands of fairness urge an even distribution of risks and benefits, and our democratic system of government demands that citizens have a voice in the decisions and policies of the government.
- *Limited resources*: Public and private funds are always in limited supply. Intellectual capital is in limited supply. Skills are in limited supply. All of these limits place constraints on how and how fast nanotechnology can develop. Of course decisions have to be made to juggle limited resources. These decisions are made in Washington; they are made by venture capitalists and by managers of large corporations with interests in nanotechnology. Researchers make choices on how to spend limited research hours and limited research funds.
- *Facts*: Facts on the ground have a kind of force. The fact that we now have carbon nanotubes, that we can produce them in good quantity, and that they

have very interesting properties, has been very important for the development of nanotechnology. Carbon nanotubes provide a partial down payment on all the grand speculative claims that have been made for nanotechnology. Facts present a kind of existence proof that is a powerful antidote to pure speculation. And *where* the facts lie has a powerful influence on *where* limited resources are expended.

- *Contextualized facts*: Who discovered carbon nanotubes? Why were they sought? Who is likely to benefit from them, and who is not? Why have we funded research to establish certain facts and not other facts? Again, who benefits and who loses? How were the decisions made to pursue certain facts rather than other facts, and who made these decisions? All of these issues play central roles in how a new technology develops, and if we want to understand and better manage the development of nanotechnology, we need to have a good handle on the contexts surrounding the emerging facts of nanotechnology.
- *Politics*: There is little doubt that governments worldwide are playing major roles in pressing the nano-research and development agenda. This means politics, and politics takes the scientific and technical facts from their safe well-controlled homes and subjects them to very different kinds of discussion and action. Hyperbole, simplification, and re-contextualization are political tools. How various publics—correctly or incorrectly—understand the facts of nanotechnology have political power in a democracy.
- *Forces of tradition*: Humans remain creatures of habit, and frequently these habits are codified in cultural, religious, or historical patterns of behavior. These habits can act as a kind of sea anchor that slows socio-technological change. All socio-technological change confronts an already existing socio-technological world. Change requires new habits and behaviors to displace old habits and behaviors, and this produces social turbulence as new and old habits struggle with each other, the more rapid the change, the more turbulent the struggle. If one adds a significant redistribution of opportunities and risks, which typically attends socio-technological change, one gets more turbulence as losers struggle to maintain position.

Metaphors

Clearly socio-technological change presents a highly complex system. It is no surprise that we do not have a good track record predicting how new technologies will fare. It would be nice if we understood or could predict—or could control—any one of these "forces" individually. It would be nice if we could perform some kind of "vector addition" to get a handle on their joint effects. It would be nice if our attempts to understand one or another of these "forces" didn't itself immediately change the "force."

Alas, social "forces" simply do not operate like physical forces. Socio-technological systems are complex and highly interactive. Humans are reflective creatures and will change their behavior in light of newly constructed conditions. Measuring humans changes humans in ways that can undermine the meaning of the measurement. This is most easily seen in our attempts to assess education. As we test, teachers learn to "teach to the test" and students learn to "learn to the test." Our attempts to assess education end up redefining what it means to be educated. If our means of assessment do not allow us access to every dimension of human knowing—and surely they cannot—then we end up with an impoverished, yet instantiated, understanding of what it means to be educated [26]. Similarly, as we come to understand a "force of technological change," insofar as that force works through reflective humans, our understanding will feedback and this will change the force.

But this talk of "force" is wrong in the first place. We are working with the wrong metaphor, imagining a field of forces, and bemoaning our inability to capture and control their joint effects. A better metaphor might see this as a complex system with ourselves as elements, interactively engaging in its development. We should think of the management of socio-technological change more like a town meeting or a political primary caucus. We come together and engage each other in a transforming exchange of information, beliefs, attitudes, hopes, fears, visions of how things could be better, or how they might be worse. Leadership matters, and this requires the ability to articulate a vision of constructive change that embraces a broad spectrum of hopes for the future, while honestly and expeditiously responding to fears. But this demands understanding, which in turn demands listening. The right, town-meeting metaphor sees socio-technological change as engaged social communication, not engaged physical manipulation.

Two aspects of socio-technological change, locus of control and process versus product, are fundamentally different viewed from this town-meeting metaphor. How we *should* think of controlling a communicative system is not the same as how we *do* think of controlling a techno-physical system. With techno-physical systems our ability to reliably predict, or even better create, the response of the system to "provocations" of the system is a central goal. This is what lies behind the empirical adequacy of "laws of nature," and this is what is necessary for some newly engineered product to "work" [26]. Fundamentally, this picture has "us" outside, attempting, in our representations and our interventions, to predict and control some part of the world that is "inside." Social systems do not work this way, and this is why interpretive and self-reflective (or "hermeneutic") methods have always been an unavoidable part of social science and humanist research. We don't control a communicative system so much as participate in it.

Since we cannot extract ourselves from the system, and since we cannot hold the defining human/social variables of the system constant, we are not in good position to conceive ideal *products* of the system. This would be to revert to the physical force metaphor. Instead we need to focus on *process*, and how different *processes* can promote or undermine fundamental human ideals. Constructive change requires giving all the stakeholders voices in the process. It requires a space where a diversity of points of view can be "put on the table" and considered for their merits. It requires sufficient time and energy for evidence to support the truth. It requires attention to perceived and/or real conflicts of interest, and it requires steps to mitigate such conflicts. At the same time, interested parties frequently are in the best position to articulate a position; advocacy, if subjected to analysis, can find truth.

This alternative metaphor should clarify why I see the obstacles that face work on the societal and ethical implications of nanotechnology as opportunities. We have become aware that we need to do something different in our dealing with socio-technological change. My diagnosis is that part of the problem here has to do with a ramification of Snow's isolated two cultures, and this has gone hand-in-glove with the wrong metaphor for understanding and controlling socio-technological change. We are at a point where a new town-meeting metaphor of engaged social communication, breaching the kinds of cultural boundaries Snow described, has an opportunity to bring about constructive change in how we as a society deal with nanotechnologies and the other technologies that will play important roles as we move into the future.

We need to create multiple and wide channels of communication. We all need to take the time to learn to understand the multiple points of view that bear on how nanotechnologies may improve or degrade our futures. As we learn to appreciate multiple points of view, we will creatively construct the directions that we take with these technologies. If we adopt a process that allows for every voice to be heard, and that promotes vision, evidence, analysis, and even advocacy, we will jointly create our better nano-socio-technological future.

Current Contexts

As we pursue this work on the societal and ethical implications of nanotechnology, developing these wide channels of communication, there are certain aspects of our current scene that are salient. While none of these are specific to nanotechnology, it is significant that nanotechnology has come on the scene at a time when these aspects are in play, and it is vital that we consider them as we begin to work on the societal and ethical implications of nanotechnology.

Universities, Research, and Technology Transfer

The relationships between the university, government, and the private sector are changing in ways that are designed to decrease the time it takes to get laboratory discoveries into the marketplace. These changes are having fundamental effects on all of the institutions involved. They are changing the nature of university research (and education), and they will have an effect on the technologies developed. For much of the half century after World War II, the United States pursued an approach to funding science and technology that was articulated by Vannevar Bush in his report to the President, *Science: The Endless Frontier* [27, 28, 29]. Bush supported substantial funding for "pure research" with the idea that technological applications would naturally follow. This idea has come into question, and in recent years we have pursued mission-oriented science aimed at technology transfer from inception [29]. The manner in which nanotechnology has been promoted and developed has certainly gone along with, if not accelerated, this trend. It is, after all, a National Nano*technology* Initiative, not a nano*science* initiative. The NNI speaks of the next *industrial* revolution, not the next *scientific* revolution [21].

Public Goods and Private Goods

Closely related to changes in the institutions that are bringing nanotechnology into being are changes in assumptions about public and private goods. What can we as a society reasonably assume should be a public good, shared by all? The distinction between public and private is especially important in respect to the increasingly important category of intellectual property, particularly the intellectual property created at institutions of higher education. When one begins to think seriously about molecular manufacturing, intellectual property becomes that much more important. These changes are having profound implications for the distribution of risks and benefits that come with new technologies.

Nanotechnology and Biotechnology

Nanotechnology follows biotechnology, particularly genetically modified foods and agricultural products, into the marketplace. Genetically modified organisms did not achieve a smooth transition, and there remains significant public resistance, particularly in Europe [30]. This has had all kinds of consequences. First, we are better aware that—and how—the market introduction of new technologies can go badly. Advocacy groups that came into being to provide an avenue for public resistance to genetically modified organisms are now well organized and easily capable of turning their attention to nanotechnology. The ETC Group is a case in point [31, 32]. ETC's early and well-produced position papers on nanotechnology demonstrate that nanotechnology is not going to get any slack as it moves from laboratory to society.

Pace of Change

There is evidence—I won't claim that it is conclusive—that the pace of technological change is itself changing, and that this change in the pace of change is reaching a critical point [33]. Humans are creatures of habit, and as such, resist many changes. Thus, if there is any credibility to this changing pace of change argument, we will have problems. If we move toward a new industrial setting where many production processes are automated, at a general level, we all gain, because of the overall productivity gains. But, this is small comfort to displaced workers. We can try to identify how such changes will impact various segments of society and to take steps to ameliorate ill effects. But an increased pace of change throws a monkey wrench into this approach. The traditionally slower, more reflective examination of societal and ethical consequences may have difficulty keeping up. Here is another reason that we need to pursue a new approach as we examine and shape the societal implications of nanotechnology.

To Do Now

We have to think carefully about time frames. The issues that matter now will not likely be the issues that will matter in five years' time. Issues of immediate concern can and should prompt a focus on specific societal and ethical contents that arise from the current situation. Longer-term, more speculative possibilities demand a focus on process more than position. We are not now in a good position to reach useful societal and ethical judgments about nanobots because we do not know what the world will be like if and when nanobots appear on the scene. But we are in a good position to establish social habits and institutions that are able to consider nanobots if and when they arrive. I am inclined to distinguish three time frames for the current consideration of the societal and ethical implications of nanotechnology: now, or near-term; medium-term, say 3–7 years out; long-term, say 10–20 years out.

Nanoparticle Toxicity and Risk

In the near term, it is nanoparticles that are coming onto the market now, and consequently nanoparticles require attention now. Are they toxic to humans or the environment? Are they appropriately regulated? How are the risks tied to these particles distributed, and how are these risks understood and communicated? Depending on what we learn about these risks, and how we communicate this to the broader publics, the toxicity of nanoparticles could ignite the first widespread resistance to nanotechnology.

We need to do this, and to publicize this work, in a way that is open, honest and engages the many interested publics as fully as possible. Suppose we start producing certain nanoparticles on a mass scale for some product or another, and we find out—after the fact—that they are highly toxic. This would create a disaster on several levels. There would be direct harm to persons and to the environment. There would be the very difficult and unpleasant problem of funding clean-up efforts. And finally there would be public backlash against nanotechnology that would have the potential to affect the whole field. Anyone concerned with the development of nanotechnology should very much want to avoid this outcome.

At the same time, as we begin to use new nanoparticles in commercial products, we should be thinking in terms of the lifetimes of these products. How will they be discarded and what will happen to them? What modes of regulation at the local, state, national, and international level will best help mitigate risks while promoting opportunities—and distribute both risks and opportunities fairly?

Thinking through Societal and Ethical Implications

We should begin serious work on the medium- and long-term ethical and societal implications of nanotechnology. In the medium term—three to seven years out—we will have smaller computers and computer components. There are likely to be more ubiquitous computing and communication capabilities, and these will raise surveillance and privacy concerns. We will have better catalysts, which will raise issues tied to industrial disruption. We will have more developed ways to interact with and manipulate the human body, raising questions about the distinction between therapy and enhancement. There will be insurance implications.

In the long term—10 to 20 years out—we are likely to have more radical human-artifact entanglement—embedded communication and sensory and effector devices. These technologies will force us to examine the question of what it means to be human. The nature of our social, cultural and national bonds will be at issue. Difficult-to-detect nano- and micro-machines will have greater functionality, with possibilities of engaging in lots of mischief, especially in a military context. We may have some form of molecular manufacturing, and this has the potential to drastically alter our understanding of manufacturing and the economic structures that are tied to it. There are likely to be problems with the fair distribution of risks and benefits.

What kinds of privacy concerns does ubiquitous computing raise, and how can we be prepared for them? How will current research impact significant industrial sectors of society—e.g., energy—and what political repercussions will be involved? How will this change the environmental discussion? If, per chance, we develop plentiful and cheap renewable sources of power, what happens to the fundamental environmental concept of sustainability? We will develop more powerful interventions with our bodies, and this will raise questions about what it means to be human, to live in a "cyborg society." We need to think about these implications. We should examine the stated goals of current nano-related research programs. For example, there is talk of "reversing the aging process" in the "NBIC" document [3]. We need to be careful what we wish for. Conquering aging would do much more than produce technical problems of over-population. It would radically impact our understanding of what it means to be human, of how individual and group interests work in concert with or at cross-purposes with each other. The reality of death drives one generation's interests forward to another generation.

While I am fascinated by all of these long-term concerns, particularly the seductive yet troubling idea of conquering aging, and while I think we must now begin to think through these matters carefully, I believe it is much more important that we focus our efforts on institutionalizing the multi-disciplinary discussion of nanotechnology.

Institutionalizing Multi-Disciplinary Engagement

We should act now to figure out how to institutionalize the examination, discussion and action with respect to the medium- and long-term social and ethical implications of nanotechnology. We need to create enduring structures that will encourage cross-disciplinary—indeed cross-cultural—exchange about the socio-technological issues that we will face in the 21st century. We need to develop the skills and habits to think and talk with each other about technology as a form of social change. Scientists and engineers need to engage their colleagues across campus in their work, but equally they need to incorporate some of the points of view of their social science and humanist colleagues into their work. Social scientists and humanists need to take greater notice of the scientific and technological developments that are coming down the pike. And we all need to do a much better job engaging the public and bringing the public's voice into these discussions. This will not be easy work. We have long habits of specialization and dissociation to combat. We won't understand each other at the outset, and we'll have to be willing to struggle in confusion as we move from talking past each other to talking to each other. This work will require sustained encouragement and support at all levels.

Institutionalizing such a multi-disciplinary discussion raises many issues. We need to think through sticky issues of conflicts of interest between funding, research, commercialization, and the ethical examination of nanotechnology. For example, should someone with significant input into how nanotechnology is funded own stock in a company that serves the nano-business sector? What about someone working on the societal and ethical implications of nanotechnology? How should work on the societal and ethical implications both be integrated with—*and yet also have an independent voice about*—scientific and engineering nanotechnology research? How can we best promote constructive discussion between and among all the stakeholders in nanotechnology without stifling voices of restraint and prudence? How can we bring advocacy groups to the table in such a way that the discussion moves towards constructive dissension, not un-engaged antagonism?

Involving Many Publics

We should begin now to develop an appropriate public education program about nanotechnology. Many people will get virtually all of their understanding about nanotechnology from the release of a movie version of Michael Crichton's *Prey*, or from other fiction sources. The entire nanotechnology community—scientists, engineers, policy makers, social scientists, lawyers, journalists, humanists, indeed the public at large—have much to gain from efforts to put into wide circulation better information about the nature of nanotechnology and what its realistic opportunities and risks are. The final discussion at the December 2003 Workshop on Societal and Ethical Implications of Nanotechnology focused almost exclusively on this topic. It clearly is critical, and perceived to be critical.

More than anything else, we need to take advantage of the opportunity presented to us to put into place institutions that will promote and support intellectual and political habits of discussion about the issues that nanotechnology raises that cross disciplinary and cultural boundaries. Only with such new habits in place will we fully tap nanotechnology's potential while avoiding its problems.

References

1. G. Stix, Little big science, *Scientific American* **285**(3), 32-37 (September 2001).
2. M. Ratner, D. Ratner, *Nanotechnology: A gentle introduction to the next big idea*, Englewood Cliffs, NJ: Prentice Hall Professional Technical Reference (2003).
3. M. C. Roco, W. S. Bainbridge, eds., Converging Technologies for Improving Human Performance*: Nanotechnology, biotechnology, information technology and cognitive science.* NSF/DOC-Sponsored Report, Arlington, VA: World Technological Evaluation Center, Inc. (2002).
4. G. Khushf, The ethics of nanotechnology: On the vision and values of a new generation of science and engineering. Unpublished manuscript (2004).
5. M. C. Roco, W. S. Bainbridge, eds., *Societal Implications of Nanoscience and Nanotechnology*, Dordrecht, Netherlands: Kluwer (2001).
6. C. P. Snow, *Two Cultures and a Second Look*, New York: New American Library (1963).
7. M. McLuhan, *Understanding Media*, New York: McGraw-Hill (1964).
8. M. McLuhan, B. Powers, *The Global Village*, Oxford: Oxford University Press (1989).

9. O. O'Neill, Lifeboat Earth, *Philosophy and Public Affairs* **4**(3), 273-292 (1975).

10. J. Butler, *Gender Trouble: Feminism and the subversion of identity*, New York: Routledge (1999).

11. W. Kymlicka, ed., *The Rights of Minority Cultures*, Oxford: Oxford University Press (1995).

12. M. Young, *Justice and the Politics of Difference*, Princeton: Princeton University Press (1990).

13. M. Young, *Inclusion and Democracy*, Oxford: Oxford University Press (2000).

14. National Science Foundation, Integrative Graduate Education Research Traineeship [IGERT] Program Solicitation, NSF 04-550. (2004). (available at http://www.nsf.gov/home/crssprgm/igert)

15. Teich, *Technology and the Future*, New York: St. Martin's Press (1990).

16. T. P. Hughes. *American Genesis:* A *century of invention and technological enthusiasm*, New York: Viking (1989).

17. Chicago Historical Society, The History Files (2004) (available at http://www.chicagohs.org/history). Material specifically pertaining to the 1933 Chicago World's Fair is at "A Century of Progress" (available at http://www.chicagohs.org/history/).

18. R. Carson, *Silent Spring*, Boston: Houghton Mifflin (1962).

19. T. P. Hughes, *Networks of Power: Electrification in western society 1880–1930*, Baltimore: Johns Hopkins University Press (1983).

20. R. Pool, *Beyond Engineering: How society shapes technology*, Oxford: Oxford University Press (1997).

21. National Science and Technology Council, Committee on Technology, Subcommittee on Nanoscale Science, Engineering and Technology, *The National Nanotechnology Initiative: The Initiative and its Implementation Plan,* Washington, DC: U. S. Government Printing Office (2000).

22. K. Smith, C. Weiner, eds., *Robert Oppenheimer: Letters and recollections*, Cambridge, MA: Harvard University Press (1980).

23. D. Baird, Organic necessity: Thinking about thinking about technology, *Techné: Journal of the Society for Philosophy and Technology* **5**(1), 1-14 (2000).

24. D. Appell, The new uncertainty principle, *Scientific American*, 18-19 (January 2001).

25. J. Tickner, C. Raffensperger, N. Myers, *The Precautionary Principle in Action: A handbook*, Lowell, MA: Science and Environment Health Network (2004) (available at http://www.biotech-info.net/precautionary .html).

26. D. Baird, *Thing Knowledge: A philosophy of scientific instruments*, Berkeley: University of California Press (2004).

27. V. Bush, *Science: The endless frontier: A report to the President on a program for postwar scientific research*, Washington, DC: U. S. Government Printing Office (1945).

28. V. Bush, *Endless Horizons*, Washington, DC: Public Affairs Press (1946).

29. Johnson, The End of Pure Science: Science policy from Bayh-Dole to the NNI. Unpublished manuscript (2004).

30. F. Francioni, ed., *Environment, human rights and international trade*, Oxford: Hart Publishers (2001).

31. ETC Group, *The big down: Atomtech—technologies converging at the nanoscale* (2003) (available at http://www.etcgroup.org/ article.asp?newsid 37 5).

32. ETC Group website (available at http://www.etcgroup.org).

33. R. Kurzweil, *The Age of Spiritual Machines*, New York: Viking (1999).

NANOTECHNOLOGY, SURVEILLANCE, AND SOCIETY: METHODOLOGICAL ISSUES AND INNOVATIONS FOR SOCIAL RESEARCH

Jeffrey M. Stanton, Syracuse University

> With the development of... technical advances which made it possible to receive and transmit simultaneously on the same instrument, private life came to an end. Every citizen, or at least every citizen important enough to be worth watching, could be kept for twenty-four hours a day under the eyes of the police and in the sound of official propaganda, with all other channels of communication closed. (George Orwell, *1984* [1, p.205])

Nanotechnology promises numerous innovations in the power, capabilities, and form factors of information and communications technologies. Some of these innovations may enable the development of mobile, networked computer devices that substantially extend audio and video recording capabilities beyond those offered by present-day camera-enabled cell phones and non-mobile, network-connected television camera systems. One might term these new devices experiential data recorders, and their application in education, business, and other areas may have the potential to significantly increase learning, productivity, and organizational effectiveness. At the same time, however, the widespread deployment of sophisticated and unobtrusive experiential data recorders also has the potential to change the nature of routine social interactions, invade personal privacy, and shift balances of power in the workplace and elsewhere.

Tens of millions of cellular telephone subscribers in the United States, Japan, and elsewhere already routinely use telephones equipped with low-resolution, full motion video cameras. Current discussions of cameras built into mobile phones suggest that a few of the issues relevant to experiential data recording have already begun to emerge [2]. In particular, the cameras have created controversy as a result of their use in capturing locker room photographs of unsuspecting subjects, but have also gained plaudits in their role assisting police in the apprehension of criminals. Mobile phone vendors have sold millions of units, and although interoperability is limited at this time, the prospect of exchanging full motion video with other phone users is not far in the future. New devices that extended the capabilities of camera phones would become possible with advances in detection, storage, and battery designs enabled by nanotechnology. Such devices might become even less obtrusive and more powerful than today's camera phones and might allow an unprecedented degree of fidelity and recording length.

The issues raised by such devices could easily serve as the basis of a substantial set of meaningful research programs on societal impacts of nanotechnology-based IT innovations. If one goal of such research is the provision of information and guidance for policy decisions about deployment, regulation, and control of nanotechnology-based innovations, then the research must occur to some extent *in advance* of widespread public distribution and acceptance of the relevant new forms of technology. As one among many variants of nanotechnology with possible, significant societal impacts, the unobtrusive experiential data recorder illustrates the methodological challenges of such prospective research. To conduct prospective research in this area, social researchers would need early access to prototype systems, technology specifications, and willing pools of volunteer participants who could realistically generate a community of interacting users. In the present paper, I use an example of a plausible experiential data recorder to explore three methodological ideas for enabling societal impacts research earlier in the nanotechnology engineering innovation process.

Enabling Nanotechnologies and the Experiential Data Recorder

Nanotechnology promises a wealth of methods for leveraging the unique mechanisms and capabilities of molecular machinery to develop powerful new products and tools. Focusing for the moment on how nanotechnology might contribute to the development of new varieties of information technology, it appears likely that order-of-magnitude advancements could occur in at least four areas:

1. substantial increases in the amount of information that one can store per unit volume through the use of three-dimensional rewritable memory devices [3]
2. computing and communication devices with vastly increased mechanical flexibility, expanded possibilities for unique, unusual form factors, and operability under a wide range of environmental conditions [4]
3. new alternatives for information transmission in addition to standard electronic and electro-magnetic techniques including improvements to photonic techniques already in widespread use [5]
4. significant reduction in the power demands of computer and communication devices based on improvements to conventional batteries and movement towards energy sources with fewer constraints than batteries [6]

The listed citations represent just the tip of the iceberg in most of these areas: Basic and applied research in nanotechnology has already resulted in thousands of research articles and presentations in these and many other areas [7]. Together, these advances will likely provide the basis for a multiplicity of ubiquitous, mobile, and networked devices that perform computing, communication, and data capture functions. Many of these devices may take physical forms entirely unlike present-day PDAs, tablets, smartphones, and wireless laptops. New devices might thus appear relatively unobtrusive, and might easily take the form of "wearable computing" devices such as those under development at MIT's media lab and elsewhere. Such devices might in some cases even be physically implanted within the human body. Simultaneously, however, this new generation of technology would offer capacities and capabilities for input, processing, and output well beyond those available in present-day devices.

To sketch an example that might be plausible within a few years, consider the lowly baseball cap worn by legions of present day undergraduates. Embed some micro-molecular devices for video and audio capture in the cap. Add a removable terabyte data storage module in the little button at the top. Include some firmware woven into the fabric that conducts real-time speech recognition. Use natural language processing to develop a narrative outline of the output of the speech recognition. Package the complete device with software drivers for automatic connection and download of data to one's laptop or desktop computer and market the package for the same price as a few pairs of high-end basketball sneakers.

The resourceful undergraduate who did not wish to take handwritten notes could wear the cap into the classroom and capture and automatically index the lecture, questions, answers, discussions, and everything else that occurred during class time. Use of the cap would not intrude in any way on the conduct of the class: A casual observer might not even notice the difference between the "data-cap" and a regular baseball cap. An enhanced version of the cap might even sense and record some of the student's key vital signs: This data stream could provide useful clues later on concerning when alertness and attention may have flagged. After class the student could share the pre-processed material with friends who couldn't make it to class. Automatic, ad hoc, proximity-based networking would make data transfer between caps simple and quick. The student could also upload the material to a laptop and use the automatically generated outline to generate a study guide and search index. One could query the search index to playback full motion video of the group discussion in which a key concept was clarified. Figure 2.1 shows a mock-up marketing display of the data cap that has been customized with logos for the Syracuse University athletics fan.

This data-cap example represents a benign, perhaps even banal application of nanotechnology to the development of an unobtrusive mobile computing device that is not in the least farfetched given current developments underway at a variety of university and industry labs. The data-cap could enable students who can afford one to improve learning, retention, and application of knowledge obtained in the classroom. Following graduation, the student could also bring the technology and the collected body of information into the workplace. Equipped with a data-cap, the new professional could treat the office and the meeting room as simply the

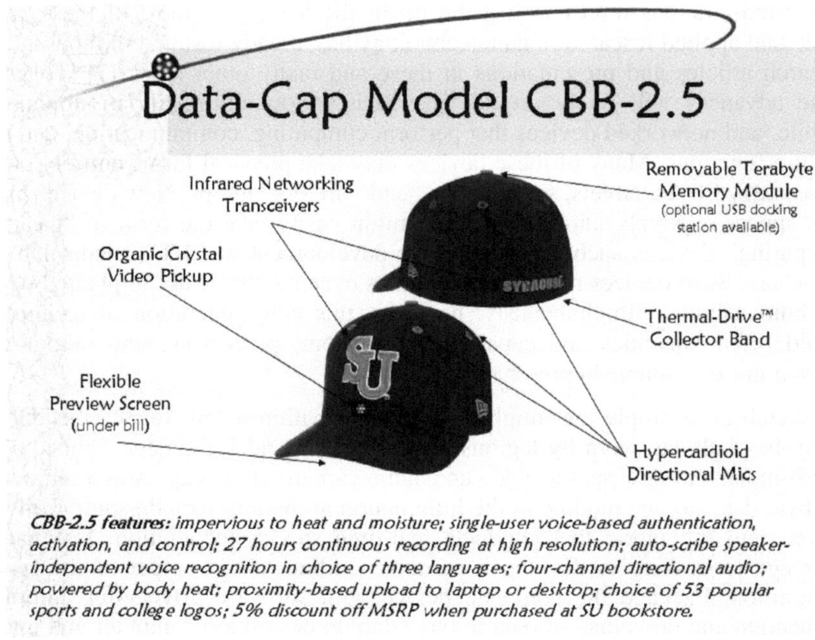

Figure 2.1. Mock marketing display for experiential data recorder.

next step in the computer-aided lifelong learning process. Every new work project would present an opportunity to extend and update a growing data library of experiential knowledge. With careful indexing and organization, one's personal knowledge base could become an important career asset, valuable to the employee, the employee's work team, the current work organization, and future work organizations. Indeed, selective sharing of indexed experiential data might provide a basis for substantially improved productivity for managerial professionals whose primary workplace roles lie in social interactions with colleagues, partners, vendors, and customers.

Let us take the next step, however, and also equip each supervisor, manager, and executive in the work organization with a data-cap (of course, these individuals would wear form factors in styles other than baseball caps). Besides increasing employee productivity, we could enhance personnel security as well. Thanks to the low cost, we can equip every security guard with a data-cap, and we can mount a variety of devices with similar capabilities throughout the workplace. Out on the street, the parking attendant is equipped with experiential data capture, as is the police officer directing traffic in and out of the parking lot. Given the proximity networking capability of each data-cap, we can also continuously monitor the security feeds at an operations center by bouncing data requests and submissions from cap to cap until they reach a cap in close enough proximity to a base station. All of the data collected at the operations center could be cross-correlated and analyzed to provide a complete, semi-permanent archive of each employee's movements throughout the workday providing possible benefits in the areas of productivity, information security, and physical security.

Possible Social and Societal Implications of the Experiential Data Recorder

Widespread deployment of nanotechnology-based experiential recorders and similar devices might provide the basis of an organically growing and shifting network of unobtrusive experiential data capture, analysis, and storage devices. This scenario comprises a simple and plausible extension of the increasingly widespread availability of wireless networks and mobile computing [8]. Research work that could lead to this scenario has already begun. For example, the "Virtual markets and wireless grids" project funded by NSF offers as its goal, "that of an adaptive network offering secure, inexpensive, and coordinated real-time access to dynamic, heterogeneous resources, potentially traversing geographic, political and cultural boundaries…" (available at http://www.wirelessgrids.net/). Just a few engineering steps lie between the present and a data-cap-enhanced future.

The data-cap could offer remarkable opportunities for increasing productivity by serving as an essentially infallible aid to personal memory. As such, the data-cap could eliminate the need for scribbled notes, meeting minutes, memos, and even some types of contractual agreements. The data-cap could facilitate project management, could assist with mentoring and career development of employees, and could help take full advantage of brainstorming and creative tasks by providing a complete and accurate record of generated ideas. In short, for individuals whose lives are replete with social interactions at work and in the home, the data-cap could enable the creation of a huge database of everyday personal interactions that would have a variety of potentially beneficial uses.

On the other side of the equation, however, the capability of making accurate, long playing, high fidelity experiential recordings of many or most personal encounters raises some weighty questions about privacy, security, confidentiality, and the foundational basis of routine human social interactions. Many aspects of human social interactions and relationships have some of their roots in the fallibility of human memory and people's extensive ability to reframe, reinterpret, and restate their past experiences in a new light. Arguably, the long-term success of human groups and societies depends in part on the ability to forget certain sequences of events or at the least to reinvent the memory of the events in a way that is more conducive to mental health and positive social relationships. Anyone who has kept a diary over a period of time may recognize the likelihood of experiencing the "what was I thinking" reaction when reading their recordings of personal experience as words on paper. Imagine the magnification of this feeling with infallible access to past interactions with coworkers, friends, and family members.

As a closing note, one needs to look no further than the present for examples of the potency of widespread recording of human experience. At the moment in Great Britain, the public and the government are already wrestling with the tradeoffs and unexpected consequences of omnipresent recording of personal behavior. Over 800 closed circuit television cameras equipped for continuous recording have been deployed through the streets of London by government authorities. Although the cameras were originally mounted as a strategy for controlling and improving traffic flow, following the events of September 11, 2001, law enforcement and government personnel began to imagine new applications of the cameras to fighting both terrorism and more typical types of crime. In combination with image processing software that can isolate and identify license plates, faces, and other image features, the cameras' data streams have become usable for a variety of tasks besides improving traffic congestion. Privacy advocates have, not unexpectedly, decried the use of the cameras for purposes other than those originally intended. But thousands more of these cameras exist in private hands beyond the easy reach of government controls, and many of these connect directly to the Internet. Further, this issue is not isolated to London: New York City also has hundreds or even thousands of cameras, as do many other U.S. and European cities.

Recently, Jay Walker, the founder of the Internet company Priceline.com, has proposed the deployment of tens of thousands of "webcams" throughout the country at sensitive locations such as reservoirs and power plants [9]. The proposal was made to the Department of Homeland Security, which, at this writing, has not publicly announced the adoption of any strategies using these ideas. Walker suggested that the widespread distribution of inexpensive devices in the hands of untrained operators (i.e., regular citizens) could serve as an effective distributed strategy for the detection and/or prevention of terrorist acts. Whether or not a national government decides to fund and adopt such an idea, it seems likely that innovations in technology and manufacturing will make the underlying technology so inexpensive and easy to use that many corporations and homeowners will adopt it for their own purposes. Given the likely widespread deployment of such technology, researchers and policymakers alike may have substantial interest in understanding the social and societal implications of such technology. In the next section I discuss the methodological issues associated with trying to understand these social and societal implications in advance of the widespread deployment of such technologies.

Methodological Issues for Societal Implications Research

Given the potential of nanotechnology-enhanced computing and communication devices to modify and enhance human capabilities, it seems evident that a variety of opportunities arise for research on the social impacts of these technologies. Technologies that affect the nature, frequency, and setting of social interactions, particularly those in the workplace, at home, and public spaces, appear to have significant potential to change the rules and boundaries of those interactions and, eventually, the essential nature of society. These changes may provide fertile ground for future social impacts research. Note, however, that some of the research that one might envision apparently must occur post facto—after the technology has been deployed in consumer versions that make its availability and market penetration widespread.

The necessity of conducting social impacts research only after a new technology appears on the public scene logically appears to limit the power of that research to influence public policy or behavioral norms in ways that are beneficial to the long-term interests of society. Although some might argue that this is as it should be, in the case of nanotechnology, counterarguments may exist for conducting a greater amount of pre-deployment social impacts research. In particular, recent human history surrounding technological developments such as the automobile and nuclear energy suggest that a greater degree of forethought in advance of widespread deployment may have the potential to offer substantial long-term societal benefits. Assuming this was the case, a need arises for alternative methodological strategies that can provide stronger prospective insights into the effects of a technology. These strategies must somehow overcome a basic set of problems with conducting social research on new technologies.

More specifically, at least three barriers exist for social researchers who wish to conduct research on implications of emerging technologies. First, researchers often lack access to cutting-edge technologies, arguably in part as a result of segregation of academic disciplines into narrow, discipline-focused departments. For instance, many sociologists work from departments in schools of arts and sciences while the cutting edge technology is developed over in the school of engineering. Second, most experimental research is conducted on individuals or small groups, rather than large interacting groups. That is, health and psychological researchers frequently conduct research on human beings where some new element (e.g., a drug or intervention) is introduced to test a hypothesis, but such tests are rarely conducted where the unit of analysis is any larger than a small interacting group. This limits researchers' ability to examine the effects of an intervention on large groups over time. Generally, those who conduct research on large groups do so from a retrospective viewpoint, by examining a group's prior history of attitudes, beliefs, and behavior. To date it has largely been impractical to conduct experimental research on large groups of individuals because of a combination of unlikely logistics and unworkable expenses. Finally, research that is prospective in nature often suffers from a lack of realism. For example, research on social judgment analysis typically requires research participants to respond to some hypothetical situations or choices [10, 11]. Although in some cases these hypothetical stimuli can shed useful light on people's likely future behavior and attitudes, the results must always be interpreted with care because of the artificiality of the research situation and the typical absence of social interaction surrounding the judgments.

To address these three issues, some innovations in social research may be necessary. In general terms it is the technology itself, and in particular the widespread availability of the World Wide Web, that may make some of these innovations cost-effective and feasible. First, to address social researchers' lack of access to emerging technologies, incentives must exist for socio-technical partnership research. Although serendipity sometimes enables fruitful partnerships between engineering researchers and social researchers, a surer way of encouraging these partnerships might focus on including a requirement for such partnerships in funding strategies. Several such partnerships exist at present, but these have primarily resulted from commercial investment and focus on a narrow research agenda. For example, NTT DoCoMo, a Japanese mobile communications company, has recently announced its "Mobile Society Research Institute." The researchers at this institute will examine "both the positive and negative aspect of mobile phones" on society. Those researchers include sociologists, technologists, and scholars from information studies, management, and a variety of other disciplines. Another example manifests in the form of the Stanford Institute for the Quantitative Study of Society (SIQSS), founded in 1999. SIQSS generally focuses on recent and present macro-societal issues by using large databases (e.g., the census database) to examine general societal trends and patterns related to information technology. SIQSS has produced worthwhile research, but intentionally avoids focusing on new types of technology currently in the lab and thus cannot provide prospective evidence about social and societal changes resulting from emerging technologies.

As an alternative to the serendipitous research opportunities that have arisen as a result of private and commercial funding, a more systematic strategy would be to provide incentives for the integration of applied engineering research and social implications research. By providing financial incentives for including psychologists, sociologists, and other social researchers on applied engineering projects, funders could provide a productive setting for social impacts research through early access to prototype devices and technologies, as well as the test facilities used to develop them. Additionally, such requirements might have the benefit of clearly communicating the criticality of social impacts research to the engineering community. The applied research resulting from such funding could achieve two integrated components: advances in the technology plus insights into how the technology may affect people if widely deployed.

Addressing the second challenge, social researchers need access to larger groups of interacting individuals to participate in research. One method of addressing this need lies in the development of "test-market" social interaction communities. Consumer and marketing researchers have long used a strategy of taking a "representative slice" of society as a testing ground for a new product or idea. The benefit of such slices lies in the ability to try out the product or idea on a representative segment of the public in a reasonably realistic social context and at a reduced cost relative to a full-scale national or international test. With the connectivity and speed offered by the Internet, many new possibilities arise for assembling communities of volunteer participants who could interact (virtually) with one another and with new technologies under experimental conditions devised by social impacts researchers.

For example, the StudyResponse project (http://www.StudyResponse.com) suggests one small step in this direction. StudyResponse maintains a database of several tens of thousands of individuals worldwide who have agreed to participate in academic research projects conducted over the Web. Although the projects conducted to date using the StudyResponse project have primarily comprised surveys, the only barrier to deploying more sophisticated research methodologies lies in obtaining appropriate funding. Although researchers have previously constructed a few specialized online research systems to explore particular questions (e.g., price setting: see, e.g., market.econ created at the University of Arizona's Economic Science Lab) these systems either do not support networks of interacting users or are impractical for widespread distribution to enable other types of studies. In short, the tools needed to create and administer communities of interacting individuals have begun to appear in isolated form, but to date few researchers and little funding have addressed the development of volunteer participant communities or their use in large-scale social experimentation.

Given the popularity and potential of the Web as a medium for making connections among large groups of individuals, this area appears to present substantial opportunities for research on societal implications of new technologies that impact social interactions. To the extent that these technologies can be represented in the form of graphical or textual stimulus materials, or audio, video, or text-based interactions among individual users, the Web provides a natural medium for experimentation and analysis [12, 13]. The capabilities of the Web for supporting interacting communities who work together to enact situations or scenarios relevant to social research await further research investment before researchers can exploit them more fully.

On a related note, the third methodological challenge for prospective social research pertains to the realism that researchers can achieve in the scenarios or situations they construct for research participants. Conventional research paradigms use so-called "paper people"—text-based descriptions of individuals and situations—or, in more sophisticated cases use videotaped vignettes in which actors model some behavior that participants must respond to or judge. While the latter is arguably more realistic than the former, both are highly contrived, and both lack the spontaneity that characterizes the events that arise in freely interacting groups. Recently, however, a new phenomenon has emerged, also linked with the burgeoning of the Web, that promises to offer a much greater degree of realism in social experimentation. In particular, the advent of high-fidelity immersive interaction gaming communities has provided a surprising new basis for experimental research on interacting communities. The successes of so-called massively multiplayer online games (MMPOGs) such as EverQuest and Ultima suggest that creating immersive, artificial environments for social interaction can engender enthusiastic participation. Indeed, social scientists have already begun to conduct research on these environments, by recruiting users who have participated in the commercial versions of games [14]. As elsewhere, however, little investment has occurred in making such environments available for social researchers. Borrowing users from existing games that have existed and evolved in unpredictable ways out of the control of researchers provides a poor substitute for a virtual world designed to examine a particular set of research propositions. In combination with the availability of volunteer research communities, the availability of immersive interaction environments for researchers would enable the development of projects that could predict, rather than simply describe the social impacts of new technologies on society at large.

The three challenges to prospective methodological research described above and the suggestions for methods to overcome them represent just the start of an extensive brainstorming process that must involve social scientists and technologists at the same table. Likewise, the numerous ethical questions raised by the prospect of intentionally designed immersive social environments require a considerable degree of further thought and exploration. Many other methodological possibilities exist in addition to those discussed here, and the specific examples proposed above primarily pertain to new types of technology that engender changes to social interactions within workplaces or the community. Other technologies that have less of a direct impact on the fabric of social interactions will inevitably require other methodological innovations. Nonetheless, the same basic problem of prospective versus post hoc research would still be operative in those cases. Effective research on societal impacts of nanotechnology requires methods that can facilitate conduct of research that anticipates some of the impacts before they occur rather than studying them *in situ* or from a historical perspective. Meaningful and credible social research conducted in advance of widespread deployment of nanotechnology-based innovations will provide a more solid basis for informed and judicious policy decisions.

References

1. G. Orwell, *Nineteen Eighty-four, a novel*, New York: Signet (1990).
2. M. Niesse, Phones play new role in crime detection; Police find cell phones' photo-taking capacity useful to track down suspects, *The Washington Post* A19 (21 March 2004).
3. M. E. Thomas, 2D area/3D volume holographic optical storage nanotechnology, *Proceedings of SPIE - The International Society for Optical Engineering* **4323**, 339-349 (2001).
4. M. Ollinger, V. Craciun, R. K. Singh, Nano-encapsulated ZnS:Ag phosphors for field emission flat panel display applications, *Materials Research Society Symposium: Proceedings* **704**, 213-218 (2002).
5. H. Suzuki, A. Yokoo, M. Notomi, Organic emissive materials and devices for photonic communication, *Polymers for Advanced Technologies* **15**(1-2), 75-80 (January/February 2004).
6. R. Tenne, Advances in the synthesis of inorganic nanotubes and fullerene-like nanoparticles, *Angewandte Chemie, International Edition* **42**(42), 5124-5132 (3 November 2003).
7. G. Stix, Little big science, *Scientific American* **285** (3), 32-37 (September 2001).
8. L. McKnight, M. Gaynor, J. Hwang, J. Park, H. Chang, A, Gupta, B. Plattner, J. Howison, P. Aravamudham, O. Uzuner, B. Chen, Wireless grid issues. Presentation at the Eighth Global Grid Forum Building Grids-Obstacles & Opportunities, June 22–24, Seattle (2003).
9. J. Scanlon, Fighting terror with webcams, *Wired* **11**(6), 11 (2003) (available at http://www.wired.com/wired/archive/11.06/start.html?pg=11).
10. G. Gigerenzer, *Adaptive Thinking: Rationality in the real world*, Oxford: Oxford University Press (2002).

11. K. Hammond, *Human Judgment and Social Policy: Irreducible uncertainty, inevitable error, unavoidable injustice.* Oxford: Oxford University Press (2000).

12. J. M. Stanton, S. G. Rogelberg, Using Internet/Intranet Web Pages to collect organizational research data, *Organizational Research Methods* **4**, 200-216 (2001).

13. J. M. Stanton, S. G. Rogelberg, Beyond online surveys: Internet research opportunities for industrial-organizational psychology, in S. G. Rogelberg, ed., *Handbook of research methods in industrial and organizational psychology*, Oxford: Blackwell (2002).

14. S. Utz, Social identification and interpersonal attraction in MUDs, *Swiss Journal of Psychology* (Special Issue: Studying the Internet—A challenge for modern psychology) **62**(2), 91-101 (2003).

NANOTECHNOLOGY AND SOCIAL TRENDS

*William Sims Bainbridge, NSF**

Science-funding agencies such as the National Science Foundation (NSF) need expert advice about what research projects to support and what research programs to establish. Here I will offer a possible framework for such advice, with the awareness that others may suggest very different approaches. My own perspective comes from a background of doing research on societal implications of other kinds of technology (notably, information and space technology) and representing two NSF directorates (first social science, then computer science) on the National Nanotechnology Initiative (NNI). It is my hope that this essay will encourage the scientific community to contribute advice for good planning and assessment of programs to study the societal implications of nanotechnology [1, 2].

The impact of nanotechnology depends not merely upon the technical capabilities of the technology, the decisions that people make in implementing it, and the relevant government investments and regulations. It also depends upon the ways in which existing social, economic, and cultural trends interact with these factors. The question is not how nanotechnology may change a stable world. Rather, we should ask how the development of nanotechnology will play into the forces that already swirl in an unstable and often chaotic world.

Thus, it will be important to identify current trends, understanding them both as forces that shape our daily lives today and as transformations that will lead to a radically different way of life in the future, even without nanotechnology. Indeed, we may find that nanotechnology could in some ways be a stabilizing influence, rather than of necessity a destabilizing one. Nanotechnology may not be directly

*The views expressed in this essay do not necessarily represent the views of the National Science Foundation or the United States Government.

relevant for some major trends, although it may have powerful indirect implications through its closer connection to other trends.

There is much room to debate the nature and meaning of current trends, and each may be defined and described in different ways. In listing trends here, I shall draw upon not only my own reading of social science, but also upon the perspectives of about 20,000 people who responded to a question I placed in a major online questionnaire study, *Survey2000*, sponsored by the National Geographic Society. It asked, "Imagine the future and try to predict how the world will change over the next century. Think about everyday life as well as major changes in society, culture, and technology." Respondents were given a text area on the Web page in which to write their thoughts, anything from a single word to several paragraphs.

To assemble the respondents' ideas into a coherent unity, I went carefully through the text, copying distinct thoughts about the future. The method of analysis was one I have used many times before, for example in surveys about the possible goals of the space program for a book I published a decade ago [3]. Naturally, some ideas were expressed in about the same language by many different respondents to *Survey2000*, and I copied only the first or best expressions I encountered. This very time-consuming process eventually gave me a new file with just over 5,000 text extracts. I then worked carefully through these 5,000 excerpts, combining and editing them into 2,000 clear statements of single ideas. Some are simple declarative statements asserting that something will be true in the year 2100. Others are more complex, suggesting what will cause a particular outcome, or combining factors to describe a general condition in the future. One byproduct was a software module, posted on my Web site and published on a CD-ROM disk, that lets the user express his or her own evaluations of the ideas. Several articles or chapters have already been published, based on these data, or are in press [4, 5, 6].

Naturally, the thousands of people who contributed ideas and evaluations were not of one mind. Indeed, they expressed several competing scenarios of the future for each of the major realms of human life. Following are listed 11 realms, with a brief statement of some of the issues expressed by the respondents, illuminated by social-scientific research, and connected to nanotechnology by theory.

Family

Survey2000 respondents had a wide variety of opinions about the future of the family. Some believed that the average family will become stronger over the coming century, with lower divorce rates and possibly a return to traditional family values. Others predicted that the long-term trend of rising divorce rates will continue, with dire consequences for children and for society as a whole. Others imagined that the family will be reinvented, and people will live in a variety of quite different but equally viable family forms. To social scientists, it is not at all clear what could reverse the trends of the 20th century, and in recent years the greatest concern has become the demographic collapse occurring in technologically advanced nations.

Table 2.1 brings together data from the 2003 *World Factbook* of the U.S. Central Intelligence Agency [7], showing the birth and death trends in 23 industrial nations, compared with data for the world as a whole. The data show that all but one of the nations are headed toward or have actually entered population decline. The one notable exception is the United States. Consider Austria, for example, a nation of about 8.2 million people, as of July 2003. The birth rate, about 9.43

Table 2.1
The Fertility Collapse in Technologically Advanced Nations, 2003

Nation	Population	Births/ 1000	Deaths/ 1000	Births/ Deaths	Growth Rate*	Fertility Rate**
Australia	19,731,984	12.55	7.31	1.72	0.93%	1.76
Austria	8,188,207	9.43	9.69	0.97	0.22%	1.41
Belgium	10,289,088	10.45	10.07	1.04	0.14%	1.62
Canada	32,207,113	10.99	7.61	1.44	0.94%	1.61
Czech Republic	10,249,216	9.01	10.74	0.84	-0.08%	1.18
Denmark	5,384,384	11.52	10.72	1.07	0.28%	1.73
France	60,180,529	12.54	9.05	1.39	0.42%	1.85
Germany	82,398,326	8.60	10.34	0.83	0.04%	1.37
Hungary	10,045,407	9.32	13.00	0.72	-0.29%	1.25
Italy	57,998,353	9.18	10.12	0.91	0.11%	1.26
Japan	127,214,499	9.61	8.55	1.12	0.11%	1.38
Korea (South)	48,289,037	12.60	6.03	2.09	0.66%	1.56
Netherlands	16,150,511	11.31	8.66	1.31	0.50%	1.65
New Zealand	3,951,307	14.14	7.54	1.88	1.09%	1.79
Norway	4,546,123	12.17	9.72	1.25	0.46%	1.80
Poland	38,622,660	10.47	9.96	1.05	0.00%	1.37
Portugal	10,102,022	11.45	10.21	1.12	0.17%	1.49
Russia	144,526,278	10.09	13.99	0.72	-0.30%	1.33
Spain	40,217,413	10.08	9.48	1.06	0.16%	1.26
Sweden	8,878,085	9.71	10.58	0.92	0.01%	1.54
Switzerland	7,318,638	9.59	8.82	1.09	0.21%	1.48
United Kingdom	60,094,648	10.99	10.21	1.08	0.30%	1.66
United States	290,342,554	14.14	8.44	1.68	0.92%	2.07
World	6,302,309,691	20.43	8.83	2.31	1.17%	2.65

*Including migration. **About 2.1 is required to sustain population stability.

babies per 1,000 population per year, is already lower than the death rate, 9.69 per thousand. The ratio of births/deaths is 0.97, and any number below 1.00 implies a shrinking population. The population of Austria is actually growing, albeit at the low rate of 0.22 percent per year, because substantial numbers of people immigrate to the country. Austria's total fertility rate is 1.41 births per woman, assuming the average woman lives through the childbearing years. A constant population requires a fertility rate of about 2.1 to offset the facts that some girls die young and that more boys than girls are born (about 105 boys worldwide per 100 girls). Except for the United States and the world as a whole, all the fertility rates in the table are well below 2.1, indicating that these nations will need substantial immigration to survive.

Three of the nations in Table 2.1 are already losing population: the Czech Republic, Hungary, and Russia. Four others have death rates higher than their birth rates: Austria, Germany, Italy, and Sweden. Demographers have long warned that birth rates of technologically advanced nations might drop significantly below replacement [8, 9, 10]. A nation that declines in population is likely also to decline in world influence [11]. Its market for goods and services may also shrink, stifling economic development. A higher proportion of its population will be elderly and dependent. A lower proportion will be young and creative.

The United Nations reports that half the world annual population growth occurs in just six nations: India (21 percent), China (12 percent), Pakistan (5 percent), Nigeria (4 percent), Bangladesh (4 percent), and Indonesia (3 percent) [12]. In the short term, their influence is likely to increase, but eventually modernization may cause their fertility rates to collapse as well [13]. We can well wonder how the demographic trends will play out, as population explosion continues in underdeveloped societies while fertility collapse threatens the long-term viability of post-industrial societies.

The mutual implications of nanotechnology and family trends for each other are not clear. However, nanotechnology could have major roles to play in medical control of fertility (whether reducing or increasing the birth rate) and in increasing the lifespan with complex implications for how older people fit into families and communities. In principle, nano-enabled medical treatments could allow infertile adults to produce children progressively later in an ever-increasing lifespan, thereby sustaining fertility at a level that keeps industrial society viable. One could imagine a time when many couples produce two generations of children, offsetting with their higher fertility those couples who produced no children, taking advantage of nano-enabled medical treatments that prolong both lifespan and the portion of it during which women are fertile. Social-scientific research is needed to help us understand whether any technical means can reverse the declining birth rate in the absence of major socio-cultural shifts.

Culture

Survey2000 respondents debated whether the "culture wars" would be settled on the basis of shared ideals and common beliefs, or if technologies like the Internet will replace the broadcast media with a babble of narrowcast ideologies and aesthetics. They worried about whether the arts and education would thrive, stimulated by wealth and communication technologies, or fall into an anti-intellectual Philistinism in which corporate greed and popular stupidity combine to prevent creativity. Nanotechnology is very relevant here, partly because it can contribute to rapid progress in technologies of computation, communication, and creativity.

Cultural trends will greatly affect the direction nanotechnology takes, in great measure because the condition of educational institutions will determine whether competent personnel will be available to develop and implement the technology, and because science education may be reformulated around principles provided and illustrated by nanoscience. Worldwide, there has been great concern that the schools are not giving many children a good introduction to mathematics and the sciences, but at the same time there is reason to be optimistic that effective methods of teaching are now at hand [14, 15, 16].

Coupled with educational reform, nanoscience could transform culture by providing a unified understanding of nature, based on the material unity of the chemical, physical, and biological world at the nanoscale. This would accomplish what Edward O. Wilson [17] called *consilience* and Mihail C. Roco and I [18] have called *convergence*: integration of the sciences and other rigorous fields of knowledge, based on shared concepts, language, and research tools. Unification of the sciences would become the basis for unification of other aspects of culture, thus providing a firm basis of knowledge not only to specialists but also to the general public.

Personality

For more than a century, social scientists have argued that technological and economic developments are indirectly but powerfully eroding the traditional basis of human personality, by creating a progressively more complex and variable social environment that forces young people to mature under highly inconsistent and individualistic influences [19, 20, 21, 22, 23, 24]. Respondents to *Survey2000* wondered whether people will continue to become increasingly sensate, hedonistic, and self-centered. Or will the pendulum swing back toward moral conservatism or social responsibility? Will people become ever more alienated, suffering anomie, neurosis, and even psychosis, or will a combination of supportive social relationships and improved psychiatric treatments successfully combat the stressful factors pushing toward collective and individual disintegration?

Nanotechnology may be directly significant for psychiatry in promising a better understanding of the brain and improved medications, but its indirect influence on mental health may be greater. We have known for a century and a half that mental disorder was strongly associated with poverty [25, 26]. As a powerful amplifier of the value of other technologies, nanotechnology can enhance individuals' abilities to attain a diversity of personal goals.

On the other hand, if human beings do become more alienated, they will be less trusting and may capriciously oppose new developments in nanotechnology because of a general suspiciousness of societal institutions rather than a careful analysis of what the technology will actually do. Thus, it is unclear whether the value of nanotechnology will be reduced by a vicious circle of psychopathology causing bad policy decisions begetting more psychopathology, or a virtuous circle of nano-enabled economic growth improving the options for policymakers and ordinary citizens alike.

Economy

The close of the 20th century gave global dominance to the economic system that its opponents called *capitalism* and its proponents called *free markets* or *private enterprise*, but people are uncertain whether the current consensus is good or bad, robust or fragile. Many respondents think this triumphant economic system will flourish and improve over the coming century, with a rising standard of living all over the world and a stable economy, compared with the booms and recessions of previous centuries. A very different view believes that large conglomerate corporations will control the lives of their workers, owning their homes and the stores where they must buy things. The living conditions and wages will be very poor worldwide, as automation replaces workers and results in lack of jobs for many people. Because they have to compete with the unemployed and with machines that have no bargaining rights, human beings will be forced to accept increasingly bad working conditions, in this view.

Much economic inequality reflects differences across nations rather than among individuals within nations [27, 28], so real improvement of economic well being must be global in scope [29]. Increased investment in current productive technologies could enable a larger fraction of the world's population to enjoy prosperity, but it is doubtful whether they could sustain universal prosperity without ruining the environment and exhausting natural resources. Thus, whatever economic system prevails, it must be fueled by technological progress, undoubtedly requiring vastly greater control of nanoscale processes and materials [30]. For example, cheap, clean, renewable sources of energy are necessary, such as nano-enabled solar power production, vehicles with nanoscale-engineered energy storage media, and vastly increased efficiency of energy use permitted by a wide range of nanotechnologies. Already, many nations have begun working to realize this hope [31].

Government

Survey2000 respondents disagree greatly about the future of government, some feeling it will become more benevolent, and others fearing it will fulfill the nightmares of *Brave New World* and *1984* [32, 33]. Perhaps the criminal justice system will grow ever more civilized, but if the public feels it is drowning in a massive crime wave, the political response could be so mindlessly punitive that it defeats its own purpose of achieving security with justice. Some fear a future big government that is a coalition of exploiters, big and small, who prey upon productive citizens. Others dream of a humane welfare state or the workers' paradise. Optimistic moderates hope that democratization and devolution will bring political renewal that can solve the problems of repressive and unresponsive government.

Government establishes the context in which nanotechnology will either develop or be stifled through establishing regulations, by investing in fundamental nanoscience research, and in supporting projects to demonstrate and evaluate particular new applications. In return, nanotechnology can provide new tools by which government can serve the public, especially in areas such as defense and communications between citizens and government, where it amplifies the value of existing technologies. Governments will have to decide how intrusive to become in nanotechnology-related issues involving other institutions of society, and how intensive a role to play in evaluating the societal impacts.

Nations

Many stresses affect international relations in migration, racism, conflict, warfare, and peace. Some nations have fragmented in recent years, such as Czechoslovakia and Yugoslavia, which experienced very different processes of division, while other nations have moved closer together, notably the member states of the European Union. The fundamental question, how varied the future will be, is partly a matter of the extent to which nations and regional blocs continue to exist. Under a single world government, it is still possible that people would be divided by non-geographic factors such as religion and ethnicity. If the world is not unified politically, however, the scope of variation and competition could be far greater. Thus it is important to know whether nations are going out of style, or are a permanent feature of human life.

Survey2000 respondents disagreed fundamentally about such issues as whether nations will consolidate into either a single world society or into a small number of competing blocs [34]. Some felt that political independence movements and the failure of governments could split many nations into even smaller units. Will the United States and allied advanced industrial societies be able to maintain their dominance, or will they fall, to be replaced by another center of power, such as China, or by world chaos? Years ago, some scholars predicted Japanese dominance, but that has not happened, so we must be prepared for the possibility that such crucial questions cannot be answered with confidence [35].

The emergence of a peaceful, progressive world society depends at the very least upon a technology that can offer abundance to a majority of people in all societies. Even in the short term, nanotechnology can be an effective multiplier of the effectiveness of many other wealth-producing technologies, and in the long term manufacturing through control of matter at the nanoscale can be one of the greatest treasures of all mankind. In a great variety of ways, nanotechnology can increase the effectiveness of military forces, so then the question becomes whether those forces will be fighting for or against civilization, freedom, and justice. In convergence with other sciences, nanoscience could become the intellectual basis for a shared, global culture, thereby helping nations to cooperate on development projects.

Religion

Survey2000 respondents identified many issues concerning the future of religion, and they proposed several competing scenarios [36, 37]. The chief issue of broad scope is whether religion will wane over the coming century, hold roughly constant, or actually increase in power. Of great importance will be the relationship between religion and science, which may be one of hostility, but cooperation is also possible [4, 38].

A unified science, based on the unity of nature at the nanoscale, could, for the first time, provide a really comprehensive explanation of the world that humans inhabit, based on concepts like the evolutionary dynamics of complex, adaptive systems. As a result, the nanoscience of the future could challenge many traditional religious beliefs, thereby leading either to conflict with religious institutions and movements, or to a synthesis in which religion adopts the truths discovered by science to offer people an even more accurate and comforting faith than they have ever known before. For understandable reasons, federal agencies appear to be reluctant to support research on religion, and many social-scientific studies of religion are supported by private foundations. However, the post–9/11 world is anxiously aware of serious social issues concerning religion, and several current policy debates bring religion into contact with science at several points. Thus, it could be very helpful to learn the current views of the scientific community about the potential role of government-supported science in this area.

Science

Many *Survey2000* respondents said that scientific research will continue to achieve significant discoveries, possibly even at an accelerating rate, but others felt we are nearing the end of the Age of Discovery. Will science soon stall, either because it has exhausted the possibilities for discovery or because society loses interest in fundamental research, or will science-based technology improve every individual's spiritual, physical, emotional, and psychological well being?

Nanotechnology offers to all of the sciences new research instruments, methodologies, and paradigms. Perhaps the greatest potential for radical transformation comes not from any one science, but from the integration of all sciences. At the nanoscale, science and technology become one, and practical applications of new laboratory discoveries can enter the market very quickly. Previously separate disciplines will blend, in a unified scientific technology that completely controls the structure and properties of manufactured objects, from the atomic level, up through the nanoscale level at which large molecules exist, on up to the scale of an entire machine, biological-mechanical hybrids, and globe-spanning information systems.

Space

Many respondents to *Survey2000* continued to have faith that a vigorous space program could lead not only to scientific discovery but also eventually to the colonization of the solar system and beyond, but where today is evidence that space technology is really moving forward? Could we be at the end of the Space

Age, rather than the beginning, as world leaders turn their backs on the universe to concentrate on mundane conflicts? Narrowly defined in terms of human travel, the Space Age lasted from December 1968 through December 1972, from the first flight of Apollo 8 around the Moon to the last flight of Apollo 17 to its surface. For more than three decades, no human being has flown much higher than 300 miles above the Earth's surface.

In economic terms, space flight is costly with relatively little return on most investments. Accomplishments to date have largely been the result of a transcendent social movement that was able to exploit international competition to support its goals, without caring very much about economic factors. A resurgence of space exploration could be launched by nanotechnology, if it simultaneously reduces the cost and increases the profit. Importantly, it may be necessary to find entirely new uses for outer space and new motivations for voyaging out into it [39, 40, 41].

Nanoscale-designed advanced materials could improve the mass ratios of launch vehicles, improve power production and efficiency for space vehicles, and potentially provide more effective ways of surviving and even thriving in extraterrestrial environments. More broadly, successes achieved in nanoscience could reinvigorate public enthusiasm for investment in a range of research and development projects, including a renaissance of the space program.

Environment

The natural environment is an extremely complex system, and *Survey2000* respondents have a very diverse range of opinions about how it may change over the coming century. Many respondents anticipate that all fossil fuels will run out, causing drastic changes in the distribution of food and manufactured goods, heavy increases in energy prices, and a halt to industrial development. Others hope that renewable energy sources will reduce pollution, appropriate technology will humanize industry, and new farming practices will allow more food to be produced. Over the coming century, the conservation movement may grow in strength, until environmentally conscious lifestyles become the norm. But there is reason to fear that the Earth will suffer massive extinction of species, terrible pollution, global warming, and exhaustion of critical natural resources. These are issues of great importance, public concern, and scientific uncertainty [42, 43].

Nanotechnology can contribute to solution of these problems through methods to prevent or remediate conventional forms of pollution, through meeting human goals by means of novel use of the most abundant and renewable resources, and by improving the efficiency of all technologies, including reducing waste of every kind. For example, nanotechnology may have an important, beneficial role to play in precision agriculture, improving the quality and quantity of food production while reducing pollution. However, some particular applications of nanoscience may threaten to pollute or to exhaust scarce resources, so we must be alert to such cases either to avoid these applications or to find ways of overcoming their environmental disadvantages.

Health

Will human health improve over the coming decades, or will problems of the elderly or of resurgent diseases grow beyond control? Are the greatest gains to be attained simply through improving people's lifestyle choices or through aggressive treatments based on genetic engineering and other radical technologies? Both the apparently slowing rate of development of effective new treatments and suggestions from demographers that we are reaching the point of diminishing returns in increasing the lifespan give cause to doubt that human health and longevity will improve significantly without a radical shift in approach [44, 45, 46, 47].

The health quality of human lifestyles can benefit from general improvements in nutrition and public health, made possible by a host of specific nanotechnologies applied to production industries and civil engineering, and from innovations such as nano-enabled microscale sensors to warn of food contamination, disease agents, and harmful pollutants. More controversial will be techniques that combine nanotechnology with genetic engineering, or that introduce nanoscale materials and devices into the human body for medical treatment. While appropriate care must be given to avoid increasing risks to human health, the policy decisions must be made in awareness of the fact that humans are already at great risk of ill health or death, and failing to explore radically new approaches could be the most inhumane choice of all.

Thus, nanoscience and nanotechnology could have significant and highly various benefits across 11 major areas of human life where powerful trends are already bringing change: family, culture, personality, economy, government, nations, religion, science, space, environment, and health. However, my comments have been based on theory more than empirical evidence. Thus, it will be important to work out appropriate research methods for examining nano-assisted change in all these realms, and to launch well-designed research projects at the earliest feasible time. Only by starting soon can we gain the tremendous scientific advantage of observing revolutionary technological change at an early stage in its development. Only by examining the societal implications of nanotechnology in many varied settings can we understand the full potential of this multifaceted world of science and engineering.

References

1. W. S. Bainbridge, Public attitudes toward nanotechnology, *Journal of Nanoparticle Research* **4**, 561-570 (2002).
2. W. S. Bainbridge, Social and ethical implications of nanotechnology. In *Springer Handbook of Nanotechnology*, ed. B. Bushan, Berlin: Springer (2004).
3. W. S. Bainbridge, *Goals in Space*, Albany, NY: SUNY Press (1991).
4. W. S. Bainbridge, Religious ethnography on the World Wide Web. In *Religion on the Internet*, ed. J. K. Hadden, D. E. Cowan, New York: JAI (2000).

5. W. S. Bainbridge, Validity of Web-based surveys. In *Computing in the Social Sciences and Humanities*, ed. O.V. Burton, Urbana: University of Illinois Press (2002).

6. W. S. Bainbridge, The future of the Internet: Cultural and individual conceptions. In *Society Online: The Internet in context*, ed. P. N. Howard, S. Jones, Thousand Oaks, CA: Sage (2004).

7. *The World Factbook*, Langley, VA: Central Iintelligence Agency (2003) (available at http://www.cia.gov/cia/publications/ factbook/ index.html).

8. K. Davis, M. S. Bernstam, R. Ricardo-Campbell, eds., *Below-replacement fertility in industrial societies: Causes, consequences, policies* (A supplement to *Population and Development Review*)(1986).

9. B. J. Wattenberg, *The Birth Dearth*, New York: Ballantine (1987).

10. K. Davis, The theory of change and response in modern demographic history, *Population Index* **29**, 345-366 (1963).

11. *Long-term global demographic trends: Reshaping the geopolitical landscape*, OTI IA 2001-045, Langley, VA: Central Intelligence Agency (2001).

12. United Nations, Population Division, Department of Economic and Social Affairs, *World population prospects: The 2000 revision*, New York: United Nations (2001).

13. N. Thomas, N. Price. The role of development in global fertility decline, *Futures* **31**, 779-802 (1999).

14. V. S. Mullis, M. O. Martin, E. J. Gonzalez, K. M. O'Connor, S. J. Chostowski, K. D. Gregory, R. A. Garden, T. A. Smith, *Mathematics benchmarking report: TIMSS 1999—Eighth grade*, Chestnut Hill, MA: International Study Center, Lynch School of Education, Boston College (2001).

15. M. O. Martin, I. V. S. Mullis, E. J. Gonzalez, K. M. O'Connor, S. J. Chostowski, K. D. Gregory, T. A. Smith, R. A. Garden. *Science benchmarking report: TIMSS 1999—Eighth grade*, Chestnut Hill, MA: International Study Center, Lynch School of Education, Boston College (2001).

16. J. D. Bransford, A. L. Brown, R. R. Cocking, eds., *How people learn: Brain, mind, experience, and school.* Washington, DC: National Academies Press (1999).

17. E. O. Wilson, *Consilience: The unity of knowledge*, New York: Random House (1998).

18. M. C. Roco, W. S. Bainbridge, *Converging Technologies for Improving Human Performance*, Dordrecht, Netherlands: Kluwer (2003).

19. F. Tönnies, *Community and Society*, East Lansing, MI: Michigan State University Press (1887; 1957 edition).

20. E. Durkheim, *The Division of Labor in Society*, New York: Free Press (1893; 1964 edition).

21. E. Durkheim, *Suicide*, New York: Free Press (1897; 1951 edition).

22. R. E. L. Faris, H. W. Dunham, *Mental Disorders in Urban Areas*, Chicago: University of Chicago Press (1939).

23. D. Riesman, *The Lonely Crowd:* A *study of the changing* American *character*, New Haven: Yale University Press (1950).

24. P. E. Slater, *The Pursuit of Loneliness: American culture at the breaking point*, Boston: Beacon Press (1970).

25. E. Jarvis, *Report on insanity and idiocy in Massachusetts*, Boston: White (1855).

26. B. Hollingshead, F. C. Redlich, *Social Class and Mental Illness*, New York: Wiley (1958).

27. G. Firebaugh, Empirics of world income inequality, *American Journal of Sociology* **104**, 1597-1630 (1999).

28. G. Firebaugh, The trend in between-nation income inequality, *Annual Review of Sociology* **26**, 323-339 (2000).

29. H. Cleveland, The global century, *Futures* **31**, 887-895 (1999).

30. M. C. Roco, R. S. Williams, P. Alivisatos, *Nanotechnology research directions*, Dordrecht, Netherlands: Kluwer (2000).

31. R. W. Siegel, E. Hu, M. C. Roco, *Nanostructure Science and Technology*, Dordrecht, Netherlands: Kluwer (1999).

32. Huxley, *Brave New World*, Garden City, NY: Doubleday, Doran (1932).

33. G. Orwell, *Nineteen Eighty-four, a novel*, New York: Harcourt, Brace (1949).

34. S. P. Huntington, *The Clash of Civilizations and the Remaking of World Order*, New York: Simon and Schuster (1996).

35. E. F. Vogel, *Japan as Number One: Lessons for America*, Cambridge, MA: Harvard University Press (1979).

36. R. Stark, W. S. Bainbridge, *The Future of Religion*, Berkeley: University of California Press (1985).

37. W. S. Bainbridge, *The Sociology of Religious Movements*, New York: Routledge (1997).

38. W. S. Bainbridge, Religion and science, *Futures* **36**, 1009-1023 (2004).

39. W. S. Bainbridge, *The Spaceflight Revolution*, New York: Wiley Interscience (1976).

40. W. S. Bainbridge, Beyond Bureaucratic Policy: The spaceflight movement. In *People in space*, ed. J. E. Katz, New Brunswick, NJ: Transaction (1985).

41. W. S. Bainbridge, The Spaceflight Revolution Revisited. In *Looking Backward, Looking Forward*, ed. S. J. Garber, Washington, DC: National Aeronautics and Space Administration (1992).

42. D. Sarewitz, R. A. Pielke, Jr., R. Byerly, Jr., *Prediction: Science, decision making, and the future of nature*, Washington, DC: Island Press (2000).

43. R. T. Watson, M. C. Zinyowera, R. C. Moss, D. J. Dokken, eds., *Climate change 2001.* Geneva, Switzerland: Intergovernmental Panel on Climate Change c/o World Meteorological Organization (2001) (available at www.ipcc.ch and www.grida.no/climate/ ipcc/regional/ index.htm).

44. S. H. Preston, *American Longevity: Past, present, and future*, Syracuse, NY: Center for Policy Research, Syracuse University (1996).

45. V. R. Fuchs, Health care for the elderly: How much? Who will pay for it? *Health Affairs* **18**, 11-21 (1999).

46. F. W. Hollmann, T. J. Mulder, J. E. Kallan. Methodology and assumptions for the population projections of the United States: 1999 to 2100, Population Division Working Paper No. 38, Washington, DC: U.S. Census Bureau (2000).

47. D. Costa, R. Steckel, Long-term trends in the health, welfare, and economic growth in the United States. In *Health and welfare during industrialization*, ed. R. H. Steckel, R. Floud, Chicago: University of Chicago Press (1997).

FIVE NANOTECH SOCIAL SCENARIOS

Robin Hanson, George Mason University

During the dotcom boom, popular discussions about the future of the Internet often included claims of dramatic social implications; the "new economy" was said to follow new rules [1]. This hype was ridiculed after the dotcom crash, but the Internet did, in fact, bring real changes, with non-trivial social implications. And early on, economists were able to play the important role of analyzing these claims, and distinguishing the hype from real changes. For example, in *Information Rules*, Carl Shapiro and Hal Varian did a great job of using economic theory to distinguish plausible from implausible claims about the Internet [2].

Popular discussions of nanotechnology have also included many claims of dramatic social implications [3, 4]. Some even point to a new "economy of abundance" [5]. Naturally, many others consider these claims to be hype. It is my hope that, as with the Internet, economic analysis might help to distinguish plausible from implausible claims about nanotechnology.

What assumptions about this technology should we base our economic analysis on? At one extreme, some think of "nanotechnology" as a new name for "materials science" and "chemistry." This view suggests that while advances in these fields will continue, there is little to say regarding nanotechnology beyond general analysis of long-term growth, and perhaps analyses of certain new enabled products, such as better surveillance. A variation on this position suggests that we are seeing the coalescence of a new research specialty, a view that may have some minor implications regarding the organization of research. At the other extreme, some think of "nanotechnology" as the technology of a new device, the "assembler," which like the computer will induce large social changes via its novel and general capabilities [6]. (Others are very skeptical about such scenarios [7].)

Economists should not choose sides in these technical disputes. While radical nanotechnology scenarios might be less likely, if realized they would have larger social implications, and we already understand the conservative scenarios reasonably well. Furthermore, it was most likely the public's interest in and concern about more radical scenarios that led Congress to request an analysis of the social implications of nanotechnology. Economists should therefore consider the social implications of radical nanotechnology scenarios. Since mild scenarios are *a priori* more likely than radical scenarios, however, we should also think about how these radical scenarios might be only partially realized.

The short informal analysis in the remainder of this paper is a very preliminary attempt at such an analysis, intended primarily to indicate what might be possible with a more careful analysis. A range of assumptions and future production costs are presented that define five economic scenarios spanning the range from conservative to radical nanotechnology scenarios. We start with the most conservative scenario, and then each scenario adds more assumptions and has additional social implications. The goal here is to identify the economic assumptions behind an imaginable radical nanotechnology scenario, as well as some milder and more likely variations.

Scenario 1: Atomic Precision

> Atom-scale manufacturing is feasible; we put some atoms where we want.

Such abilities may eventually allow many new products, such as cheaper and smaller computers and sensors, and perhaps tiny medical implants. Exactly which products are feasible would depend on exactly which assembly contexts allow such precision, and at what cost. Economic growth could go far before hitting limits. Particular products may have particular implications.

Scenario 2: General Plants

> General-purpose manufacturing plants, using fewer kinds of inputs, displace most special purpose plants, as general-purpose computers have displaced most special-purpose signal processors (This is mature "3D printing" or "direct manufacturing"). [8]

Computers displaced special-purpose signal processors because of scale economies in computer production (the more we made, the cheaper they got) and because it was usually easier to program a general computer for a particular task than to design and build a special-purpose device for that task. These advantages usually outweighed the fact that special-purpose devices could produce the same results faster, with fewer transistors, and with less energy.

Similarly, for more general plants to dominate, the scale economies of making them, and the reduced cost of designing a production plant and retooling for it, would have to overcome the efficiency advantages (e.g., time, energy, inputs) of specialized plants. Generality is relative, of course; a "general" plant might make most consumer goods, most kinds of household furniture, or just most kinds of mattresses. More general plants should use fewer kinds of inputs, have production costs that depend on fewer design details, and cost more themselves to design. The skills of manufacturing workers would be less specialized to particular kinds of products.

The reduced design and retooling costs of more general plants should allow more product differentiation, and more rapid product evolution. When transport costs matter, production should be at the general plants of the relevant type nearest to each customer.

Scenario 3: Local Production

> Small general plants, located in or near homes, dominate manufacturing.

This requires a high level of plant generality, and requires that production processes, including diagnosis and repair of problems, be almost fully automated, with human intervention rare. Such high levels of generality and automation are harder to design. But once such devices existed, they would allow hobbyist designers of products (and production plans), as PCs allowed hobbyist programmers. As with PCs today, open-source product design and sharing of stolen product designs could become issues. Any companies who owned the standards for such devices, such as Intel and Microsoft do today for PCs, would hold a commanding position in the economy.

In this scenario the costs of transportation of products and of labor for their production have been mostly eliminated. What remain are marginal costs of energy, inputs, waste disposal, plant rental, marketing (people learning about product price, quality, and features), and regulation (such as of use externalities), and fixed costs of design, plant setup, marketing, and regulation.

Software and cable TV companies now offer a few large packages of diverse products in order to better price-discriminate. (This works when marginal costs are low and item values are not too positively correlated [9].) Future consumers might similarly be offered a few lifestyle packages costing most of their income and entitling them to use a wide range of product designs (e.g., clothes, furniture, food, cars) at their local plant (if the customer pays for inputs, energy, etc.). This would require a lot of coordination by (or concentration of) sellers of consumer-good designs, and deterrence of sharing between buyers of different packages. Geographic separation, which now lets cruise ships and resorts offer all-you-can-eat-and-play deals, might help deter sharing.

Scenario 4: Over-Capacity

> Local general plants are so fast and cheap that they are usually idle, like PCs now.

Most home PCs today are usually idle (or doing very low value work); they are as capable as they are because they are cheap, and on occasion we want that much capacity. Similarly, if it is cheap enough to create local general plants, they might usually be basically idle, and be only rarely used at capacity. If so, the relevant cost of capital would usually be very low; the marginal costs of most products would be inputs, energy, waste disposal, marketing, and regulation. Fixed costs of design, regulation, and marketing would usually dominate total costs, as with software and music today. In this scenario, "information economics" would describe most consumer goods [2].

Scenario 5: Self-Reproduction

> A local manufacturing plant can create a copy of itself in much less than a year.

Once we add in this assumption, we reach the most radical nanotechnology scenarios [3, 4], where plants built to atomic-precision can reproduce themselves (as all life forms do today), and in addition be programmed to produce other products (as a few life forms can now do in some limited ways). The problem of designing such entities seems very hard, but solving this design problem is one possible route to achieving over-capacity of local general plants, and perhaps soon.

Self-reproduction could give a large and sudden, and hence destabilizing, cost advantage to the commercial or military power that first achieves this ability. How large an advantage depends on just-prior costs, and how sudden an advantage depends on self-reproduction time, the development lead held by the first successful group, and the availability of product designs taking advantage of this reduced cost. Self-reproducing military or terrorist weapons might be a concern if these entities were small enough and reproduced fast enough.

Conclusion

The five social scenarios presented above, with their matching economic assumptions and social implications, span the range between conservative and radical nanotechnology. While the conservative scenarios are more likely, the radical scenarios have larger social implications, and so are worth considering. The preliminary analysis here of these scenarios is intended primarily to indicate what might be possible with a more careful analysis.

References

1. K. Kelly, *New Rules for the New Economy: 10 radical strategies for a connected world*, New York: Viking Press (1998).
2. C. Shapiro, H. R. Varian, *Information Rules: A strategic guide to the network economy,* Boston: Harvard Business School Press (1999).

3. E. Drexler, C. Peterson, G. Pergamit, *Unbounding the Future: The nanotechnology revolution*, New York: William Morrow and Company (1991).

4. N. Stephenson, *Diamond Age: Or a young lady's illustrated primer*, New York: Bantam Books (1995).

5. B. Bruns, Nanotechnology and the commons: Implications of open source abundance in millennial quasi-commons, Technical report (2000) (available at http://dlc.dlib.indiana.edu/archive/00001051/).

6. K. E. Drexler, *NanoSystems*, New York: John Wiley and Sons (1992).

7. R. Baum, Nanotechnology: Drexler and Smalley make the case for and against "molecular assemblers," *Chemical and Engineering News* **81**(48), 37-42 (2003).

8. Imato, Instant manufacturing, *Technology Review* (November 2003).

9. Y. Bakos, E. Brynjolfsson, Bundling information goods: Pricing, profits, and efficiency, *Management Science* **45**(12), 1613-1630 (1999).

TECHNOLOGICAL REVOLUTIONS AND THE LIMITS OF ETHICS IN AN AGE OF COMMERCIALIZATION

John Trumpbour, Harvard Law School

Nanotechnology has raised vital questions about the social impacts of new technologies. This essay will try to supply some historical perspectives on technological revolutions by assessing (1) their contributions to economic growth, and (2) the ensuing stresses from the commercialization of research institutions. In the desire to reap part of the rewards of ongoing technological innovation, academic institutions in particular will have to be vigilant about guarding their independence. Otherwise they may find themselves less capable of imparting a social and moral dimension to policy making about new technologies, as well as being hampered in delivering scientific knowledge in the public interest.

Technological Revolutions and Economic Growth

It is frequently asserted that nanotechnology will contribute to significant bursts in economic growth and productivity, eventually culminating in an age of post-scarcity. Similar claims were advanced for the IT revolution. In "A Magna Carta for the Knowledge Age," Esther Dyson, George Gilder, Jay Keyworth, and Alvin Toffler heralded the primacy of IT and declared that "The central event of the 20th century is the overthrow of matter. In technology, economics, and the politics of nations, wealth—in the form of physical resources—has been losing value and significance. The powers of mind are everywhere ascendant over the brute force of things." They later make clear in cadences reminiscent of Frederick Jackson Turner on the North American frontier: "Cyberspace is the land of knowledge, and the exploration of that land can be a civilization's truest, highest calling [1 pp.26 and 28]."

When it comes to overall productivity, the IT revolution may, in hindsight, have provided more sobering lessons. IT contributed to the economic dynamism of the U.S. economy in the 1990s, though growth rates still lagged behind many previous decades (i.e., the 1960s saw a 3.3 percent leap in GDP per capita in the United States, compared to 2.35 percent in the supposedly roaring 1990s. Japan grew 9 percent in GDP per capita during the 1960s compared to 1.1 percent in the 1990s) [2, p.47].

In an article for *Challenge*, economist Jeff Madrick provided a longer frame for observing computerization and productivity in the United States: "Business has increased its investment in computers by more than 30 percent a year since the early 1970s, but the rate of growth in productivity has fallen from about 2.85 percent a year between 1947 and 1973 to about 1.1 percent a year since 1973 [3, p.50]." In the late 1990s and even throughout the recession of 2001 and its aftermath, productivity growth has sometimes been majestic (6 percent in the second half of 2003). But even *Business Week* concedes: "...this recovery has left the unemployed and poor behind while mainly helping owners of assets such as stock and homes....Indeed the only group whose wage gains are significantly outpacing prices are managers and executives [4, p.45, 5]."

Under President Bill Clinton, the ratio of CEO wages to the average U.S. worker skyrocketed from 113 to 1 in 1991 to 449 to 1 at the end of his Presidency [6, p.9]. While that result ought not to be blamed on IT, it should raise questions for those who expect technological marvels to assure that prosperity is shared throughout society. The Congressional Budget Office did a detailed survey of U.S. income from 1979–1997 and concluded that the bottom quintile declined significantly, the middle quintile stagnated, and the top 1 flourished, as household income (in 1997 dollars) soared for the latter from $256,400 in 1979 to $644,300 in 1997 [7].

On a global scale, the IT era has brought a boost to India and China, countries that comprise a significant proportion of the world's population. Still the richest 10 percent of nations in 1980 had 77 times higher median income than the poorest 10 percent; by 1999, it had grown to 122 times [8]. After total growth of 34 percent between 1960 and 1980, sub-Saharan Africa actually declined by 15 percent in per capita GDP between 1980 and 2000. Latin America grew only 7 percent between 1980 and 2000 in contrast to the 75 percent total growth in per capita GDP from 1960 to 1980 [9]. The IT revolution delivered little to these regions.

So there are those who raise the specter of a digital divide spiraling into a further nanotechnology divide. For many advanced industrial nations, nanotechnology is seen as a means of assuring or re-capturing a lead role in the global economy. During the 1970s, Germany had been the "hot" economy, but then stumbled in the 1980s. Japan then became the center of attention in the 1980s, but fell ignominiously in the latter half of the 1990s. The United States became for many the center of dynamism in the 1990s, but it too started the 21st century with lackluster performance. Through its Ministry of Economy, Trade, and Industry (METI), Japan clearly regards nanotechnology as a component of achieving restoration as the pacesetter in the world economy. The United States hopes that biotechnology and nanotechnology will help it overcome sluggishness lingering from the dotcom meltdown. In *Small Times* magazine (November/December 2003), former Israeli Prime Minister Shimon Peres discusses how Nano-Zionism will lead his tiny nation to "economic growth," which will expand the "scope for cooperating with neighboring states." Foreseeing a world of pacific wonder, he believes nanotechnology could provide Israel with an "increased standard of living for its citizens, and this is one of the best ways to promote a lasting and equitable peace [10]."

When it comes to social stratification or global inequality, it is more likely that nanotechnology will accentuate divisions, in spite of Peres's own view that it could deliver regional prosperity to the Middle East. Those who are optimists about the technology and its contribution to increased productivity should probably refrain from claiming that part of its promise includes widely distributed prosperity. It will take reforms in the economic and political spheres to deliver those kinds of results.

Historians such as Duke University's Alex Roland suggest that the marketing of scientific agendas requires extravagant claims; otherwise politicians and the science bureaucracies would balk at their exorbitant costs [11]. Many science boosters are sincere in their theological devotion to earthly heaven in the making; that is, if society could find the resources to realize technology's unbound potential.

Denise Caruso, founding editor of Digital Media, has raised the specter that the need for venture capital encourages something more sinister, what might be characterized as an exuberant techno-hucksterism: "The newly minted 'visionary' asserts he has found the equivalent of the Holy Grail—if not in terms of the actual technology or concept, at least in terms of his ability to induce religious ecstasy in investors." She then explains the ensuing carnage after the visionary and imitators take their companies public: "The venture capitalists and founders sell off a pile of stock as soon as possible. The folks they sell it to lose their shirts when the stock starts hemorrhaging, as it inevitably does. The news media start a backlash to the mania, which they were complicit in creating [12]."

For the many sincere techno-optimists, there is a belief that the quadruple revolution of nanotechnology, biotechnology, information technology, and cognitive science (NBIC) will produce a concatenation of growth transforming the world of the 21st century. The rightist ideologue Dinesh D'Souza characterizes it this way: "A world without scarcity is one in which income or wealth differentials should cease to have much effect [13, p.101, 14]." Reflecting on the vision of later nineteenth century political economists and social theorists who held that the industrial revolution would eventually bring triumph over scarcity, John Gray of the London School of Economics counters that proponents of the latest wave of technologies may be making a similar monumental error. He counsels caution:

> The trouble with the cult of technology is not that it exaggerates the power that comes with the practical application of scientific knowledge. It is that it forgets the unregenerate human beings

> who use it. Developments in genetics may allow the eradication of inherited defects that produce painful or disabling diseases; but they also allow for the creation of genetically selective bioweapons. New types of energy may be invented which greatly reduce the role of hydrocarbons in industrial economies. If so, we can be sure they will soon be turned to military uses. [15]

For the anti-Enlightenment social critic John Gray, this is not the dawning of the Golden Age. The vast diffusion of scientific knowledge combined with the proliferation of nation-states should increase the likelihood that some will pursue agendas of deadly consequence, he argues. In the most extreme cases of the contemporary age, politicians, warlords, and rapacious corporations are converting lands with plenty into zones of famine and human misery. Gray's world is far removed from that of the *Ladies Home Journal*, which in 1953 foresaw that nuclear energy would soon deliver a planet where "there is no disease.... Where hunger is unknown... where food never rots and crops never spoil.... Where 'dirt' is an old fashioned word.... Where the air is everywhere as fresh as on a mountain top and the breeze from a factory as sweet as from a rose [16, p.4, 17]."

Like most path-breaking technologies, nanotechnology has produced its share of utopian and dystopian scenarios. The dialectic of doom greets the party of hope, who are warned that Yeats's "blood-dimmed tide" will be loosed, and "everywhere the ceremony of innocence is drowned." Nanotechnology partisans have grown weary of dire predictions of a planet converted into "grey goo" or some similar soup-like concoction.

For those in search of a message less apocalyptic and more practical, there are a few lessons that might help in putting this technology's great potential in proper perspective. First, scarcity is likely to remain a part of the human condition, so forecasts of a Golden Age should be tempered with a better appreciation of technology's limits, especially under the constraints of the current social order.

Second, many technologies have brought great transformations, but their contribution to broad-based prosperity has been something short of stellar. In the case of the IT revolution, there is a fabulously wealthy software aristocracy and a well-remunerated managerial stratum who continue to sing of its wonders and the promise that the best is yet to come. But even Microsoft, by many measures the most successful firm of the IT revolution, has a total workforce of 50,500 people in 2003, "just one-seventh those of both Ford and General Motors," observes former *Financial Times* journalist Eamonn Fingleton, who adds that "the latter two companies as of 2003 were a fraction of their former selves in employment terms" [18 pp.42-43]. In the early 1980s, GM indeed had over 850,000 employees and Ford, approximately 500,000.

Finally, for some U.S. partisans, nanotechnology is seen as a "silver bullet" that might rescue the lackluster sectors of the national economy, particularly manufacturing. It should be recalled that the United States has often had technological supremacy over capitalist rivals, but quickly squanders the lead. Manufacturing entails a lot more than just having the best technology, and it requires the proper orchestration of many elements. Fingleton tells the story of steppers, which are "the mainstays of the semiconductor equipment business—a business that in a good year generates close to $50 billion in revenues [18, p.131]." Steppers are the machinery putting the complex patterns onto microchips. Nikon, Canon, and ASM Lithography (Netherlands) dominate this industry, which in the early 1980s was largely monopolized by the U.S.-based CGA Corporation. CGA invented the equipment, but it was soon crushed by Japanese superiority on a number of fronts: better optics, a highly-trained workforce whose lifetime job security rendered them less likely to take skills to rival firms, and wells of long-term capital supplied by Mitsubishi and other firms less afflicted with the short-term stock market bonanza horizons of the U.S. managerial elite.

Nanotechnology will have a substantial role in the production processes of the future. But similar to IT and many remarkable technologies, nanotechnology should not be expected to fulfill the Machine Messiah prophecies of its most devoted adherents.

Technology, Ethics, and the Commercialization of Higher Education

From its very inception, the history of technology as a field of intellectual inquiry carried concerns about the balance of power between machines and humans. In his essay "The Drama of the Machines," Lewis Mumford asserted that the machine "has conquered us. Now our turn has come, not to fight back, but to absorb our conqueror." [19, 20] The mechanized slaughter of World War I had contributed to the sense that the machine could now overwhelm the human capacity for resistance.

While Mumford believed that better historical understanding of technology and a societal vision of its goals might put the machine back under human control, the 21st century response has been a call for a strengthened regime of ethics in the scientific curriculum and in the training of corporate leaders. It should not be forgotten, however, that in the late 1980s an enormous amount of resources went into the expansion of ethics for aspiring corporate leaders. In 1987, outgoing Securities and Exchange Commission chairman John Shad gave the Harvard Business School (HBS) $20 million to set up professorships in business ethics, for as he told the *Boston Globe*: "I've been very disturbed by the great number of leading business and law school graduates becoming felons [21]." Alluding to SEC prosecutions, Shad elaborated: "Some of those we're bringing cases against are Baker Scholars, Rhodes Scholars, Phi Beta Kappas. It's the cream of the crop, and that's what is so shocking [22]."

What then were the results of Shad's extraordinary commitment of resources to redressing elite moral erosion? During the 1990s, Enron executive Jeff Skilling (HBS class of 1979) boasted of hiring 250 MBAs per year, many of whom received training from Shad's corps of ethics faculty [23]. Few appeared to have any scruples concerning Enron's business model, which rewarded fraudsters and mountebanks with the evident approval of "independent" financial analysts.

When it comes to the age of nanotechnology and biotechnology, Michael West, the founder of Geron and later CEO of Advanced Cell Technology, seems to admit that bioethicists play little role in shaping company policy because it may be easy to select ethicists who are already in accord with your practices. "In the field of ethics," he told Stephen S. Hall [24, p.323],"there are no ground rules, so it's just one ethicist's opinion versus another ethicist's opinion.... You're not getting whether something is right or wrong, because it all depends on who you pick." Bioethicists on occasion argue that they have been more successful at establishing standards than business ethicists, but they are under considerable challenge today.

There has been admirable concern that the project of nanotechnology is informed by ethical and social science insight. Contributors to the STEM Pathways workshop held at NSF in Fall 2003 argued that some of NASA's recent spectacular failings could have been addressed and possibly remedied by a better understanding of the social science on organizations and institutional dynamics. A favorite of New Labour in Britain and organizational theorists in many business schools, Charles Leadbeater puts it this way:

> Creative work will increasingly involve people working in teams which combine members with different skills and backgrounds. These teams are most effective when people can trust fellow team members to play their part. In low-trust organizations, people will tend to hoard knowledge and only share ideas formally through memos and when requested. In high-trust organizations, people are more likely to bestow their knowledge on one another and develop joint understandings of problems and their solutions. Trust and co-operation will be vital to the work cultures of the future. [25, p.153]

The pursuit of "pure science" is not enough, especially in enterprises that increasingly require massive teamwork to accomplish distant goals. Nevertheless, it may be necessary to assess candidly the limits, the failings, as well as the triumphs of ethical and social scientific modes of inquiry.

Universities are under acute pressure to develop structures and institutions that can better promote the new sciences, including biotechnology and nanotechnology. Critics of biotechnology have long argued that it is a driving force behind convergence between university science and corporate science. The commercialization of university laboratories ignited social criticism; but corporate strength continues to grow, perhaps augmented by the fiscal crisis of the state. As state budgets decline for universities in the early 21st century, corporate interests and resources receive a warmer welcome from beleaguered administrators. Nanotechnology is likely to strengthen calls for corporate-university collaborations. One of the engineers of greater corporate-university ties in the 1980s, former Harvard president Derek Curtis Bok, warns today that the pendulum has swung too far and that society will pay a serious cost if commercialization continues to progress without challenge [26].

It might be useful to reflect on some of the dangers to the university if corporate interests have too dominant a role in shaping science. Disclosure restrictions appear to be a growing feature of university-industry research centers. A survey conducted by Carnegie Mellon University in the early 1990s indicated that over half of the centers allow corporate partners to delay publication, and a third permit industry to delete information from papers slated for publication. Dianne Rahm's survey of the top-100 R&D-performing universities confirmed the CMU findings. In Richard Florida's summary of her findings, "79 percent of technology managers and 53 percent of faculty members reported that firms had asked that certain research findings be delayed or kept from publication [27, p.70]." Restrictive covenants are contributing to public cynicism about the hidden agendas of academic science. UCSF scientists came under enormous pressure to suppress research showing the efficacy of a drug that could save patients with thyroid problems $365 million per annum. Sheldon Krimsky of Tufts University has cited this example as the workings of science in the private interest, rather than the public interest [28].

Academic institutions want to become part of for-profit ventures; but getting in on venture capital action has had baleful results. Placing itself "in the venture capital game, a high-stakes contest where they don't belong," reflects Richard Florida, "Boston University, for instance, lost tens of millions of dollars on its ill-fated investment in Seragen. These activities do little to advance knowledge per se and certainly don't attract top people. They simply tend to distract the university from its core missions of conducting research and generating talent." In fairness to corporations, it may well be the case that university administrators and academic entrepreneurs among the faculty are driving this phenomenon. In the CMU survey, "Some 73 percent of the university-industry research centers indicated that the main impetus for their formation came from university faculty and administrators. Only 11 percent reported that their main impetus came from industry [27, p.69]."

Those committed to the progress of nanotechnology will need to give important consideration to the independence of research universities in the decades ahead. It will take vibrant societal structures and institutions to ensure that technological advances benefit the broader society, rather than merely enriching the favored few.

This essay has sought to provide perspective on technology's contributions to economic growth and the role of societal institutions in establishing ethical frameworks for knowledge. In the search for greater efficiency and wealth, many institutions have succumbed to privatization and what is regarded as "the commercialization of everything." The university's independence may well have been exaggerated in the past, just another form of belief in the Golden Age. However, if there are to be spaces where technology can be evaluated without fear or favor, it will require public support for institutions that can look out for the general well being. Otherwise the social impacts of new technology may fall short of the promise of its visionaries, and public disenchantment will likely create a stormier landscape for future scientific endeavors.

References

1. E. Dyson, G. Gilder, J. Keyworth, A. Toffler, A Magna Carta for the knowledge age, *New Perspectives Quarterly* (Fall 1994). Several versions of this text circulate in cyberspace.

2. R. Brenner, The boom and the bubble: The U.S. in the world economy, New York: Verso (2002).

3. J. Madrick, Computers: Waiting for revolution, *Challenge* (July-August 1998).

4. M. J. Mandel, Productivity: Who wins, who loses; The U.S. is reaping big—but uneven—gains from its highly efficient workforce, *Business Week* **45** (22 March 2004).

5. J. C. Cooper, The price of efficiency: Stop blaming outsourcing. The drive for productivity gains is the real culprit behind anemic job growth, *Business Week* **45** (22 March 2004).

6. R. Pollin, Contours of descent: U.S. economic fractures and the landscape of global austerity, New York: Verso (2003).

7. K. Phillips, How wealth defines power; the politics of the new gilded age, *The American Prospect* (May 2003 supplement).

8. C. E. Weller, A. Hersh, Free markets and poverty, *The American Prospect,* A13 (Winter 2002 supplement).

9. M. Weisbrot, The mirage of progress, *The American Prospect,* A10 (Winter 2002 supplement).

10. S. Peres, quoted by Juan de la Roca, A nation's architect finds blueprints for the future: Shimon Peres sees nanotech as Israel's missing link between a pioneer past and technological progress, *Small Times* **3** (7) (November/December 2003), 12-13 (available at http://www.smalltimes.com).

11. Roland, P. Shiman, Strategic computing: DARPA and the quest for machine intelligence, 1983–1993, Cambridge, MA: MIT Press (2002).

12. D. Caruso, Technology: Digital commerce; Contributing to the bumpy ride for Internet stocks is a familiar overload of eager hangers-on, *New York Times* (16 August 1999).

13. D. D'Souza, The moral limits of wealth, *Forbes* (9 October 2000).

14. D. D'Souza, The virtue of prosperity: Finding values in an age of techno-affluence, New York: Free Press (2000).

15. J. Gray, World at war; Terror on the Tube and plans to invade Iraq are just the beginning; Industrialisation, it was thought, would end human conflict; in fact, it leads to a fiercer battle for scarce resources, *New Statesman* (25 November 2002).

16. Ladies Home Journal quoted in V. Mosco, The digital sublime: Myth, power, and cyberspace, Cambridge, MA: MIT Press (2004).

17. S. L. Del Sesto, Wasn't the future of nuclear energy wonderful? In *Imagining tomorrow: History, technology, and the American future*, ed. J. J. Corn, Cambridge, MA: MIT Press (1986).

18. E. Fingleton, *Unsustainable: How economic dogma is destroying American prosperity*, New York: Thunder's Mouth Press/Nation Books (2003).
19. L. Mumford, The drama of the machines, *Scribner's Magazine* **82**(2), 150-161 (August 1930). See discussion of this text in W. Solomon, *Literature, amusement, and technology in the Great Depression*, Cambridge: Cambridge University Press (2002).
20. D. L. Miller, *Lewis Mumford: A life*, New York: Weidenfeld & Nicolson (1989).
21. R. Lenzner, SEC Chairman to donate $20 million to Harvard for business ethics program, *Boston Globe* (31 March 1987).
22. A. L. Cowan, Harvard to get $30 million ethics gift, *New York Times* (31 March 1987).
23. J. Doward, Business & media: The firm that built the house of Enron: McKinsey refocused the energy firm; Now it fears collateral damage from the collapse, *The Observer* (24 March 2002).
24. S. S. Hall, Merchants of immortality: Chasing the dream of human life extension, Boston: Houghton Mifflin (2003).
25. C. Leadbeater, *Living on Thin Air: The new economy*, New York: Viking (1999).
26. D. Bok, Universities in the marketplace: The commercialization of higher education, Princeton: Princeton University Press (2003).
27. R. Florida, The role of the university: Leveraging talent, not technology, *Issues in Science and Technology* (Summer 1999) (available at http://www.issues.org).
28. S. Krimsky, Science in the private interest: Has the lure of profits corrupted biomedical research? Lanham, MD: Rowman & Littlefield (2003).

NANOREVOLUTION: IMPLICATIONS FOR THE ARTIST

Tanwin Chang, National Bureau of Economic Research

Art is both a reflection of society, and a driver of its evolution. As interpreters of cultural change, artists need to be attuned to new developments in human knowledge as a source of ideas and inspiration [1]. Artists have a history of exploiting technological advances in the exploration of aesthetic experience, thereby forming a conduit through which new technology impacts the general public. When examining the societal implications of nanotechnology, it is useful to consider nanotechnology's effect on artists.

The invention of photography provides an example of the synergistic power of art and science to transform culture. In the early 1800s, Daguerre was a respected stage designer seeking a method of recording an image, which he believed would be useful as a tool for painting his signature dioramas. Collaboration with a gentleman inventor Niépce, who was educated in physics and chemistry, provided the initial technology. After a decade of further experimentation, the invention of the "daguerreotype," the first commercially available photographic process, was announced in 1839. Imagine modern life without photography in its innumerable forms: no photo albums, no glossy magazines, and no film industry. Nanotechnology may similarly have extensive consequences on the world of art, and therefore society at large.

Already artists are commenting on—and indeed educating the public about—nanotechnology. For instance, the Los Angeles County Museum of Art is running an exhibition through September 2004 called "Nano" that makes "nanoscience visible, tangible, and experiential." While artists certainly have much more to contribute in this manner, what is more interesting is the potential effect of nanotechnology on artists, rather than the converse. In other words, let's focus not on artistic *depictions* of nanotechnology. Instead focus on how nanotechnology can *be* art, or otherwise *create* or *enable* art.

Here are three predictions regarding nanotechnology and artists:

1. Nanomaterials will provide a wealth of new media that will be adopted by artists.
2. An increasing percentage of the artist population will be scientifically trained.
3. Artists will explore a new "sensory space" engendered by nanosensor technology.

Nanomaterials as Experimental Artistic Media

The use of nanomaterials to enhance beauty and aesthetics has already begun.

The vibrant colors of opals and peacock feathers are captured in the form of iridescent paints, which is usually a liquid medium carrying a suspension of tiny microscale flakes. Each flake is about 1 micron thick and 12 microns wide, and is composed of several layers of transparent material [2] designed to exploit the fact that light can interfere with itself. Each layer is typically in the range of 100 nm to 350 nm thick, and the precise thickness of each layer determines the way the color changes depending on viewing angle. For instance, the color of a car coated with iridescent paint will appear to shift in hue as it passes by. Iridescent paints are already commercially available at art supply stores, and for automobiles at specialty body shops.

Nanotechnology is finding application in fashion. Stain-resistant pants have already been the subject of a splashy marketing campaign since they were introduced a couple of years ago. Each fiber in the fabric is coated with a covering of nanoscale whiskers, which causes water to bead-up and roll off the fabric to prevent a stain [3]. Other research is directed towards nano-fibers comprised of a variety of materials in order to engineer the optical, electrical, and/or chemical properties of the fabric to create so called "multifunctional textiles [4]." Eventually, nano-enabled smart clothing will respond to the wearer in ways inspired by the ideas of haute couture fashion designers. For example exercise clothing might broadcast one's dedication to fitness by vibrantly changing color as a reaction to perspiration or increased body heat.

Nanocapsules or "nanosomes" are used in cosmetics to increase the effectiveness of skin creams by encapsulating and transporting active ingredients beneath the outer layers of the skin [5]. Presumably, these nanosomes, which will eventually be designed to implant temporary pigments into the skin, could be effectively used by make-up and tattoo artists to further their craft. Socially conscious performance artists may choose to alter skin color to comment on race relations. Extrapolating to the more fantastic, science fiction writers have written about societies where body art is applied and removed daily like clothing to confer mood, circumstance, or status [6].

Carbon nanotubes possess at least 30 times the tensile strength of steel at one-sixth the weight. Once they can be reliably braided into cords and cables, advances in the field of architecture and sculpture will follow. The design of suspension bridges, certainly, will undergo a transformation. Carbon nanotube cables should also invigorate the sub field of architecture that is concerned with tensile structures, which are presently used for enclosures that span a large area, such as in a sports stadium, airport, or train station.

Future chefs will laud the rise of nanotechnology: "Food may also be designed to suit individual profiles, and even selected tastes and textures. Nanotechnology is being applied to studies into improved flavor deliver, encapsulating flavor particles in nanoparticles to protect them from the environment until they are released, thereby maintaining freshness [7, p.21]." Also, there is an intriguing possibility to create photonic crystals constructed from edible materials as a benign additive to alter the appearance of food [8].

Artists seek in their choice of materials a spectrum of properties that may be combined, mixed or otherwise worked to reflect an aesthetic vision. Nanomaterials offer the promise of media with combinations of sensory properties that have never been experienced before. The examples above are the proverbial tip of the iceberg.

The Return of the Scientist to the Artist Community

Some argue that the creativity of the Italian Renaissance was in part due to the intermingling of art and science at that time [9]. The discovery of perspective drawing and the desire of visual artists to render more realistic paintings facilitated innovations in science and engineering—particularly in architecture and anatomy. Those investigators proficient at the sciences were also accomplished artists, because it was precisely through artistic renderings that scientific knowledge was transmitted.

The artist and the scientist are similar beings. This is evidenced by the examples of scientists/artists living among us. The National Academy of Sciences *Beauty of Phenomenon* exhibition (2003) features the work of eight prominent scientists/artists. The curator's statement:

> Since the 19th century a belief has flourished that science and art are two mutually exclusive cultures. However, both scientist and artist are on similar quests: to explore and express our understanding of the world around us.

As a group, scientists have an inherent affinity for art and the artist's lifestyle. This may be attributed to the romanticized aspiration of many scientists to be the maverick who upsets conventional wisdom and works feverishly in a basement laboratory validating his or her own theories. In reality, much of science and technology is done in large groups, with several levels of hierarchy, and where most researchers must serve the vision of the group leader. For these researchers, who are disillusioned with the process of big science, the artist's lifestyle is particularly inviting; a lifestyle which generally rewards vivid imagination without demanding intellectual compromise. Presently the journal *Leonardo*, published by the International Society of Arts, Science, and Technology, serves as a forum for artists and scientists to exchange ideas. The Art and Science Collaboration, Inc (www.asci.org) publishes a Web site fostering joint projects between artists and scientists.

Nanotechnology is fertile ground for scientists and art to meet. The book *NanoCulture: The New Technoscience and its Implications for Literature, Art, and Society* attests to the power of nanotechnology to stimulate the artistic mind [10]. Nanoscience researchers have already created art that is enabled by nanotechnology. In 1996, researchers at IBM's Zurich division constructed a nanoscale abacus, which could arguably be considered one of the smallest sculptures ever, as a vivid demonstration of the capabilities of scanning tunneling microscopes. Eric Heller, a physicist and participant at Harvard's Nanoscale Science and Engineering center, has produced a collection of digital images derived from the transport of electrons in nanostructures. Several of his works are featured in the National Nanotechnology Initiative's 2004 budgetary documents [11].

Moreover, the advent of nanotechnology will enable scientists to take up artistic pursuits as never before. The simplest reason is that scientists will have the most access and best understanding of new nanomaterials. Secondly, the process of nanomanufacturing has a strong component of design. That is, nanoengineering projects will tend to offer many degrees of freedom, so engineers may increasingly find themselves guided by aesthetics and essentially following in the tradition of architects. Also, nanotechnology will eventually increase the productivity of the society as a whole, so the community of scientists and engineers will have greater means to take up artistic affairs. Presently, many potential artists are discouraged by the difficulty of making a living through their artwork. As productivity increases, goods become more affordable, and it becomes easier for people to take up artistic activities that are not necessarily remunerative. Nanotechnology workers, in particular, may enjoy "sabbaticals" between employment that can be devoted to the arts. That's because nanotechnology is really a collection of different technologies that will cause the typical nanotechnology worker to change employers fairly regularly as one segment of nanotechnology rises economically over other segments.

Novel Sensory Spaces

Miniaturization of "bionic" equipment is underway. Several companies offer implantable hearing aids. NASA has funded projects to develop artificial retinas. Direct silicon/nerve interfaces are being developed to assist amputees control prosthetic devices. Eventually entire suites of sensors may be placed in one's body to monitor health. More dramatically, sensors may be placed in other objects, but electronically linked to a nerve implant, so that one could "feel" what it's like to be a car, or a bat [12], or even another person.

This is the ultimate canvas for the artist. If this type of technology passes from science fiction to reality, the sensory space available for artists to explore will be enormous. As an example, consider Figure 2.2, the art project "What does it feel like to be a bridge?" Bridge safety and maintenance is an important issue that some claim can be addressed by inserting fiber optic sensors into the load- bearing elements of the structure. These sensors will be designed to measure the forces on the bridge due to weather and traffic. Presumably a computerized system will monitor the data from the fiber optic sensors to keep track of the "health" of the bridge. In the age of implantable nanosensors, it is conceivable that a suite of nerve stimulators could be implanted in a person's hand. Then the sensor data from the bridge could be wirelessly downloaded and used to stimulate one's palm to give a sensation that is directly correlated with the state of the bridge.

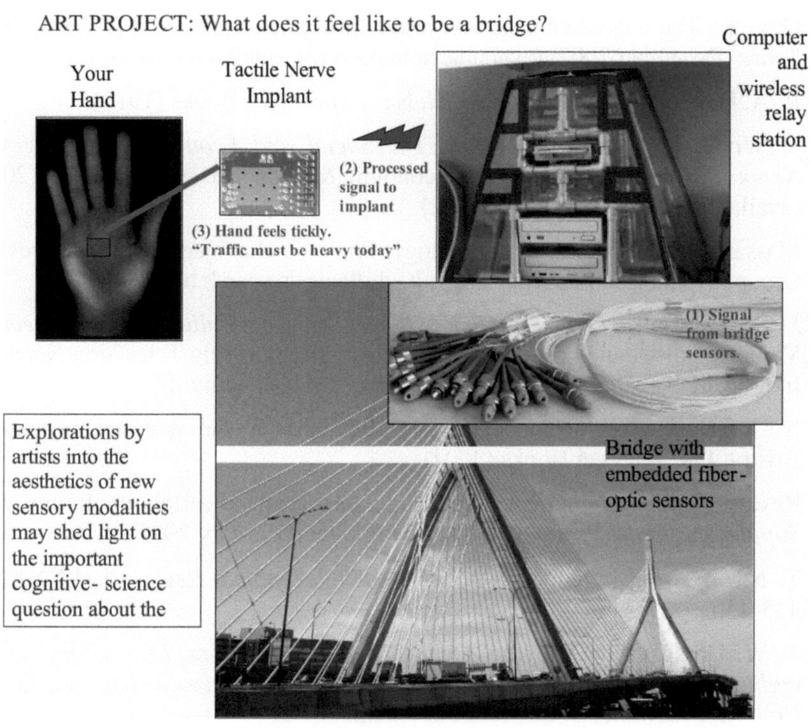

Figure 2.2. Feeling something outside yourself (Photographs by the author).

Artistic interest in projects such as this would contribute to science as well as aesthetics. Cognitive scientists are interested in the problem of qualia: "Qualia is a term which is applied to indivisible primary feelings such as the feeling of seeing red in an apple or the feeling of pain in one's foot. Traditionally qualia are considered to be purely first-person or subjective in nature [13, p.76]." But when it becomes possible to "feel" another person's body as if it were one's own, the difference between subjective and objective may be a gray area indeed.

Nano-enabled art that examines this gray zone, made by artists fascinated with subjective/objective experience, will undoubtedly serve the cognitive sciences as surely as the study of anatomy was informed by Renaissance artists who were fascinated with realistic depictions of the human body.

References

1. S. Wilson, Research and development as a source of inspiration and ideas for artists, *Leonardo* **24**(3) (1991).
2. Thin-film interference color, *ChromaFlair Technical Bulletin* No.1 (May 2000) (available at http:// www.colorshift.com).
3. Get ready for new "nano" products, *Wired News* (July 2003).
4. Lightening the load on soldiers, firefighters, *Results: Research and Graduate Studies at North Carolina State University* **2**(2) (2002).
5. G. Oger, The nanoscience behind beauty is serious business at L'Oreal, *Small Times* (December 2002) (available at http://www.smalltimes.com).
6. R. A. Heinlein, *I Will Fear No Evil*, New York: Ace Books (1970).
7. S. Wood, R. Jones, A. Geldart, *The Social and Economic Challenges of Nanotechnology*, Swindon, UK: Economic & Social Research Council (2003) (available at http://www.esrc.ac.uk).
8. Massachusetts Institute of Technology: Photonic Bandgap Fibers and Devices Group (available at http://mit-pbg.mit.edu/Pages/research.html).
9. J. W. Dauben, *The Art of Renaissance Science: Galileo and perspective.* Videotape distributed by SETN (Science and Engineering Television Network, Inc) (available at http://www.crs4.it/Ars/arshtml/arstoc.html).
10. N. K. Hayes, ed., Nanoculture: Implications of the new technoscience, Bristol, UK: Intellect Books (2004).
11. Research and development supporting the next industrial revolution, *NNI Supplement to the President's FY2004 Budget* (February 2003).
12. T. Nagel, What is it like to be a bat?, *Philosophical Review* **4** (LXXXIII), 435-450.
13. T. W. Griffith, M. D. Byrne, Qualia: The hard problem. *Proceedings of the eighteenth annual conference of the Cognitive Science Society*, ed. G. W. Cottrell, Mahwah, NJ: Lawrence Erlbaum, 76-79.

VISION, INNOVATION, AND POLICY

William Tolles, Consultant

Innovation begins with a gleam in the eye ("vision"). Technologies emerge with successful innovation. Characteristics of successful technologies (those that displace existing ones) have been examined extensively by Clay Christenson [1]. A "disruptive technology" emerges first by the establishment of a new product in a niche market. Thus, by examining the appearance of newly established niche markets, the path of change may be anticipated. Appropriate policies may then be introduced to encourage or discourage various aspects of the anticipated path. In the absence of newly established niche markets, examining emerging patterns in patents may give a clue as to where niche markets might possibly be established. For successful examples, the emergence from niche market to dominance takes typically 10–20 years. Thus, development of policy typically has that period of time after niche markets have been established to consider revising or introducing policies affecting a new technology.

By examining the frequency of patents that refer to nanometer structures (these patents have had an order-of-magnitude growth over the past five years), a measure of emphasis is obtained (unpublished data). For patents in the United States, by far the largest number of patents in this category falls in the area of pharmaceuticals, antibiotics, medical diagnostics and treatments. Many new patents have emerged that make use of nanoparticles in plastics and composites, as well as food and agriculture. The greatest impact of nanotechnology appears to be emerging in these areas presently. The path of the semiconductor industry continues with the next most frequent number of patents, followed by glasses, ceramics, laboratory test equipment, and image-producing devices.

The emergence of a "vision" constitutes only the first step toward impact of an innovation. Developing a stimulating vision is important, even critical, for stimulating and integrating many individual efforts toward desired innovative goals. Nanotechnology emerged with a wide variety of visions enunciated by a diversity of groups. The issues emerging for nanotechnology represent a complex set of interactions between competing visions (some real, some imaginary) and societal concerns. It is important to distinguish the real from the imaginary and to develop a vision that is credible, achievable within a lifetime, and that has a positive impact on society.

The field of nanotechnology was stimulated by the emergence of new tools that provided information about the behavior of materials at dimensions previously not possible. Macroscopic behavior is frequently dependent on material properties at nanometer dimensions. The goals of nanotechnology were readily stated in terms of societal impact, imagined or real. A positive aspect of the emergence of the field is that a number of good academic research programs are crafted about goals to which society can relate.

An Appropriate Vision

To this end, recognize that there are a number of detractors to a positive image of what is achievable through nanotechnology. The extent to which nanotechnology has been misrepresented is illustrated by many examples. In China there has been a clamor for truth after vendors advertised "nanotech washing machines and refrigerators" based on the fact that some paint may have been used to line a part of an appliance [2]. Claims that a microliter of salt solution containing DNA has the capacity to carry out 66 billion operations per second stretch the truth [3]. This leads the public to feel that computational capabilities at the molecular scale are many orders-of-magnitude greater than today's computers. This avoids the issue that the path of particles is not equivalent to controllable computations in a computer. There are many examples of exaggerated or irrelevant claims about the potential of nanotechnology. Public criticism of nanotechnology is feeding on such claims:

> Nanotechnology is being presented through the media as a vehicle for scientific fantasy rather than of community advancement [4].

Pronouncements extolling the possible impact of nanotechnology to be far beyond what is possible have provoked criticisms of the science in proportion to the exaggerated visions enunciated. One of the first of these was the treatise of Bill Joy [5]. The book *Prey* by Michael Crichton [6] has based its theme on the highly unlikely event of an intelligent swarm of nanoparticles that replicate with nourishment from flesh. Visions of self-replicating nanobots developed by a country that can overwhelm another country are emerging faster than any new real self-replicating life forms [7]. The Action Group on Erosion, Technology, and Concentration (ETC) has issued proclamations about the need to control research in this field [8, 9]. Prince Charles has reacted to such documents [10], setting off a number of studies in the United Kingdom [11]. Ignoring the signs of "irrational intelligentsia" can be debilitating; witness the impact of sentiment against genetically modified foods, which resulted from inattention by scientists [12]. What is badly needed is a vision that is credible, realistic, and shared by the public as well as the scientific community [13]. "Gray goo" as a threat has been recognized by an understanding Congress to this point [14]:

> As many people here know, the most extravagant fear about nanotechnology is that it will yield nanobots that will turn the world into "gray goo." That's not a fear I share, but I do worry that the debate about nanotechnology could turn into "gray goo"—with its own deleterious consequences.

Fundamentals

Fortunately, science builds on a firm base of an exceedingly complex and interlinked set of hypotheses, proofs, and experimental observations that emerge with amazing consistency. The fundamentals of thermodynamics are critical to understanding, for example, why perpetual motion is not possible as a source of energy. Such fundamentals also place limits on the manner in which matter may be transformed from one state to another. The behavior of particles as waves has gained extraordinary insight with the tools of quantum mechanics. Electronics and information theory have emerged with many fundamental truths associated with the creation, storage, and distribution of bits of information. These fundamentals form a basis for refuting many of the irrational visions being propagated today.

At the same time, concerns about the interaction of nanoparticles with living species is a valid one, and laboratory tests, incomplete at present, are an essential prelude to introducing large quantities of products into the public marketplace.

It is in the best interests of the scientific community to develop a rational basis for visions and limits to those visions based on scientific fundamentals, particularly those behaviors that are associated with the irrational fears about the vision of nanotechnology today. Thermodynamics, chemical dynamics, biology and biochemistry contain the concepts that will reveal what is possible with nanobots and self-replicating species. When it comes to information technology, vast resources in industry have responded to commercial opportunities over the past several decades, and have developed an amazingly capable complex of researchers knowledgeable about what is possible with today's materials and treatments involving information technology. Any academic community concerned with information technology (for example, molecular computers) would benefit greatly by joining forces with the intellectual components of this industry to search for the true limits of what may be realistically expected.

Policy, patent law, and social acceptance will follow rational visions, suitably developed in a reasonable time frame. It is imperative that credible scientific discourse lead the visions and actions of those shaping social, legal, and policy frameworks in the future.

References

1. C. Christensen, *The Innovator's Dilemma: When new technologies cause great firms to fail,* Boston: Harvard Business School Press (1997).
2. Financial Times Information Global News Wire, China Online (12 April 2001); FT Asia Intelligence Wire, China Daily (22 June 2001).
3. N. Easen, Is life the key to new tech? CNN.com International (22 September 2003). (available at http://edition.cnn.com/2003/TECH/09/19/wow.tech.life.computing/index.html).
4. P. Binks, Questions loom large in nanotech's tiny world, *The Age* (21 October 21, 2003). (available at http://www.theage.com.au/articles/2003/10/20/1066631346170.html) B. Joy, Why the future doesn't need us, *Wired* (April 2000). (available at http://www.wired.com/wired/archive/8.04/joy.html).
5. M. Crichton, *Prey*, New York: HarperCollins (2002).
6. L. Navrozov, Nanotechnology—Part II, WorldTribune.com (15 August 2003). (available at http://216.26.163.62/2003/lev8_15.html).
7. ETC (Action Group on Erosion, Technology, and Concentration), *The big down*, Winnipeg: Author (2003). (available at http://www.etcgroup.org/documents/TheBigDown.pdf).

8. ETC, *No small matter II: The case for a global moratorium*, Winnipeg: Author. (available at http://www.etcgroup.org/documents/Occ.Paper_Nanosafety.pdf)

9. Prince Charles whips up a nanotech storm in the UK, *Trends in Nanotechnology Weekly* (5 June 2003). (available via free subscription at http://www.cientifica.info/html/TNT/tnt_weekly/archive_2003/issue_5.htm#_Toc40082455).

10. Royal Society, *Nanotechnology: Views of scientists and engineers*, London: Author (2003). (available at http://www.nanotec.org.uk/workshopOct03.htm).

11. D. Brown, Perception may be nano's biggest enemy, leaders tell Congress, *Small Times* (10 April 2003). (available at http://www.smalltimes.com/document_display.cfm?document_id=5809).

12. J. R. Von Ehr, Testimony of Mr. James R. Von Ehr II, Chief Executive Officer, Zyvex Corporation, before the U.S. Senate Committee on Commerce, Science, and Transportation (1 May 2003). (available at http://commerce.senate.gov/hearings/testimony.cfm?id=745&wit_id=2015).

13. S. Boehlert, Opening statement for hearing on nanotechnology consequences, U.S. House Committee on Science (9 April 2003). (available at http://www.house.gov/science/hearings/full03/apr09/boehlert.htm).

3. CONVERGING TECHNOLOGIES

NANOTECHNOLOGY'S IMPLICATIONS FOR THE QUALITY OF LIFE

Carlo Montemagno, UCLA

It is important to consider nanotechnology developments in the context of the future and the potential implications that they may have. Some coming opportunities may not be immediately obvious, based upon what we have seen in the past. In his 1959 speech, "There's Plenty of Room at the Bottom," Richard Feynman said: "The biological example of writing information on a small scale has inspired me to think of something that should be possible....Consider the possibility that we too can make a thing very small, which does what we want—that we can manufacture an object that maneuvers at that level!" [1, p.29] Why did it take 40 years for this to occur?

In the last four decades, a number of major advances have enabled the technology to move ahead, creating micromachine devices that enable us to start working with molecules at the single scale. We are starting to understand how cells work and how to manipulate matter on the scale of single molecules. We have been able to minimize and store huge quantities of data and have access to these data. Together, those developments form an enabling event that generated immense vigor around this area. As a result, many new advances have occurred, particularly in diagnosis and medical treatment, involving precise manipulation of matter.

Examples include laboratories on a chip such as the LabChip systems produced by Caliper Life Sciences Corporation, the micro-tweezers produced by MEMS Precision Instruments, and the long-standing work on neural microprobes at the University of Michigan's Center for Neural Communication Technology. Especially notable are the microcages, pneumatic manipulators, and DNA microarray injection pins created by C. J. Kim's Micromanufacturing Laboratory at UCLA [2, 3]. At Rice University, Jennifer West's laboratory has been developing medically useful nanoshells, a new type of nanoparticle with tunable optical properties, including gold "nanobullets" that could be employed in cancer therapy where they would be guided by magnetic resonance and then generate heat to destroy tumors [4].

The ability to use machines to manipulate matter a single molecule at a time renders many things possible that were impossible before. For example, this ability has enabled our team at UCLA to make a device that is a motor being run by a single molecule [5, 6, 7, 8, 9]. More recently, it has allowed us to begin engineering some basic motility devices that use the concerted reaction of a number of molecules to allow motility to occur. These things altogether have fostered what we call nanotechnology.

To get to the nanoscale, one of the things that we have to be able to do is interface with matter at larger scales, because those are the scales at which we work. One of the goals people talk about is making bio-bots, robots that are small, comparable

M.C. Roco and W.S. Bainbridge (eds.),
Nanotechnology: Societal Implications — Individual Perspectives, 131–168.

to microscopic organisms. The strategy, of course, is to manipulate the chemistry and topology of the structures, creating the devices using MEMS technology, and have them self-assemble and make our own bio-bots. Our team has created a robotic device about 160 microns in length that can walk across a surface, small enough to fit easily inside the zero on the head of a penny. Admittedly, this is still microtechnology, but it shows that by using nanotechnology, we can work on that interface between the two.

A Paradigm Shift

The question then becomes: How do we go to larger scales? That requires a paradigm shift in understanding what is really required to make this new next-generation type of materials. Consider the work done in the research laboratory of Robert H. Singer at the Albert Einstein College of Medicine in New York, studying the reproductive behavior of yeast cells [10, 11]. Yeast cells are very interesting organisms in their own right that reproduce by budding. Only a female yeast cell can bud, and whether it buds or not depends on the presence of a single protein. The problem is how a female yeast cell can produce a male offspring. If she produces the protein during reproduction, she becomes a he, and the budding process does not work any more.

Here is how it does work. Messenger RNA is produced, put onto carrier molecules, then onto a kinesin onto a microtubule, and then transported down to the budding cell where the protein is actually produced. There is a purposeful transport of information, occurring through the stochastic interactions of basically seven molecules. This is a higher order of behavior that sends information down from the mother cell to her son, emanating only through molecular interactions.

So from this, we have come up with the tenet that one should look at the stochastic interactions in single molecular systems. They have higher orders of scales of observation, in which emergent properties come through as it goes to scale—from single molecules up to groups of large clusters of molecules up to the scale of change. Higher-order functionality emerges from stochastic non-linear interactions at a lower level that is observable at a higher level. Figure 3.1 suggests that this rule may apply at many nested levels.

The implication here is that the living system does not work like clockwork, but comes about as an emergent result of the fact that there is randomness associated with the processes. To elicit this higher order of behavior requires us to incorporate this robustness and the randomness associated with it.

Integrative Technology

A key part of the paradigm shift is *integrative technology*, the fusion of biotechnology, nanotechnology, and informatics. We look at nanotechnology for the precision engineering of the matter. We look at biotechnology providing the fundamental building blocks for our function. And then we look at informatics for understanding and controlling how the flow goes between those entities.

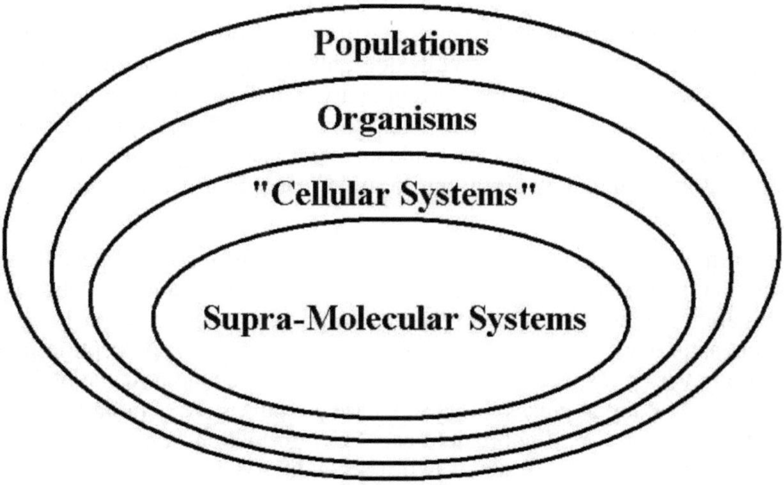

Figure 3.1. Emergence of functionality from lower to higher levels.

So how do you try and incorporate that? Through nanoscale structures. A living cell is not just a bag of chemicals, but a structure of membranes. Information is incorporated in the structure associated with cells, and that structure provides delineations of functionality that occurs inside the cells. So if you create a biomimetic membrane, a membrane that allows you to insert functional building blocks of living systems, you should be able to get higher-order behaviors.

We have looked at aquaporin, a protein that transports water across membranes, for systems that purify water. We have experimented with energy transduction systems going from light to chemical energy, and light to electricity. We have valves. All these are building blocks that are found in living systems, and 80 percent of the proteins in the human genome are membrane-bound proteins. All those functional building blocks are there for us to use like Legos, which can be assembled to create artificial organelles performing unit functions. These are chemical factories, unit operations that function as a result of them working together.

Combined in the right manner, these methods can achieve higher-order functionality. Right now, we produce straight linear operations. We filter water. We pump protons. But those are all things that we expect out of the process. One way to achieve a higher order of behavior is to incorporate ion-gated channels. These are the same type of channels that are used in cells to communicate with one another, such as nerve cells for processing information.

Mathematical analysis indicates that a higher order of behavior can be elicited with vesicles about 150 to 200 nanometers in size. They can be made to operate like a monostable multivibrator. This may not seem like much, but it would be an unexpected, emergent property from the system. And it is also a fundamental property that we find in living systems. Notably it is part of the process that we find within the sinoatrial node of the heart that triggers your heart to beat.

This is the very beginning of an induction path. If you have a system in which you incorporate these artificial nodes that self-stimulate, you should be able to mimic the same function that a group of cells does, a higher-order type of function. Can you do this with neuronal cells? Current analysis indicates that the action potentials may be slower than neuronal cells, so some work in engineering will be required. Success with that process could enable transporting information between neurons, taking stimulation from one location, cascading it and stimulating it to another neuron.

Consider the way our brain works when we process information, especially spatial information that has structure. In cortical sensory integration, spatiotemporal patterns are integrated over time to produce responses selective to specific patterns and their natural variants. The fundamental basis of sensory perception involves integration over both space and time. It is difficult to decode how the brain works merely from looking at pulses, because where the pulses are coming from and the associated structure are important pieces of information. This information all comes over a time period of about half a second. Given how slowly the neurons respond, nobody should ever be able to hit a baseball, because there is not enough time for the process.

Human beings do vary rapid pattern recognition based upon their interpretation of reality. This insight provides a framework where we can make environmentally aware materials and systems. These systems would incorporate other biological molecules with chemo receptors or mechanical receptors that stimulate them inside these processes.

In the human brain, there are several neural operations that are fundamental computational building blocks. Detecting transiently similar firing rates is naturally executed by networks of spiking neurons. The collective synchronization event is a "MANY ARE CURRENTLY APPROXIMATELY EQUAL" operation. Integrate-and-fire neurons implement a fuzzy "AND" and a fuzzy "OR." Neurons with transiently similar firing rates can transiently synchronize, and this is how they process banks of information.

We have been experimenting with vesicles on the order of 130 nanometers in diameter that could perform the same logical operations, given appropriate changes in parameters such as diameter, the number of molecules, and perhaps other properties. Adjusting these parameters impacts coordination number, the trigger threshold, and the duration and quality of current pulse and recovery period. From a system of such vesicles, it should be possible to elicit a higher order of behavior. In principle, the result would be a system that processes information based upon interactions among the vesicles. They transmit information between themselves by the communication of where the molecules are. The result could be an engineered bio-computational intrinsically smart system, three-dimensional like the human brain, rather than two-dimensional like computer chips.

Social Implications

Capabilities like those described above require us to start thinking about the social implications of where things are going. In a very few years, DNA-based diagnosis may determine ultrasensitive and accurate pathology of disease through DNA hybridization. This suggests the possibility of replacing traditional doctors, and it raises implications on how medical insurance will be affected. Will machines and automated health care represent peace of mind for patients?

Similar DNA technology may offer disease prediction, forecasting the future of one's health. This suggests the possibility for termination of health care insurance based on results, raising serious issues of legality and ethics. Does withholding results represent illegal action? What are societal ramifications of the technological crystal ball?

The implications that are associated with this level of technology in terms of human privacy and preserving health are issues that have to be reserved. When we have nanotechnology, and we are able to acquire all this information and have it accessible, the impact that it will have on people's lives can be quite substantial unless proper safeguards are addressed.

Many other potential technological outcomes can be imagined. One is a radically new approach to manufacturing, *broadcast architecture.* Nature dictates self-replication by delegating replication codes from a central "server," and a similar assembly approach can be developed for manufactured goods. It will be based on replicating codes of information, and the information available will be transmitted and will be indigenous to the materials and devices that you manufacture. Engineered systems can address the regulation of autonomous activity and undesirable replication.

Another direction for development is unparalleled biomimetics, integrating natural with inorganic systems and engineering hybrid technologies. Biomimetic systems bring the advantages, without the weaknesses, of enhancing stability and robustness while harnessing nature's most potent processes (i.e., energy transduction enzymes) in synthetic materials. There will be a whole suite of new sorts of biofunctional materials, which take the building blocks that living systems have. One would be something that pumps a proton and transforms it into an electron. Put many of them together, and you get a bio solar cell. Take something that pumps a proton, and it is gated to a valve. Now you have a valve that automatically self-regulates pH. You have unit operations in the classical sense of the way that we build chemical factories.

Spying on the nanoscale has been touched upon numerous times, as it is a very real issue. Nanotechnology operates at the scale that current devices cannot reach, enabling us to observe what we cannot see with our eyes. One concept that must be addressed is bio TTL—biological tagging, tracking and locating individuals. We need to make sure that we find the right balance between our own national security needs and our rights for privacy.

We may even have a responsibility for stopping certain developments, but any nanotechnology prohibitions must be based upon correct information rather than rumors. One of my students asks, "Are we prepared for uncontrollable self-assembly processes?" The argument goes like this: Functional nanotechnology relies on self-assembly to mimic natural systems. The process of self-assembly can represent the hazards associated with nanoscience. Current nanobiotechnology concepts include devices based on autonomous behavior of nanoscale-engineered systems. *Are* we prepared for uncontrollable self-assembly processes? What are the environmental and medical-related ramifications? Can we address issues associated with autonomous malfunction of devices we can't see? Will the environmental, military, and public health risks posed by nanotechnology outweigh the proposed benefits?

In fact, I do not think we are going to have uncontrolled self-assembly processes. However, educated students who are working in this area are influenced by a lot of the media to believe this is possible. This is very, very telling, showing that the media are very far-reaching in their influence. We scientists have to make sure we deal with this misconception in a very aggressive manner, so that we don't get ourselves caught in the same type of dilemma that the people who do stem cell research got themselves caught into.

More realistic concerns relate to nanopharmaceuticals and nanorobotics. Those are areas in which I think we're going to have fundamental impacting technologies, and risks really will be associated with them. In the case of nanopharmaceuticals, can uncontrollable aggregation result in catastrophic effects? Therapeutics can react against healthy cells, resulting in new problems. We may see challenged acceptance of proposed technology based on ingesting bacterial materials (i.e., pore proteins) as drug delivery vehicles.

Nanorobotics may be designed to scavenge hazardous materials. Can the robots then serve as mechanisms to re-deliver the scavenged chemicals? If they exhibit autonomous behavior, will autonomous malfunction create destructive functionality? Whether rightly or wrongly, because devices are invisible, people may fear that virus-like replication could result in uncontrollable problems. Will they be misused for bioterrorism?

We need to face the new frontier. This requires a true partnership, including agencies like NIH, DARPA, and NSF, with the legislative branch of government, industry sectors, society, and the world at large. We have a responsibility for educating the public and demystifying the impossible. We also must deal aggressively with our lawmakers and with our policymakers, keeping them educated and in the loop, and jointly consider appropriate regulatory mechanisms, such as laws, commissions, and standards.

I am very upbeat about the technology, and I think an enormous amount of things can happen. Health care will benefit from nano-bullets to kill tumors and form artificial organs. Electric batteries structured at the nanoscale will deliver cleaner energy. Nanoelectronics will enable cheaper and more efficient consumer electronics, computers, communications, and information systems. The civil infrastructure will have stronger structures, using materials with properties controlled at the nanoscale. The environment will benefit from nano-enabled recycling and waste management, and by real-time toxin detection and cleanup. We should be cautious about imagined military and surveillance applications, such as dissembler weapons, self-replicating nanomachines, and the invasion of privacy.

In general, the possibilities of this fusion of technologies are very real and very positive. We will be able to do things that we never dreamed possible. The next 10 to 15 years are going to be quite extraordinary!

References

1. R. Feynman, There's plenty of room at the bottom, *Engineering and Science* **23**, 22-36 (1960).
2. J. Ok, M. Chu, C.-J. Kim, Pneumatically driven microcage for micro-objects in biological liquid. *Proceedings of the IEEE micro electro mechanical systems workshop*, Piscataway, NJ: Institute of Electrical and Electronics Engineers, 459-463 (1999).
3. J. Tsai, C-J. Kim, A silicon micromachined pin for contact droplet printing. In *Proceedings of the second joint EMBS/BMES conference,* Piscataway, NJ: Institute of Electrical and Electronics Engineers (2002).
4. L. R. Hirsch, R. J. Stafford, J. A. Bankston, S. R. Sershen, B. Rivera, R. E. Price, J. D. Hazle, N. J. Halas, J. L. West, Nanoshell-mediated near-infrared thermal therapy of tumors under magnetic resonance guidance. In *Proceedings of the National Academy of Sciences* **100**, 13549-13554 (2003).
5. C. D. Montemagno, Nanomachines: A roadmap for realizing the vision, *Journal of Nanoparticle Research* **3**(1), 1-3 (2001).
6. G. D. Bachand, C. D. Montemagno, Constructing organic/inorganic NEMS devices powered by biomolecular motors, *Biomedical Microdevices* **2**, 179-184 (2000).
7. R. K. Soong, G. D. Bachand, H. P. Neves, A. G. Olkhovets, H. G. Craighead, C. D. Montemagno, Powering an inorganic nanodevice with a biomolecular motor, *Science* **290**, 1555-1558 (2000).
8. C. D. Montemagno, G. D. Bachand, Constructing nanomechanical devices powered by biomolecular motors, *Nanotechnology* **10**, 225-331 (1999).
9. R. K. Soong, H. P. Neves, J. J. Schmidt, C. D. Montemagno, Engineering issues in the fabrication of a hybrid nano-propeller system powered by F-ATPase, *Biomedical Microdevices* **3**(1), 71-73 (2001).
10. P. Chartrand, X. H. Meng, S. Huttelmaier, D. Donato, R. H. Singer, Asymmetric sorting of ash1p in yeast results from inhibition of translation by localization elements in the mRNA, *Molecular Cell* **10**(6), 1319-1330 (2002).
11. X. Darzacq, E. Powrie, W. Gu, R. H. Singer, D. Zenklusen, RNA asymmetric distribution and daughter/mother differentiation in yeast, *Current Opinion in Microbiology* **6**(6), 614-620 (2003).

MANAGEMENT OF INNOVATION FOR CONVERGENT TECHNOLOGIES

Errol Arkilick, NSF

There is no shortage of description for the potential of convergent technologies to radically alter the industrial landscape and bring vast improvement to society. In particular, the convergence of nanotechnology, biotechnology and information technology along with cognitive science has been described as having the potential to significantly influence a society's ability to project economic power in the 21st century [1]. Recognizing the potential impact on national security, nations are positioning their science and technology public investments and policies to establish leadership in these convergent fields. This potential provides a vision of our future for which we reach; now it us up to us to develop the plans and vehicles to get us there. To do so, I believe it instructive to study and classify the innovations we envision with an eye toward learning from the past as we invent the future.

In order to develop a tactical plan to bring about our convergent futures, I believe it instructive to study the innovations we envision by applying known best practices of innovation management. Many innovations are actually extensions of current paradigms. For instance, in the future we will be able to remove more heat from a processor (nano/IT convergence), or develop immuno-neutral coatings to decrease the body's rejection of implanted medical devices (nano/bio convergence), or develop new structures that enhance the ability of targeted pharmaceutical agents to pass through the various epithelium (nano/bio convergence). These types of innovations will likely entrench or conserve the market/customer linkages of current market leaders, but may make obsolete technologies or production capacities of these same leaders.

Consider a case where current paradigms will be extended. For example, for an immuno-neutral hip replacement joint, instead of prowess in the ability to machine and market unique components out of titanium for hip replacement joints, the ability to quickly deliver a personally specific immuno-neutral cast carbon-filled joint will be rewarded. (We are assuming that it is this capacity and not some other aspect of business such as customer service or deep relationship with the reimbursement/insurance industry that is rewarded in the marketplace for this example.) To the extent that the convergent-enabled innovations are embraced by the market, the industry-dominant players that don't provide the capability will be demanded by their customers to do so. And, depending upon the business landscape (including intellectual property), the leaders will respond and acquire the demanded capability. In the "transilience framework" of Abernathy and Clark, this type of innovation—one that entrenches market/customer linkages while simultaneously obsoleting technology or production capacity—is classified as revolutionary [2].

The market risk with revolutionary innovations can be relatively small and most innovations of this type fall into the "better, faster, cheaper" category. To be sure, there are vast opportunities for these innovations brought about by convergence, and I am not downplaying the significance of wealth creation enabled by entrenching market/customer linkages. But, revolutionary (as defined in the framework of Abernathy and Clark) innovations typically do not "change the world." They can, however, be used by knowledgeable entrepreneurs to build enterprises that are then acquired by market leaders (or often the number two player) transferring new capability to the market.

For world-changing innovations enabled by convergence, think of innovations that simultaneously act to enable completely new market/customer linkages as well as obsolete technology or production capacity. In these cases, the leading firms will be threatened on two fronts and it is here that *huge* shifts in the fortunes of enterprises (and people) can occur. New industries are formed, new customers are served and brave new worlds are created—but in a good way. Think of personal computing and the fortunes of such venerable companies as Digital Electronics Corporation (DEC) or Wang.

With DEC and the personal computer, problems arose because management did not understand the significance of the difference in the dynamics of the minicomputer verses the personal computer market. They did what all companies in search of excellence do; they listened intently to their customers. What they found was that desktop computers did not suit the needs of DEC's "best" customers. The problem was, a new paradigm for value was emerging with the new personal computer and the capability that DEC possessed simply did not address the new value placed on inexpensive lower-performing (as compared to the mini-computer) microprocessor-based desktop platforms. So, both the customer/market linkages possessed by DEC as well as the technical capability to build high performance minicomputers in the face of the new value landscape of the personal computer ceased to be advantages and actually became liabilities for DEC.

In the framework of Abernathy and Clark, innovations that simultaneously make obsolete market/customer linkages as well as technical or production capacity are termed architectural. And it is within this sector of the "transilience framework" where the most exciting opportunities for convergence lie and where the entrepreneurial grand-slams will be made.

Innovation type classification can be valuable because the fields upon which innovation-enabled enterprises play are uneven and knowledge of macro forces can be used to advantage. Classification can help define the appropriate strategies that should be engaged as we invent our futures. Said another way, it is important to consider classification because different innovation classes represent different opportunities/threats to the enterprise. In particular, the resources and skill sets required to bring successful innovation to market differ for architectural and revolutionary innovations.

Using the concept of gap analysis, namely looking to bridge the gaps between where we are today and where we want to be tomorrow, and using the transilience framework of Abernathy and Clark, we can develop appropriate strategies to increase the effective use of resources to build our convergent futures.

Revolutionary and architectural innovations enabled by technology convergence have different gaps to fill and will require different management techniques and resources for success. Revolutionary innovations are mainly technical in nature, market risk is minimal, and those with established market/customer relationships are in the best position to bring the innovation to market. This realization has important implications when it comes to resource investment and enterprise positioning.

With architectural innovations, market risk and technical risk are involved, and often the entrenched customer relationship, as well as technical capability, actually hinder an enterprise's ability to recognize the potentially disruptive nature of the innovation [3]. In this case, new entrants may be better suited to bring the innovation to market.

In the United States, we are fortunate to possess a wide range of institutions where both basic and applied research are conducted—research that leads to revolutionary and architectural innovations. We have an industrial base that can readily market revolutionary innovations as well as a large pool of aggressive entrepreneurs backed by a mature venture capital industry. This combination can build the architectural innovations into the industries of the future. In addition, our society supports innovation and commercialization of nascent technology through programs such as the Small Business Innovation Research (SBIR), Small Business Technology Transfer (STTR) and the Department of Commerce's Advanced Technology Program (ATP). In particular, The National Science Foundation SBIR/STTR program, in the Engineering Directorate under the Division of Manufacturing and Industrial Innovation, supported a program specifically in convergence (nano-, bio- and information technology) for Security Technologies.

The prominence of the American market over the past 60 years, may have given American enterprises an edge in bringing new innovation to market. However, the emergence of new global markets, particularly in Asia and Europe, and the recognition that the capabilities outlined above are not uniquely American may change this. In order to compete effectively and develop the tactics that will allow us to capture the value of the convergent technologies, America will have to continuously hone its innovation management and technology development tools to embrace best practices. Note, I am not calling for a central planning body or roadmap-developing committee for planning our tactical efforts to reach the future—these rarely succeed. I am suggesting that each of us study the gaps that lie between us and the convergent future with an eye toward understanding the nuances associated with the differences in revolutionary and architectural innovation. By identifying how we and our enterprises can contribute to filling those gaps I emphatically believe there is much potential for wealth creation.

References

1. M. C. Roco, W. S. Bainbridge, eds., *Converging Technologies for Improving Human Performance*, Dordrecht, Netherlands: Kluwer (2003).
2. W. J. Abernathy, K. B. Clark, Innovation: Mapping the winds of creative destruction, *Research Policy* **14**, 3-22 (1985).
3. C. M. Christensen, *The Innovator's Dilemma*: *When new technologies cause great firms to fail*, Boston: Harvard Business School Press (1997).

THE "INTEGRATION/PENETRATION MODEL:" SOCIAL IMPACTS OF NANOBIOTECHNOLOGY ISSUES

Linda Goldenberg, University of Calgary

Social scientists are firmly committed to including the public voice in defining societal impacts of nanotechnology. Accomplishing this requires effectively identifying, articulating, and communicating issues to a variety of "publics" with diverse perspectives, backgrounds, interests, and levels of education, and adjusting communication level, style, and content to particular contexts. It is essential for social scientists to have a working knowledge of the science and technology of nanotechnology, because roots of significant emergent issues lie there. While many societal impacts lay nested within biotechnology issues, unique issues are also arising from nanobiotechnology itself.

In response to these observations, I have developed a series of models for the practical purpose of creating organizational frameworks for the public consultation process, and a set of tools to explain nanotechnology and societal impacts in understandable language. The Integration/Penetration Model serves this purpose for nanobiotechnology. The model is a simple framework that can be totally or partially expanded without compromising its basic integrity, which provides flexibility to accommodate needs and complexity levels of social scientists and the public in diverse contexts and for varying purposes.

The Integration/Penetration Model is also an organizational strategy of inquiry for social scientists to navigate through large, primarily scientific and technical nanobiotechnology bodies of literature. Nanobiotechnology is a fast-paced and emergent field of study, and its academic literature is following the development path of scientific disciplines outlined by John Ziman [1]. In *Reliable Knowledge*, John Ziman outlines the development of scientific discourse, where scientists undergo processes of "consensibility" and "consensus," referring to a pattern of development of an initial foundational layer of primary scientific literature, followed by a secondary layer of scientific literature commenting, challenging, examining, and referring to the primary layer. Nanotechnology is currently at the stage of development of the primary layer.

There are hundreds of thousands of academic citations on the science of nanobiotechnology, derived from research activity clustered in major research centers. In the United States the primary hub is the National Nanotechnology Initiative (NNI), and the departments funded through NNI. It is too new to be an orderly, organized, and mature body of literature, and discourse is currently from industry, government, and academic perspectives. In addition to mastering the science and technology dimensions, these perspectives need to be familiar to social scientists as well because no single perspective will capture the multiple dimensions from which issues are emerging. Nanobiotechnology is a dynamic, fast-paced science, with significant breakthroughs occurring on a daily basis, and keeping abreast of the field requires familiarity with a variety of perspectives, flexible thinking, and sufficient understanding to pick up nuances and subtleties and read these through a societal impacts perspective.

This paper presents the Integration/Penetration Model for nanobiotechnology, and identifies and organizes societal impacts of several technologies within its framework.

The Integration/Penetration Model

In the Integration/Penetration Model, the nanobiotechnology developmental trajectory is conceptualized as a two-dimensional continuum. The model is simple, and reflects the composition and path of the nanobiotechnology developmental trajectory continuum on two significant dimensions:

- *Integration Dimension*: Compositionally, the trajectory comprises two components, "wet" and "dry" strands, initially separate and apart (in the *Interface Range*), which converge and intertwine until ultimately the "dry" is totally assimilated into the "wet" at the level of genetic material (in the *Integration Range*).
- *Penetration Dimension*: Dynamically, the trajectory's path penetrates three concentric circles representing an individual's social and physical space and genetic space: (1) in the Interface Range, it penetrates the social world; (2) in the Integration Range, it pierces an individual's skin and enters the body; and (3) ultimately the genetic material is penetrated, with total wet/dry assimilation.

The two-dimensional continuum concept provides the means to examine the dynamic interaction of the trajectory's wet/dry composition with its path through an individual's social and physical spaces, and to organize, locate and articulate significant societal impacts and issues along the continuum.

The Integration/Penetration Model is presented in simple schematic form in Figure 3.2. The balance of this paper positions a number of nanobiotechnologies on two ranges of the nanobiotechnology developmental trajectory, in the Interface Range and Integration Range, and discusses relevant social impact issues.

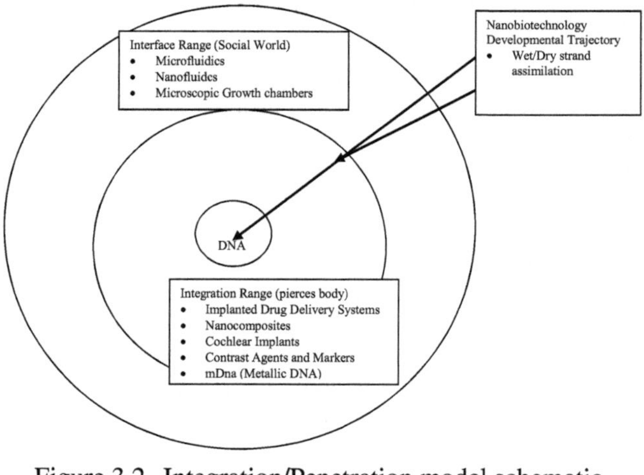

Figure 3.2. Integration/Penetration model schematic.

Nanobiotechnology: The Science Dimension

Nanobiotechnology refers to merging the living and non-living, referred to as the wet/dry interface, where "wet" refers to biology and "dry" refers to engineered nanomaterials [2]. Nanomaterials are foreign to biological entities, so the term "nanobiotechnology" includes integrating the wet and dry through material development and processes to bridge the two dimensions.

The exemplary technologies for the Integration/Penetration Model are derived primarily from Cornell University's NBTC (Nanobiotechnology Center) definition:

> Nanobiotechnology is an emerging area of scientific and technical opportunities. Nanobiotechnology applies the tools and processes of nano/microfabrication to build for studying biosystems. Researchers also learn from biology how to create better nanoscale devices. .... [3]

The scope and diversity of research, potential uses, and applications are wide, and five themes will provide a basic technical anchor and form a fundamental set of areas for the model:

1. *Analytical and diagnostic tools*: Testing devices to profile metabolism and biochemistry, for example to measure neurotransmitter function and drug toxicity. Key devices include microfluidic and nanofluidic technologies, commonly referred to as "lab-on-a-chip" and "mouse-on-a-chip," used in drug development and clinical diagnosis.
2. *Highly functional and bioselective surfaces*: Molecular templates, such as self-assembled monolayers, to research and understand spatial relationships involved in and mediating allergen recognition, and defining and exploiting surface topographical and chemical characteristics that influence biological system behavior.
3. *Rare or sparse cell isolation*: Devices and processes for isolating and preserving particular cells for subsequent research, including cancer cells, fetal cells, and lymphocytes from different biological sources.
4. *Molecular motors*: Nanoscale devices capable of rotational or linear movement, with application to fluidic systems requiring pumping, such as implantable drug delivery devices.
5. *Markers and contrast agents*: Nanoparticle-enhanced optical imaging technologies designed to target DNA, RNA, or other specific biological particles.

These areas provide the initial technical anchor and exemplary technologies for the Integration/Penetration Model.

The Integration/Penetration Model: The Interface Range

The Interface Range is the first of two ranges examined on the trajectory continuum. In the compositional dimension, the wet/dry strands of the trajectory remain separate with no integration, and in the penetration dimension, the trajectory penetrates an individual's social world.

Penetration into the Social World: Microfluidics, Nanofluidics, and Microscopic Growth Chambers

Microfluidic platform technologies, or "chips," are tiny platforms containing networks of microscale flow channels and reservoirs down which biological material is sent and sorted according to size or other defined criteria. These devices can be platforms, or three-dimensional chambers with discrete geometries and dimensions designed to identify and sort specific biological materials. Microfluidic devices can sort a wide variety of biological material, including fluids, proteins, enzymes, and chemicals. Research in this area continues at a rapid pace, such as adding probes and sensors to the channels, which increases complexity and expands the range of testing capabilities. Microfluidic devices can be organized into arrays, creating Microsystems, and entire systems custom-designed for specific purposes. Their small size and portability have earned them names such as "lab-on-a-chip" and, in the case of pharmaceutical development, "mouse-on-a-chip." Basic miniaturization processes continue in *nanofluidics*, which have application for sorting at the genetic level.

There are broad application opportunities for these devices. Direct applications to humans include clinical and nonclinical diagnosis, and compressed time from patient diagnosis to therapeutic intervention, with an increased range of diagnostic capabilities for pathogen and toxin identification, cancer detection, and genetic analysis. Portable devices, such as "lab-on-a-chip," enable these applications to occur outside a clinical or hospital setting, and facilitate quick response to epidemics and disease outbreaks.

Microfluidic devices could drastically reduce the time to market in drug discovery and development by *externalizing* testing processes. For example, biological material could be removed from the body and processed through a microfluidic or nanofluidic device, and reaction and interaction with specific drugs or chemicals could be assessed outside the body. Benefits of externalization include the ability to test biological reaction to a limitless variety of pharmaceuticals and chemicals, and virtually eliminate animal testing.

In a related vein, microfluidic and nanofluidic devices will advance personalized medicine, which involves using the devices to analyze personal genetic and biological makeup, and interaction with various forms of proteins, vitamins, and pharmaceuticals for the purpose of proactive, personalized intervention therapy to prevent health problems and optimize personal health.

Military applications include detection of toxic substances involved in chemical and biological warfare, and environmental applications include air, soil, and water testing and remediation processes.

Privacy, Security, and Tracking and Surveillance Systems

Microfluidic devices and applications raise legal and ethical questions regarding privacy, security, ownership, and control of personal information similar to the issues raised around biotechnology, genetic testing, and information and communication technologies with respect to positioning individuals in existing social matrices. Given the expansive range of uses and social drivers, such as commercial opportunities, fiscal responsibility required of governments, needs for economic efficiencies and excellence in health care, and consumer demand for personalized medicine, these devices will find easy acceptance. Technological advancement of nanofluidics will further extend and amplify these devices.

It is noteworthy while the devices will fill some needs, they will create other issues, such as a need for disposal systems for the biological and toxic material in the devices. This is an important but undiscussed point in environmental impacts and safety concerns.

Accuracy and Technical Complexity in Fluidic Environments

Social context provides one set of issues, and another set derives from considering the accuracy and reliability of the devices in respect of physical and interpretive factors. A closer examination of challenges in fluidics lends insight into other issues.

Externalized testing is not a new concept or practice; it is routinely performed in a wide variety of simple static tests where a sample is placed in solution, and results are generally considered reliable. In contrast, testing and sorting in a *dynamic* fluidic environment presents a myriad of challenges because the system must be capable of

- taking in the sample
- preparing and isolating the sample by filtering out dust or other contaminants
- mixing particles with fluid reagents
- pumping and transporting the mixture down channels
- detecting and identifying pathogens by immunoassay
- data analysis and reporting

Each of these capabilities is a complex system in itself, and together, they create a complex interrelated technical system. Challenges in the fluidic testing and sorting environment are articulated in *Handling Fluids in Microsensors*, a report by the U.S Department of Energy and Department of Defense [4], on projects focused on developing instruments and technologies for sensing material, separating particles (fractionation), and identifying particles based on dielectric properties; dielectric materials permit the passage of an electrostatic field but do not conduct current. The dynamic dimension presents challenges, and continued miniaturization ultimately to the nanoscale compounds the complexity exponentially in ways not known. Some challenges have been identified and solutions proposed:

- Acoustic mixing as an alternative to physical mixing, where small channel dimensions make creating turbulence difficult
- Electronic filtering to capture and suspend particles to remove contaminants before fractionation, through utilizing dielectrophoretic forces to manipulate particles according to their dielectric properties

- Magnetohydrodynamic (MHD) pumps as an alternative to pneumatic power for moving fluids through the sensor. The MHD pump consists of an electromagnet and a series of metal electrodes. Multiple pumps on the same chip can be driven independently by varying their electrode current amplitude and phase relative to the electromagnet, thereby enabling routing in complex integrated systems.

Micro and nanofluidic environments are complex environments and systems, and complexity is increased by properties of biological material and processes. While relatively simple descriptions of these devices are adequate for discussions related to social context application issues, they do not adequately capture two other categories of issues.

One issue is accuracy and reliability, both of the physical devices themselves and the extent to which the externalized environment accurately replicates the body environment. With respect to the former, even if a mechanical challenge is identified and addressed, physical construction of the fluidic device may be problematic. For instance, a flow channel may be too small to accommodate certain molecules, fluid viscosity may depend on channel size, chemical composition of the device may interact with the fluids or particles, or physical manipulation such as using pressure to drive a viscous fluid or particle through a small orifice could result in particle deformation and potentially compromise its integrity. Channels can be damaged or obstructed in the process of making a device, rendering it inaccurate. In terms of replicating the body environment, all potentially significant interacting or influential variables may not be present in the externalized environment, and processes and reactions may be time-, temperature-, and/or pH-sensitive.

Clearly, the fluidic environment is a complex one, increased by characteristics and properties of biological material and processes themselves. In addition to mechanical challenges, living cells are themselves complex, dynamic systems forming part of a larger body system, and bring in another set of fluidic properties.

This raises important questions about reliability and accuracy, and the extent to which applications in areas such as drug development will be reliable, given the range of potential physical challenges. Questions to be confronted include: What are acceptable levels of accuracy? Who will define them? When these devices roll out of research labs and into industrial settings for mass production, the risks of inaccuracy and/or damage during production will likewise increase.

New Forms of Knowledge: Computer Simulations and Modeling

To address the complexity of microfluidics and nanofluidics and the biological environment, computer simulations and modeling are increasingly seen as a solution, which raises questions relating to the "knowledge" created, and the extent to which modeling and simulation actually depict and reflect the "reality" and processes they are intended to. A series of questions is confronted: How accurate, reliable, and complete is this methodology? What has been left out? Are these "illusions of knowledge"? An important question is: What are the acceptable boundaries for relying on computer generated information or simulations?

Epistemological Dimensions Associated with Methodologies: Substantial Equivalence of Knowledge?

New sciences such as nanotechnology and nanobiotechnology are inexact and speculative bodies of research findings derived from a variety of mediated forms of inquiry, including experimental, theoretical, and computer simulation and modeling. There is not and cannot be unmediated, direct human experience at the molecular level. The biotechnology concept of "substantial equivalence," which refers to considering only outcome and not processes in product development, can also be applied to knowledge. This poses a central epistemological, ethical, and practical question, as depicted in Figure 3.3: *Are all forms of knowledge equivalent?*

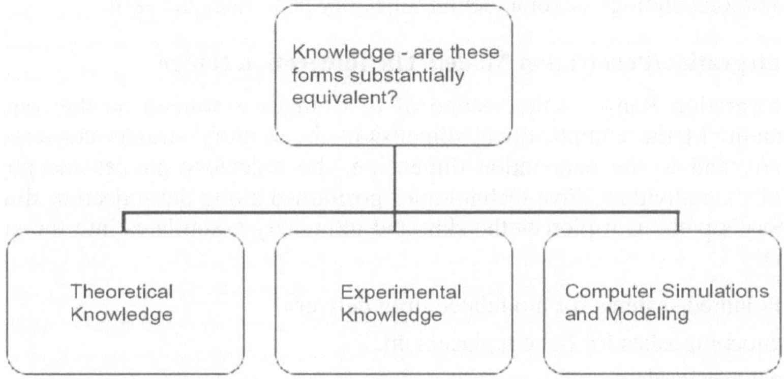

Figure 3.3. The forms of knowledge.

An equally important question is: *When does it matter?*

These and other key questions are important in contexts and applications with substantial consequences, such as drug development and pharmaceutical approvals, where both economic and health impact stakes are high:

- Are all types of knowledge—theoretical, experimental, simulated—*substantially equivalent* and interchangeable?
- Is knowledge derived from externalizing processes through microfluidics, nanofluidics, and microscopic growth chambers *substantially equivalent* to knowledge derived from drug trials on humans?
- What are acceptable boundaries and circumstances for relying on different types of knowledge, and who should determine these?

Ethical issues associated with pharmaceutical approval surge to the forefront, and raise questions such as:

- Do similar arguments apply to labeling drugs as to labeling genetically altered food?
- Should drugs be labeled or characterized according to the type of knowledge utilized in development and regulatory approval?

- Should there be an economic relationship between types of knowledge and the cost of pharmaceuticals?
- Where is responsibility for adverse events—in regulatory bodies or manufacturers?

It becomes clear that technological advancement creates significant societal implications on a number of levels, including industrial, regulatory, epistemological, and ethical. Issues such as those drawn out should be brought into the public consultation process to establish acceptable regulatory positions in line with public values.

All issues raised in the Interface Range transfer to the Integration Range, where the nanobiotechnology developmental trajectory penetrates the skin.

The Integration/Penetration Model: The Integration Range

The Integration Range is the second of two ranges examined on the trajectory continuum. In the compositional dimension, the wet/dry strands converge and integrate, and in the penetration dimension, the trajectory pierces the physical body of the individual. Five technologies positioned along the trajectory illustrate trajectory's path, as it pierces the skin and ultimately assimilates into the genetic material:

- implanted systems for modulated drug delivery
- nanocomposites for bone replacement
- cochlear implants
- contrast agents and markers
- mDNA (metallic DNA)

The five examples raise different issues along the developmental trajectory.

Penetration into the Physical Body: Implanted Drug Delivery Devices and Nanocomposites

Implanted drug delivery devices and nanocomposites for bone replacement pierce the skin of the individual and reside inside an individual's physical body. They raise potential health and safety issues at the individual level, primarily complex technical issues required to be approached and resolved at that level (allergy, reaction, rejection, etc.).

At the Center for Biological and Environmental Nanotechnology (CBEN), for example, composites of thermally sensitive hydrogels and optically active nanoparticles are under development for biomedical applications, including photothermally modulated drug delivery and optically controlled microfluidic systems [5]. Implanted drug delivery devices for applications such as insulin delivery need to take into account that temperature requirements for storing insulin within the narrow range safe and acceptable for use are lower than the temperature of the human body. An implanted device for this purpose requires a refrigeration component, increasing complexity of the microsystem by adding another dynamic system. CBEN is addressing challenges related to temperature by developing systems incorporating gold nanoshells that absorb light into hydrogels to trigger a temperature change in response. Optically responsive composites have numerous applications, including posts in microfluidic channels that collapse in response to particular wavelengths, permitting flow in the channel. This represents increased complexity in microfluidic devices as well.

In terms of issues, personal decisions regarding medical procedures and interventions are between patient and physician. From implanted drug dispensing devices to nanocomposites for bone replacement, matters squarely enter the domain of risk assessment and management for health and safety, and reliability and accuracy of the technologies and knowledge relied upon in approval and regulatory processes are important components of the decision making process. Clearly, interested parties benefit from accurate information regarding body reactions such as rejection, allergic reaction, or blood clotting.

Cochlear Implants

Cochlear implants exemplify "normalization" issues [6, 7] and illustrate ethical issues such as the debate over "fixing" the human body, and so can be offered as an alternative to scientific and technical issues (which may be beyond the grasp or interest level of some people), and make public consultation more inclusive. Normalizing technologies raise questions such as: How far will "normalizing" go, once virtually every aspect of a human body becomes transformable, right down to the genetic level? Technologies such as cochlear implants resonate with ethical issues associated with biotechnology (designer babies, playing "God"), and illustrate the many issues yet to be confronted by nanobiotechnology currently nested and rooted in biotechnology issues.

Markers and Contrast Agents

Midpoint in the Integration Range are nanoparticle-enhanced optical imaging technologies, which include innovative contrast agents and biological markers designed to target DNA, RNA, or other specific biological particles. Contrast agents and markers involve nanoparticles varying in scattering and absorption properties over an engineered broadband of visible and infrared spectral regions. Reflectance, absorption, scattering, and concentration are a function of the optical properties of the nanoparticles. At CBEN, researchers (West, Halas, and Drezek) are leveraging advances in nanoparticle technologies to develop innovative nanoparticle optical contrast agents that bind to specific cells, which can be targeted to specific molecular signatures of disease. The nanoparticles have ideal optical and chemical properties for optical imaging, and it is anticipated nanoparticle-enhanced optical imaging technologies have potential to provide significant breakthroughs in detection of precancerous conditions.

An even closer step to complete convergence of the wet/dry strands and penetration into genetic material occurs in fluorescent imaging, as illustrated by emission enhancement of intrinsic fluorescence DNA. According to Lakowicz et al. [8], high-sensitivity DNA detection is essential for genomics, but the very weak intrinsic fluorescence of DNA necessitates reliance on extrinsic fluorescent probes. They have developed a procedure for enhancing intrinsic fluorescent emission through controlling and manipulating spatial proximity of the DNA to "silver islands," which essentially harnesses excited-state fluorophores interactions with metallic surfaces. In this case, silver particles in close proximity amplify local electric fields, and stimulate increased excitation in nearby flurophores. By creating "islands" of particles, and through examining the emission spectrum of DNA in relation to the proximity to these "islands," researchers noted an 80-fold increase in emission intensity near the metal islands. Investigation continues into this "region of enhancement," as researchers explore the reasons for the enhanced emission, and potential uses for it within nanobiotechnology.

Fluorescent imaging raises issues of privacy, testing, health, and safety, and the public voice needs to be heard in determining application and uses.

mDNA – Metallic DNA

The final point on the Integration Range of the trajectory continuum involves total integration of the wet/dry strands, and penetration of the trajectory into the genetic material. Lee and Aich [9] accomplished this with a process for creating mDNA (metallic DNA) by inserting conducting metal ions such as zinc, cobalt, or nickel into the center of the DNA helix, creating a semi-conductor. In a similar vein, researchers at the Sony Materials Science Laboratories, ATCS, Sony International (Europe) [10] are investigating the electrical conductivity of metallicized DNA molecules forming nanowires for applications in future electronics, using gold nanoparticles. Crossing this threshold into the genetic material blurs the boundaries between the living and the engineered. These advances will open up new discussions and issues extending far beyond the scope of this paper, but are destined to emerge in time.

This section has outlined the nanobiotechnology development trajectory path to complete assimilation of the wet/dry strands at the genetic level, thereby achieving total penetration of the trajectory into an individual's social, physical, and genetic existence.

Regulatory Structures and Society

A central purpose of the public consultation process is regulation and an integral part of the process is to include a consideration of the constraining or enabling influence of regulatory structures. Significantly, there is a fundamental misfit between static regulatory structures and dynamic society, which provides a rationale for public participation in the regulatory process.

Inasmuch as society creates its structures, society and these structures should reflect and be mirror images of each other. However, social evolution far outpaces structural evolution so this is not the case. Society is constantly in a "cultural lag" [11], where legal and regulatory structures are a palimpsest of past decisions and needs, with layer upon layer of irrelevant past systems and frameworks often impeding progress. Societal mechanisms for structural change are reactionary, backwards-oriented and precedent-based, relying on formal authority and amendment through mechanisms such as lengthy governmental procedures.

Regulatory structures are not future-oriented or equipped to proactively manage future or potential problems, particularly when problems are uncertain—indeed a problem must appear before it can be regulated or, in the case of the judicial system, become a precedent. Frequently, to invoke the legal or regulatory systems, it is necessary to dig through layer upon layer of irrelevant material, and make comparisons of concepts never intended to serve the analytical and judgment purposes they are being called on to support. A prime example is copyright, and its inability to adequately be effective in the digital environment of the Internet. The inadequacy of a constraining, backward-oriented system is striking, and it is neither equipped nor adequate to handle the challenges of rapidly emerging nanobiotechnology issues, many of which are characterized by uncertainty or simply unknown.

A forward-looking system of regulation is more relevant to a dynamic, fast-paced society. Public consultation can identify issues of concern to the public, and resources channeled into issues prioritized by the public.

Conclusion

When considering the nanobiotechnology developmental trajectory and its future path, concepts and technologies can be tied to larger societal and ethical issues, including extending and furthering issues arising from biotechnology and genetic engineering. However, the science and technology of nanobiotechnology also opens a new range of issues, many of which I have touched on in this paper, including normative issues, rights, security, privacy, and control, as well as epistemological, ethical, regulatory and practical issues.

Societal impacts and issues are varied and can be complex, and the Integration/Penetration Model is a comprehensive, flexible organizational framework and tool for the public consultation process. It can accommodate multiple frameworks placed on the same technology, and contribute to interdisciplinarity and understanding a variety of perspectives. For example, each nanobiotechnology discussed exists at the intersection of science, technology, medicine, philosophy, law, humanities, social sciences, ethics and communication.

Issues like autonomy, state intervention, privacy, and the ethics of digging into and attempting to uncover the secrets of life itself, within the context of humankind's relationship to nature, the universe, the meaning of life—and God himself—will take on renewed importance. These issues can also be easily positioned and contextualized within the model's framework.

References

1. J. Ziman, *Reliable Knowledge,* Cambridge, UK: Cambridge University Press (1978).
2. CNST (Center for Nanoscale Science and Technology, Rice University) (available at http://cnst.rice.edu/cnst.cfm?doc_id=1209).
3. NBTC (Nanobiotechnology Center, Cornell University) (available at http://www.nbtc.cornell.edu/).

4. Handling fluids in microsensors, *Science & Technology Review* (November 1999) (available at http://www.llnl.gov/str/Miles.html).

5. CBEN (Center for Biological and Environmental Nanotechnology, Rice University) (available at http://www.ruf.rice.edu/~cben/).

6. A. Dreger, When medicine goes too far in the pursuit of normality, *Bioethics Bulletin* **10** (1), (1999) (available at http://www.phen.ab.ca/ materials/ het/het10-01a.html).

7. G. Wolbring, Science and technology and the triple D (disease, disability, defect). In *Converging Technologies for Improving Human Performance*, ed. M. C. Roco, W. S. Bainbridge, Dordrecht, Netherlands: Kluwer (2002).

8. J. R. Lakowicz, B. Shen, Z. Gryczynski, S. D'Auria, I. Gryczynski, Intrinsic fluorescence from DNA can be enhanced by metallic particle, *Biochemical and Biophysical Research Communication* **286**, 875–879 (2001) (available at http://www.idealibrary.com).

9. J. Lee, P. Aich, Developers of a "molecular wire" win innovation award, *University of Saskatchewan News* (15 May 2002) (available at http://www.usask.ca/events/news/articles/20020515-2.html).

10. O. Harnack, W. Ford, F. Scholz, A. Yasuda, J. Wessels, Electrical conductivity of self-assembled metallic DNA nanowires, Presentation made at the Ninth Foresight Conference on Molecular Nanotechnology, 9–11 November, Santa Clara, CA (2002) (available at http://www.foresight.org/ Conferences/MNT9/Abstracts/Harnack/).

11. W. F. Ogburn, *Social Change*, New York: B.W. Huebsch (1922).

THE USE OF ANALOGIES FOR INTERDISCIPLINARY RESEARCH IN THE CONVERGENCE OF NANO-, BIO-, AND INFORMATION TECHNOLOGY

Judith Klein-Seetharaman, University of Pittsburgh and Carnegie Mellon University

Modern research would be unthinkable without interdisciplinary approaches. For example, many of the biological disciplines provide proof by their names: biochemistry and biological chemistry, biophysics, biotechnology, biomaterials, biostatistics, bioinformatics, and computational biology all encompass aspects of biology that are influenced by at least one other scientific discipline. How do converging technologies fit into this existing, rich interdisciplinary framework of modern biological research? Technologies are based on science that has found a particular application area, where the driving force is a practical outcome. Different application areas can have the same scientific principles and approaches in common and only differ by the details of their implementation. Thus, in principle, the combination of two scientific disciplines that form the basis for two technologies in different application areas should yield the same benefit as the combination of the two technologies. While this is true in principle, in practice the convergence of technologies provides complementary benefits to existing interdisciplinary research. Technology in one application area is transferred to another application area by way of *analogy*. The use of analogies provides a link to the general public since analogies allow people to grasp concepts immediately when they can relate them to something they know. Thus, converging technologies have tremendous societal impact because they provide a direct means of communicating interdisciplinary research using analogies.

We will demonstrate this concept using our ongoing Biological Language Modeling Project, which exploits an analogy between language—familiar to everyone—and biology—familiar in detail only to experts. We expect that the same principles will hold true for other examples with more direct relevance to nanotechnology such as the manufacturing of miniature machines in analogy to biological machines.

The Language–Biology Analogy in Convergence of Information Technology and Biotechnology

The Biological Language Modeling project is a cross-disciplinary collaboration that converges Human Language Technologies with Biological Chemistry, in an effort to provide novel approaches to the mapping of biological sequences to the structure, function, and dynamics of proteins and, ultimately, of biological systems. Complex biological systems are built from cells that have differentiated to perform specialized functions. This differentiation is achieved through a complicated network of interacting biological molecules. The main action is carried out by proteins, which may be viewed as nano-sized biological machines (see below) that are composed of strings of characteristic sequences of the 20 amino acid building blocks. The sequences of the strings are encoded in their entirety in the genome. The linear strings of amino acids contain in principle all the information needed to fold a protein into a 3-D shape capable of executing its designated function. With the advent of whole-genome sequencing projects, we now have complete lists of all the protein sequences that define the complex function carried out by several organisms that have been sequenced hundreds to thousands in bacteria and tens of thousands in humans [1]. Individual proteins and functions have been studied for decades at various levels, atomic to macroscopic. Most recently, a new field has evolved, that of proteomics, which looks at all the proteins in a cell simultaneously. This multitude of data provides new opportunities for the applicability of statistical methods to yield practical answers in terms of likelihood for biological phenomena to occur.

The availability of enormous amounts of data has also transformed linguistics. The analogy between biology and language is shown in Figure 3.4. In language, instead of genome sequences, raw text stored in databases, Web sites, and libraries maps to the meaning of words, phrases, sentences, and paragraphs as compared to protein structure and function. After decoding, we can extract knowledge about a topic from the raw text. In language, success in this process has been demonstrated by the ability to retrieve, summarize, and translate text. Examples include powerful speech recognition systems, fast Web document search engines,

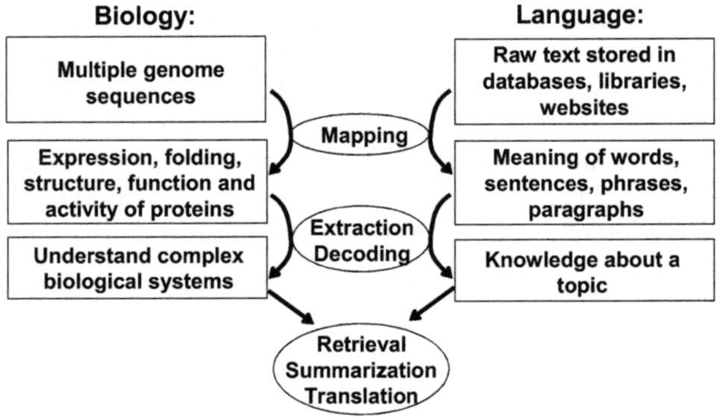

Figure 3.4. Concept of analogy between language and biology.

and computer-generated sentences that are preferred by human evaluators in their grammatical accuracy and elegance over sentences that humans build naturally. The transformation of linguistics through data availability has allowed convergence of linguistics with computer science and information technology. Thus, even though a deep fundamental understanding of language is still lacking, e.g., a gene for speech has only been discovered recently, data availability has allowed us to obtain practical answers that fundamentally affect our lives.

In direct analogy, transformation of biological chemistry by data availability opens the door to convergence with computer science and information technology. This convergence between biological chemistry and computer science and information technology has already happened in the past—namely, when the disciplines of bioinformatics and computational biology emerged. In fact, human language is, just like biological chemistry—a domain of application for statistical approaches. What can the analogy to language provide that is not already provided by the previous, more general convergence of disciplines? The answer lies in the implementation details. For example, classification of protein sequences into families and subfamilies based on their functional and evolutionary relatedness is a typical problem in bioinformatics and has been studied extensively previously. Typical classification methods such as k-Nearest Neighbor approaches, Hidden Markov Models, Support Vector Machines, and Neural Networks have all been applied to this problem without reference to language technologies. At the same time, all of these methods are also used extensively in the language technologies domain. Previously, it had been concluded that it was necessary to apply classifiers of very high complexity, i.e. Support Vector Machines, to the protein classification problem to achieve high accuracy. However, application of the much simpler Naïve Bayes methods to the same problem, in combination with feature selection using chi-square that has been found to be a very successful combination in the language domain [2], has been shown to outperform the complex Support Vector Machines in this task also [3]. The reason for this success lies in the deeper analogy between language and biology that impacts methodical details, the feature selection process.

The principle for feature selection in the language domain as compared to the biology domain is illustrated in Figure 3.5. The left-hand side illustrates feature selection in language. The bags of words in three related (ball games) but different documents are shown. Some of the words identify the relatedness of the documents ("ball"), while some identify the differences ("hoop–basket", "touchdown–tackle", "glove–bat"), while some are irrelevant for distinguishing these texts from each other and from unrelated texts ("a–to–the"). The right-hand side of the figure shows feature selection in biology. Word equivalents used in protein sequence language are short stretches of amino acids, here shown mapped onto a secondary structure representation of a G protein coupled receptor. Only some amino acid positions (labeled) are useful in distinguishing different subtypes of G protein coupled receptors, while the helices (center) are common to all G protein coupled receptors, and other areas cannot be distinguished from any other protein. This example demonstrates that experts in language technologies are able to augment current interdisciplinary research in bioinformatics, although the set of tools in principle is in common.

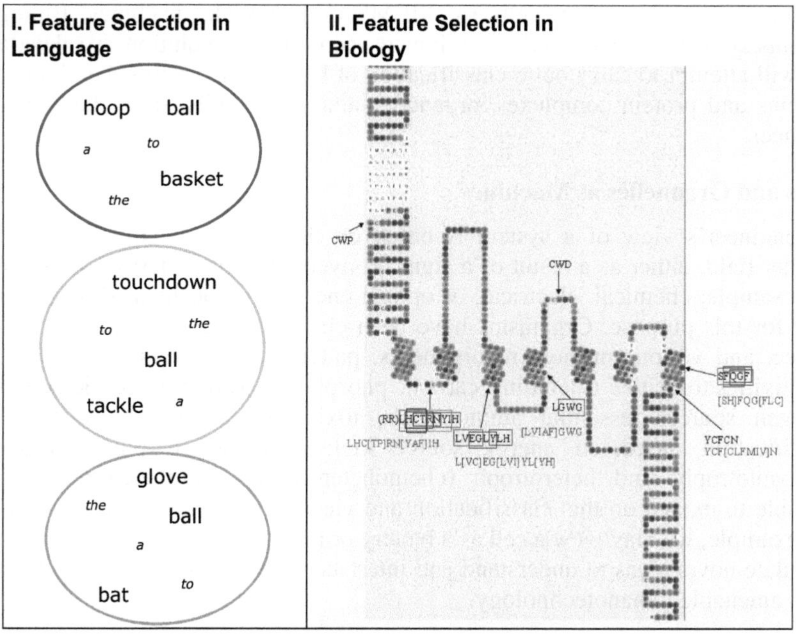

Figure 3.5. Example for the utility of details in language for biological questions.

Analogy Miniature Machines and Functional Biological Systems in Convergence of Nanotechnology and Biotechnology

Inasmuch as the application of language technologies to biological chemistry provides a novel view on old research problems, the application of other technologies may do the same. One particularly promising example is nanotechnology. The ability to control structures at a nanometer-scale level is a necessity for improving magnetic, electric, catalytic, or mechanical properties of materials for the next generation. Nanoscale cantilevers are being designed for applications in various future devices such as actuators for "smart gates" in drug-delivery systems or for data storage [4].

Methods traditionally employed in the nanotechnologies field such as Atomic Force Spectroscopy [5] and Laser Tweezers have found important applications in the protein science field, such as protein folding [6], pharmacology [7], and membrane protein structure studies [8]. However, because the field of nanotechnology is much newer than biological chemistry, there is great promise in the reverse process: the application of knowledge and principles from biological chemistry for nanotechnology. Proteins, cells, and organisms are machines of increasing complexity and size that carry out specific functions. They respond to an imposed energy field by an output that is often in the form of a different field. For example, certain organisms respond electrochemically to chemical fields with an electrical signal; others respond with a structural and thus mechanical response to an optical perturbation. Biological systems range widely in sizes, ranging from nanoscale to microscale and can be potentially viewed as micro/nano-scaled actuators, sensors, or fuel cells for MEMS or nanoscale machines. Biological systems are usually classified by biologists according to evolutionary relatedness. We will attempt to carry out a classification of biological systems, i.e., individual proteins and protein complexes, organelles and single cells, from the eye of an engineer.

Cells and Organelles as Machines

An engineer's view of a system is based on the conversion of one field into another field, either as a result of a signal conversion or for energy requirements. For example, chemical, electrical, or optical energy and mechanical work can be used for this purpose. Organisms have been classified according to their energy source, and various nutrient requirements, patterns, or conversion abilities and sensitivities/toxicities (including carbon, phosphorus, sulfur, trace elements and nitrogen sources, essential amino acids, toxic metals). Some examples for classification based on energy source include auxotroph (chemoautotroph, photoautotroph) and heterotroph (chemoheterotroph, photoheterotroph). It is possible to extend on this classification and view a cell based on its capabilities. For example, we may view a cell as a battery or an actuator switch. This view may stimulate novel ideas to understand and interface with biological systems to make them amenable to nanotechnology.

Proteins and Protein Complexes as Machines

Molecular studies of protein mechanisms have provided detailed understandings of the molecular mechanisms underlying functions in individual proteins. Three examples are shown in Figure 3.6. motor, switch and pump. On the left is an image of a kinesin microtubule motor, downloaded from the Protein Data Bank (PDB) [9] (PDB ID 2NCD [10]). In the middle is a proton pump bacteriorhodopsin (PDB ID 1C3W [11]). And on the right is a light switch rhodopsin (PDB ID 1L9H [12]). For example, the light-sensitive molecule rhodopsin converts light-energy into chemical energy by way of conformational changes in the protein [13, 14]. The nature of these conformational changes has recently been shown to be very much mechanical [15], and is therefore directly analogous to human machines. Testing such hypotheses may further increase our understanding of this molecular process, but in turn it may be possible to mimic this mechanism invented by nature within the nanotechnology arena. While it is not possible to build proteins man-made, and even less so cells and organisms, due to the high complexity, it seems more feasible that one can use the natural systems within a nano-technology context. The details of this process are a major challenge for current and future convergence of nano- and biotechnology.

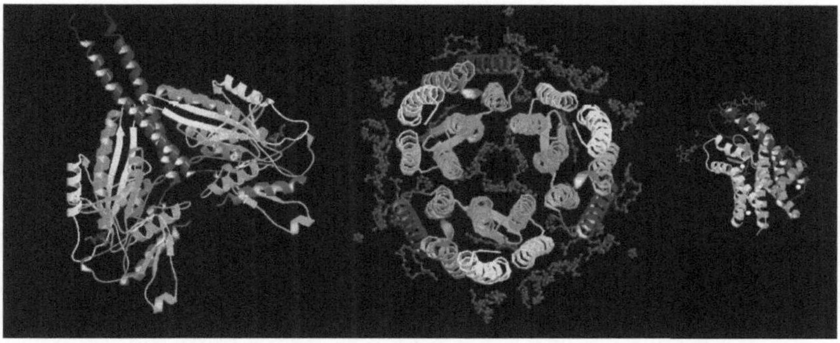

Figure 3.6. Examples of molecules representing engineering functions: motor, pump, switch (figures courtesy Protein Data Bank; also [right figure] *Nature* [11], [middle figure] *Journal of Molecular Biology* [12]).

References

1. J. C. Venter, et al., The sequence of the human genome, *Science* **291**(5507), 1304-1351 (2001).
2. Y. Yang, J. T. Pedersen. A comparative study on feature selection in text categorization. In Proceedings of the Fourteenth International Conference on Machine Learning, ed. D. Fisher, 8-12 July, Nashville, TN (1997).
3. B. Cheng, J. Carbonell, J. Klein-Seetharaman, Protein Classification based on Text Document Classification Techniques, *Proteins - Structure, Function and Bioinformatics* **58**(4), 955-70 (2005).
4. A. Hellemans, Cantilever tales, *Science* **290**(1529) (2000).
5. G. Binnig, C. F. Quate, C. Gerber, Atomic force microscope, *Physical Review Letters* **56**(9), 930-933 (1986).
6. F. Oesterhelt, et al., Unfolding pathways of individual bacteriorhodopsins, *Science* **288**(5463), 143-146 (2000).
7. R. S. Pereira, Atomic force microscopy as a novel pharmacological tool, *Biochemical Pharmacology* **62**(8), 975-983 (2001).
8. D. Leckband, Force as a probe of membrane protein structure and function, *Current Opinion in Structural Biology* **11**(4), 433-439 (2001).

9. Research Collaboratory for Structural Bioinformatics, Protein Data Bank (available at http://www.rcsb.org/index.html).

10. E. P. Sablin, Direction determination in the minus-end-directed kinesin motor ncd, *Nature* **395**(6704), 813-816 (1998).

11. H. Luecke, et al., Structure of bacteriorhodopsin at 1.55 A resolution, *Journal of Molecular Biology* **291**(4), 899-911 (1999).

12. T. Okada, et al., Functional role of internal water molecules in rhodopsin revealed by X-ray crystallography, *Proceedings of the National Academy of Sciences USA* **99**(9), 5982-59877 (2002).

13. W. L. Hubbell, et al., Rhodopsin structure, dynamics, and activation: A perspective from crystallography, site-directed spin labeling, sulfhydryl reactivity, and disulfide cross-linking, *Advances in Protein Chemistry* **63**, 243-290 (2003).

14. J. Klein-Seetharaman, Dynamics in rhodopsin, *ChemBioChem* **3**(10), 981-986 (2002).

15. B. Isin, et al., Predisposition of the Three-Dimensional Dark State Structure of Rhodopsin for Functional Conformational Changes. In preparation.

CONVERGING TECHNOLOGIES: INNOVATION, LEGAL RISKS, AND SOCIETY

Sonia E. Miller, Converging Technologies Bar Association

> I am going to have the Society outlawed, every member removed from any responsible post. And all executive and technical positions, henceforward, can be filled only by applicants signing a non-Society oath. (Isaac Asimov, *I, Robot* [1, p.268])

Imagine if we could have the opt-in or opt-out choice of freely joining society. What if with the simple wave of a magic wand and certain special wizardry, we could continuously adjust, modify, or alter our societal conditions to improve our current human state? If only such magic existed…

While society does not offer us a legal opt-out provision, converging technologies do present us with magical prerogatives for unprecedented improvements to our human performance. Through the convergence of four science and technology provinces—nanoscience and nanotechnology, biotechnology and biomedicine, information technology, and cognitive science and cognitive neuroscience, commonly referred to as NBIC [2]—which also can be combined with other related sciences and technologies, the recipients of such advancements will experience revolutionary betterment heretofore unknown to mankind.

However, this multidisciplinary blending of NBIC confronts us not only with auspicious enrichment, but also with analogous potential adverse implications and risks. Risk accompanies innovation and change [3]. Pinpointing the impact of latent risks to these miraculous benefits could take years [4]. But, we must begin the precautionary dialogues now to safeguard society from the possibilities of unforeseeable harm in the future.

What are some phantom risks, whose very existence are yet unproven and perhaps unprovable, but raise true concerns at the interface of science and the law, arouse public fears, and may lead to public controversy and litigation [5]? How will the risk trade-offs between the innovation derived from these sciences-in-progress and the personal risks inherent in their use be communicated?

In addressing the societal implications of the convergence of NBIC, it is integral to explore and identify probable, atypical, and outlying legal risks, as well as a range of different conjunctions of risk [6]. While developing risk management strategies across sectors such as health care, pharmaceuticals, agriculture, energy, defense, and the environment might sound plausible, the task presupposes that future risks can be understood, measured, and to some extent predicted [7]. Additionally, global interdependence coupled with lack of harmonization in regulations and standards make risk management even more complex.

While the scientists and engineers are the practitioners of innovation, it is the role of the law and legal institutions to deal with the risks, uncertainties, and unintended deleterious effects that may pervade society. These risks may be moral, physical, or economic, obvious or ordinary. Attorneys are trained to identify or mitigate legal risks [8]—the element of uncertainty in an undertaking with the chance of injury, damage, or loss.

Yet, through what legal processes, formulations, and techniques will risk be controlled within converging technologies when it transcends several areas of law? How will the judicial systems assess and adjudicate risk-laden issues through the courts, not just in the United States, but internationally, when the courts lack expertise in risk analysis? What standards of proof will correspond to scales of certainty or uncertainty that constitute acceptable basis for legal decisions in a variety of practical contexts? How can policy be separated from science in risk assessments [9]? What policies will be implemented related to risk?

Bear in mind that in the past three decades, U.S. liability laws have undermined innovation in the courtroom [10]. Determining the responsibility for societal costs arising from the amount of risk people should assume in using products has been a contentious issue; at times, calls have been issued for tort reform and limitations to compensatory and punitive damages. Even when old risks are brought under control, new risks may evolve. Issues for consideration regarding a new product or process development include the following:

- Professional judgments from design to end use can be introduced as evidence.
- Liability can exist even when products are built according to accepted industry standards; the only proof required is that it was possible to produce a safer product, even if production costs were so prohibitive that the product became unmarketable.

- Manufacturers are increasingly being held responsible for human error and poor judgment in product use.
- Nonscientific juries are deciding highly complex science, engineering, and technology issues.
- Sometimes the technology itself goes on trial, and the more unfamiliar and innovative it is, the more harshly it is judged.

As innovation results in more complex products and processes, and perhaps greater risk, it is imperative that judges and juries be well informed about the aspects on which they are being asked to deliberate. Because oftentimes members of the same technical, engineering, or scientific disciplines disagree, what becomes admissible as scientific evidence is called into question.

Through multidisciplinary strategic alliance-building, we can begin to identify attendant legal risks, remove barriers to innovation, and educate the stakeholders while seeking socio-cultural, socio-economic, and socio-political parties, groups and institutions to convey their values and concerns about the intangible uncertainties of converging technologies. However, for effective collaboration, it will be necessary to resolve the cognitive differences found in communication styles among attorneys, scientists, engineers, and the respective societal sectors, the degree of cooperation needed, clear understanding of the questions, the perception of risk, truth, expertise, clarity of explanation, and expectations.

Time is of the essence. At stake is the well being and safety of society, U.S. economic competitiveness, the future of innovation, and new unimaginable discoveries to improve human performance. There is no opting-out of Society in this Age of Convergence.

References

1. Asimov, *I, Robot*, New York: Bantam Books (1950).
2. M. C. Roco, W. S. Bainbridge, *Converging Technologies for Improving Human Performance*, Dordrecht, Netherlands: Kluwer (2003).
3. P. L. Bernstein, *Against the Gods: The remarkable story of risk*, New York: John Wiley & Sons, Inc. (1996).
4. P. D. Shaw, *Managing legal and security risks in computing and communications*, Burlington, MA: Butterworth-Heinemann (1998).
5. K. R. Foster, D. E. Bernstein, P. W. Huber, *Phantom Risk: Scientific inference and the law*, Cambridge, MA: MIT Press (1999).
6. R. Baldwin, P. Cane, W. G. Hart, *Law and Uncertainty: Risk and legal processes*, Dordrecht, Netherlands: Kluwer Law International (1996).
7. T. A. Caulfield, B. Williams-Jones, *The Commercialization of Genetic Research: Ethical, legal & policy issues*, Dordrecht, Netherlands: Kluwer (1999).
8. M. Whittaker, Lawyers as risk managers, Butterworths Journal of International, Banking and Finance Law (January 2003).

9. P. Slovic, *The Perception of Risk: Risk, society and policy series*, London: Earthscan Publications, Ltd. (2000).

10. R. M. Morrow, Technology issues and product liability. *In Product Liability and Innovation: Managing risk in an uncertain environment*, ed. J. Hunziker, T. O. Jones, Washington, DC: National Academies Press (1994).

SHORT-TERM IMPLICATIONS OF CONVERGENCE FOR SCIENTIFIC AND ENGINEERING DISCIPLINES

Cyrus Mody, Cornell University

Much of the rhetoric and organization of nanotechnology emphasizes the extraordinary interdisciplinarity of this venture, and highlights the need to accelerate an already unavoidable convergence of disciplines under the nanotechnology umbrella. For instance, a review of the National Nanotechnology Initiative stated: "Nanoscale science and technology are leading researchers along pathways formed by the convergence of many different disciplines—biology, physics, chemistry, materials science, mechanical engineering, electrical engineering, and others. . . . It is expected that the number of interdisciplinary groups will grow as it becomes evident that an interdisciplinary approach is necessary to tackle the interesting and complex problems that are part of nanoscale science and technology [1, p.1]."

Indeed, much of the early money for nanotechnology has been spent specifically in order to foster interdisciplinarity—through, for example, the building of nanocenters that are structured (both physically and organizationally) in ways that are meant to promote their use by researchers from a wide variety of disciplines [2, 3]. This emphasis on interdisciplinarity can (and in the long term probably will) be one of nanotechnology's great strengths. The traditional disciplines are contingent products of history, rather than conformal mappings of organizational structure onto the structures of the real world [4, 5, 6, 7]. The societies of the late 19th century in which the disciplines formed and professionalized no longer exist; today's societies may demand newer and more integrated ways of organizing research. Finding ways around barriers created by the boundaries of the traditional scientific and engineering disciplines can generate new paths of innovation and foster a wider sense of ownership of nanotechnology throughout the scientific and engineering community.

I would like, however, to point to some potential difficulties that may be generated in the short and medium term by overly optimistic assessments of nanotechnology's revolutionary impact on the traditional disciplines. Interestingly, I would recommend that research to understand and alleviate these difficulties is itself best undertaken from an interdisciplinary perspective by combining insights from the history, sociology, and anthropology of science and technology. Historically, the interdisciplinary goals of nanotechnology reproduce trends in a number of postwar science and engineering movements. Some of these movements, such as materials science, succeeded in coalescing into new, interdisciplinary constellations of research and development (though disciplinary frictions within materials science departments are still common) [8]. Other movements, such as cybernetics, did less well at enrolling different disciplines into a common project [9]. Yet others, such as information science, are still in the process of formation, though they show great promise in constructing a coherent rubric that their participants can acknowledge as guiding their common project. Nanotechnology can only benefit from a better historical understanding of these "umbrella disciplines" and their struggles to build local convergences.

Sociologically and anthropologically (and from a policy perspective), there is a need to better understand (and be mindful of) how researchers actually understand, react to, and implement or resist interdisciplinarity in nanotechnology and the other NBIC convergence streams. If we avoid thinking only in terms of the implications of nanotechnology for "society" writ large, and concentrate on looking at the diverse implications of nanotechnology and convergence for various pockets within society, we can see that the subcultures most immediately affected by nanotechnology are communities of scientists and engineers. Unless environmental risks become a hot button issue (as they have for another umbrella discipline, biotechnology), in the short and medium term, it will probably matter very little to the general public that various products are being developed by "nanotechnologists" rather than, say, chemists or materials scientists. But for scientists and engineers in relevant disciplines, nanotechnology is presently and profoundly influencing their daily routines, institutional affiliations, disciplinary loyalties, and material resources.

If one talks to researchers in these disciplines, one finds the full spectrum of orientations to nano—from outright hostility to full-fledged enthusiasm. In between those two extremes, there is a wide array of locally defined understandings of nano. Some cheerily think of nanotechnology as a way to continue things they've been working on already, but in conjunction with new kinds of friends; some think nanotechnology is great for generating money but not much else; some have seen their old subdisciplines decline and are, perhaps reluctantly, retooling themselves as nanoscientists; and some see nanotechnology as a great source of money, but not a coherent or long-lasting vision of how to organize research. Even for people who agree with the general aims of the NNI or the visions of individual enthusiasts, it is by no means self-evident why they should change what they are doing in order to fit with the nanotechnology rubric. Indeed, those most committed to nanotechnology may be most prone to disputes about which disciplinary languages to use, which disciplines should take the lead in nanotechnology research, and how nano-interdisciplinarity should be fostered.

Nanotechnology is the site of what sociologists of science call "boundary work"—the construction and reconstruction of delineations between communities. Like any redrawing of borders, we can expect some people to disagree with the new map [10]. Fostering interdisciplinarity requires instituting "trading zones" to promote interaction and building linguistic "pidgins" to enable communication between different kinds of practitioners; but, as the history of colonialism shows, pidgins usually emerge in an environment of asymmetric influence [11]. If one discipline or community is seen as imposing its vision, then other communities may become disaffected with the project of nanotechnology. Importantly, many of the constituent communities of nanotechnology research have seen fierce debates about disciplinary autonomy before, and nanotechnology organizers should be aware that the history of those debates will influence how communities react to nanotechnology [12, 13, 14].

Therefore, I recommend that we look at these heterogeneities in the communities that are being co-opted into nanotechnology and that we try to understand those technical subcultures as we would understand any subculture that is being coopted into a large, new social movement. It is entirely natural in those circumstances for there to be a complex weave of resistance and enthusiasm and opportunism. We need to acknowledge that nanotechnology is not a self-evident way of organizing research for everyone, and look at the sources of varying attitudes toward nano. I think what we will find is that perceptions of nanotechnology have much to do with the histories of various subcommunities within nano, and the histories of particular individuals in those communities. In the probe microscopy field, for instance, I have found that what a person thinks of nanotechnology has a lot to do with whether he/she just started buying microscopes recently, or was building them in the 1980s, or was an early adopter of the technology in various disciplines like biophysics or electrochemistry or surface science [15].

If we think of "nanotechnology" writ large as a tool for organizing research, then I think we will find that it is successful at coordinating scientists and engineers in new ways in some places, and not so successful in others—and if we acknowledge that different attitudes toward nanotechnology have their own history and their own logic within various communities, I think we can go a long way toward preventing those attitudes from hardening. We need to look carefully at what it is that makes some people very attracted to nano, and makes other researchers cautious or even hostile—and if we understand that logic of enthusiasm, resistance, and opportunism, we can offer more subtle, institutional and cultural rather than monetary, inducements to make nanotechnology a more sustainable way of organizing science and engineering.

References

1. NRC (National Research Council), Small Wonders, Endless Frontiers: A review of the National Nanotechnology Initiative, Washington, DC: National Academies Press (2002).

2. T. F. Gieryn, Two faces on science: Building identities for molecular biology. In *The Architecture of Science*, ed. P. Galison, E. Thompson, Cambridge, MA: MIT Press (1999).

3. P. Galison, C. Jones, Factory, laboratory, studio: Dispersing sites of production. In *The Architecture of Science*, Cambridge, MA: MIT Press (1999).

4. D. J. Kevles, *The Physicists: The history of a scientific community in modern America*, Cambridge, MA: Harvard University Press (1995).

5. E. T. Layton, Jr., *The Revolt of the Engineers: Social responsibility and the American engineering profession,* Cleveland: The Press of Case Western Reserve University (1971).

6. R. Kohler, *From Medical Chemistry to Biochemistry: The making of a biomedical discipline*, Cambridge, UK: Cambridge University Press (1982).

7. M. Rudwick, *The Great Devonian Controversy: The shaping of scientific knowledge among gentlemanly specialists*, Chicago: University of Chicago Press (1985).

8. B. Bensaude-Vincent, The construction of a discipline: Materials science in the United States, *Historical Studies in the Physical and Biological Sciences* **31**, 223-248 (2001).

9. Pickering, Cyborg history and the World War II regime, *Perspectives on Science: Historical, Philosophical, Social* **3**, 1–48 (1995).

10. T. F. Gieryn, *Cultural Boundaries of Science: Credibility on the line*, Chicago: University of Chicago Press (1999).

11. P. Galison, *Image and Logic: A material culture of microphysics*, Chicago: University of Chicago Press (1997).

12. R. Kline, Construing "technology" as "applied science:" Public rhetoric of scientists and engineers in the United States, *Isis* **86**, 194-221 (1995).

13. R. Kline, Facing a (perceived) paradox of "engineering science:" Epistemology and American engineering education in the early Cold War. In Annual Meeting of the Society for the History of Technology, Charlottesville, VA (1995).

14. E. T. Layton, Jr., American ideologies of science and engineering, *Technology and Culture* **17**, 688–701 (1976).

15. C. Mody, From the topografiner to the STM to the AFM: What probe microscopy can tell us about nanoscience instrumentation. Paper presented at Discovering the Nanoscale International Conference, 20–23 March, Columbia, SC (2003).

CONVERGING TECHNOLOGIES AND THEIR SOCIETAL IMPLICATIONS

Nora Savage, Environmental Protection Agency

The convergence of nano-, bio-, info- and cogno-technologies will have a major effect upon the various federal agencies. In particular, for the Environmental Protection Agency (EPA), whose mission is to protect human health and safeguard the environment, these impacts may alter the very nature, scope, and method of conducting business. The basic principles by which the Agency regulates industries, monitors the environment, and seeks to ensure a clean, healthy ecosystem will undergo major transformations as developed technologies enable novel capabilities that will depend upon vastly different premises and assessments.

In addition, the extent to which the duties of agencies with related legislative authorities will merge or conflict may increase significantly. Much of this will result directly from technological advances as traditional disciplinary lines and boundaries are blurred through interdisciplinary research and product development. Some will be the result of evolving agency priorities and mandates.

Accordingly, the EPA, in anticipation of the significant impacts upon the manner in which its mission is accomplished, is currently engaging in a variety of activities. These activities include sponsoring research and product development on the potential applications and implications of nanotechnology for the environment; coordinating and participating in strategic research planning meetings concerning the potential role(s) for emerging technologies with respect to environmental protection; and providing information at a variety of conferences and workshops composed of academic, industrial, government, and nongovernmental organization (NGO) representatives dealing with possible societal and environmental impacts of novel technologies.

One potential application of the convergence of these technologies to the environment is the development of rapid, accurate, miniature environmental sensing and monitoring devices that will eliminate much of the risk from hazardous compounds to ecosystems and the public. The incorporation of "smart" technology, i.e. the ability to perform some specified action depending upon the nature, concentration, and location of the detected substance(s) and to self-repair in the event of failure or trauma, will enhance this capability. Alerts can subsequently be issued immediately both to the general public and to Agency personnel concerning toxic substances in the atmosphere, water supplies, and subsurface areas so that both avoidance by the public and treatment by the Agency can occur quickly. Anticipated capabilities would include the ability to detect minute concentrations of a variety of compounds simultaneously, to record the collected data, to transmit the data to a central facility in a user-friendly format, to provide warnings for dangerous concentrations of toxic compounds, to isolate these compounds in order to minimize environmental and human impacts, and to remediate potentially harmful compounds. Accordingly, the technology would allow for the detection of minute concentrations of multiple compounds simultaneously and for the implementation of rapid responses to safeguard people and the environment. The work of Wan Shih at Drexel University involves the use of highly piezoelectric microcantilever arrays for in-situ, rapid, and simultaneous multiple pathogen quantification in source water with the ability to detect pathogens [1]. Additionally, the work of Joseph Wang and associates involves remote sensing using DNA recognition for the development of electrochemical sensing devices [2].

Treatment and remediation techniques can also be greatly improved through the development of technologies based upon the convergence. The potential exists to develop inexpensive remediation and treatment technologies that enable the rapid and effective cleanup of recalcitrant compounds, especially those located in inaccessible areas. Currently, many of the methods that the Agency employs to remove toxic contaminants from the environment involve laborious, time-consuming, expensive techniques that often require pre-treatment processes and removal of portions of the surrounding environment. The development of technologies that can perform *in situ* remediation—and by their ability to reach into crevices, below aquifers, and into other areas that have proven difficult to access, eliminate the necessity for costly pump-and-treat operations that can damage the existing ecosystems—would greatly facilitate the remediation of many contaminated sites, especially those on the Superfund list. The work of Angela Belcher and her lab associates at the Massachusetts Institute of Technology in inducing tiny benign viruses to produce inorganic materials with semiconducting or magnetic properties [3] shows promise as a novel method to remove heavy metals. Also, the work of Wilfred Chen at the University of California, Riverside [4] involves the development of tunable biopolymers with metal-binding properties for the removal of heavy metals, such as cadmium, mercury, and arsenic. Traditional techniques for remediation of these compounds have proven to be very costly due in large part to their location in difficult-to-access areas and to the characteristics of the surrounding environment.

In addition, environmentally benign manufacturing and processing methods will result in the elimination of toxic wastes and by-products and enable bottom-up chemical and industrial manufacturing that utilizes "green" processes. Such "green" or environmentally benign manufacturing processes will not only eliminate waste streams (via precise manufacturing) as a resultant product, but also will reduce risks associated with the use of hazardous reactants and solvents through the utilization of non-hazardous starting materials. Such processes also have the potential to significantly reduce the consumption of energy and to make commercially viable many of the alternative clean energy sources (i.e., solar, fuel cells). This will obviate the need to constantly play "catch-up" as the Agency currently seeks to treat an ever-increasing number of polluted and contaminated sites throughout the country.

The convergence of these four technologies will also provide a stimulus for the various federal agencies to exert a more coordinated effort in terms of their respective regulatory duties and sponsored research opportunities. It will enable a more concerted effort among the various agencies in the development and implementation of various strategies utilized while performing public service. It can also result in a restructuring of the regulations under which the agencies act in order to provide more efficient and effective benefits and to offer enhanced protection and utility to the general public.

Conversely, the convergence of these technologies may raise serious environmental, public health, and societal issues. Such issues include environmental risk due to the intended or accidental release of materials that may occur as a result of the processing and manufacture, the usage and application, and the disposal or disassembly of products and materials resulting from the converging technologies.

There may also be an alteration in the nature and extent of personal privacy due to increased and more penetrative monitoring capabilities. Changes in personal privacy may also result as personal medical information becomes more readily accessible and more widely available. The possession by employers, government officials, and others of in-depth personal medical information including predisposition to certain physical and mental afflictions, will cause changes in the level of privacy experienced. Furthermore, the placement of various monitors that utilize face-recognition software for such purposes as security and adherence to various laws may be viewed as intolerable intrusions upon individual rights. As personal privacy changes in both level and scope, increased communication between scientists, engineers, industrial representatives, government personnel, and the general public is vital. Also important is increased research funding for the social and ethical issues that will arise from the development of products as a result of the convergence of these technologies.

Convergence may also give rise to enhanced socio-economic disparities due to higher costs of state-of-the-art technologies. Providing comfort and security may well depend upon expensive new technologies that may be beyond the reach of those in the lower socio-economic groups. In addition, the prevalent use of such technologies may result in the alienation within our society of groups of people as these technologies become commonplace and infiltrate various aspects of daily life. Alternatively, a decrease in the cost of these technologies may also occur due to the use of inexpensive components and processes, thereby closing the gap between socio-economic groups. This may allow for better protection of health and improved living conditions for many. It is unclear what the overall impact will be.

Legal issues such as tort complaints may arise from the unintended and unexpected consequences of these technologies as the products come to market and enter everyday use. Intellectual property disputes may also develop as patent areas merge and overlap and as those applying for patents seek broad application and protection for ideas and devices.

Environmental concerns include issues of the potential persistence and possible synergistic effects of these compounds with other contaminants or naturally occurring compounds in the environment. How reactive such compounds may be in the environment, the reactions involved as they degrade, and the types of compounds that result are crucial, yet largely unknown questions. There is also the issue of the potential bioavailiability, bioaccumulation, and biotransformation capacities of novel compounds developed as a result of this convergence. The capacity of bio-nanotechnology products to accumulate in certain organs, or other components of living systems in various species, needs to be explored, along with the metabolic alteration of these substances and subsequent effects upon living systems. Additionally, knowledge about the transport of nanomaterials that reach the environment is important, and yet currently unknown, information. How these materials move from one medium to another, from one organism or ecosystem to another, and from organisms to the environment and vice versa will be critical for understanding and implementing proper manufacture, use and disposal options. In order to effectively assess these impacts a full life-cycle analysis of the various constituents, products, and compounds must be undertaken—a look at the product from the accumulation of starting materials to the development, manufacture, use, and eventual disposal or reuse of the item or portions thereof.

Finally, there are concerns regarding the interaction, overlap, and oversight of the regulatory authorities of certain federal agencies that must be considered. Confusion may result over jurisdictional issues for agencies with oversight over similar matters. Regulatory purviews may begin to cover identical topical areas or may overlook and omit other areas. Consequently, important issues may be resolved with differing, conflicting conclusions by different agencies, causing confusion among the regulated industries and the general public. For example, upon the convergence of nano-, info-, cogno-, and bio-technologies, the purview of the EPA may overlap and intersect with those of the Food and Drug Administration (FDA), the National Institute for Occupational Safety and Health (NIOSH), the Occupational Safety and Health Administration (OSHA), and the Consumer Product Safety Commission (CPSC). The collaboration of duties and coordination of activities among these regulatory agencies will be critical. This will be increasingly important as products with nano-, bio-, info-, and cogno-materials or using converging technological processes begin to be manufactured on a mass scale and enter the marketplace.

References

1. W. Y. Jeong, W. Y. Shih, R. Mutharasan, W-H. Shih, *In situ* cell detection using piezoelectric lead zirconate titanate-stainless steel cantilevers, *Journal of Applied Physics* **93**(1), 619 (1 January 2003).

2. J. Wang, et. al., Magnetically Induced Electrical DNA Detection, *Journal of the American Chemical Society* **124**, 4208 (2002).

3. C. E. Flynn, S-W. Lee, B. R. Peelle, A. M. Belcher, Viruses as vehicles for growth, organization and assembly of materials, *Acta Materialia* **51**(19), 5867-5880 (25 November 2003).

4. W. Chen, J. Kostal, J-Y. Kim, A. Mulchandani, Elastin Biopolymers for environmental remediation and sensing. Paper 203d presented at AIChE Annual Meeting, 3-8 November, Indianapolis (2002).

4. ETHICS AND LAW

ETHICAL ISSUES IN NANOSCIENCE AND NANOTECHNOLOGY: REFLECTIONS AND SUGGESTIONS

Robert E. McGinn, Stanford University

One of my main scholarly interests is the study of ethical issues raised by developments in science and technology. Exploring such ethical issues as may be raised by developments in nanoscience and nanotechnology (NST) is something I regard as intellectually challenging and socially important. It was with this interest and orientation that I attended the Workshop on Societal Implications of Nanoscience and Nanotechnology. Extensive attention was devoted at the workshop to discussion of societal issues likely to be engendered by emerging and future developments in NST. Some of that attention was directed at a particular subset of societal implications of NST, namely, those held likely to raise ethical issues, properly so-called. The first day of the workshop included two ethics-related presentations: "Ethics and Nanotechnology: A Survey" and "Navigating Nanotechnology through Society." On the second day, one of the 10 small two-hour-forty-minute group discussions was devoted to "Ethical, historical, governance, philosophical implications, risk and uncertainty," a domain of concerns as diverse in content as it is enormous in scope.

The above-mentioned presentations were excellent and the small group discussion was stimulating. However, as regards ethical issues likely to be raised by NST, after the workshop I found it difficult to believe that, at least at this point in time, academic ethicists are much more knowledgeable than the modest fraction of the public that has read Bill Joy's essay in *Wired* [1], with its ominous references to molecular-level replicators and gray goo, or the presumably slightly larger fraction that has read Michael Crichton's *Prey* [2], with its unpredictable swarms of programmed, learning-capable nanobots. To date, the scholarly literature about ethics and NST is thin. A few writers have suggested in general terms that developments in NST may well engender important ethical issues of privacy, intellectual property, distributive justice, and public safety [3]. It is useful to underscore such concerns. But I discern little reliable knowledge about what ethical issues NST will actually engender, at what junctures they will emerge, no indication of whether any will prove unique to this dynamic field of inquiry, and little searching analysis of their structure, stakes, or resolutions.

In light of the foregoing, I regard "NST ethics" as essentially an unopened Black Box, for scholars and thoughtful segments of the general public alike. How might one respond to this situation? Four general programmatic suggestions follow, ones I hope to heed in my own effort over the next few years to identify and analyze the key ethical issues in and around NST. I offer these suggestions as candidate parameters for any inquiry into NST ethical issues that aspires to avoid the irresponsible polar extremes of facile optimism (that NST will not raise any noteworthy ethical issues) and resolute pessimism (that NST will raise ethical

M.C. Roco and W.S. Bainbridge (eds.),
Nanotechnology: Societal Implications — Individual Perspectives, 169–205.

issues so grave as to call for terminating the endeavor permanently, as with persistent calls by some parties for doing precisely that with human therapeutic cloning research). My general suggestions are as follows:

1. Strive to assure that the activities and practices of all relevant groups of actors involved with NST are scrutinized from an ethical point of view.
2. Strive to assure that all phases of NST product life cycles are scrutinized from an ethical point of view.
3. Undertake ethics-related empirical inquiry in communities of NST practitioners.
4. Combat myths that prevent or impede groups of practitioners involved with NST from recognizing and coming to grips with ethical issues that arise in their work.

Regarding 1, while I am especially interested in exploring ethical issues that pertain to scientists and engineers working on NST, I came away from the workshop with the realization that scientists and engineers are far from being the only groups of actors whose members are likely to be confronted by NST-related ethical issues and decisions. I am thinking not only of manufacturing workers, end users of NST-enabled products, and those affected by such use, but also of NST-related groups like entrepreneurs, government regulators, educators, legislators, judges, media professionals, and even members of the general public. Whether they currently realize it or not, members of each of these groups are likely to be ethically obliged or entitled to act or to refrain from acting in various ways as a function of, among other things, the kinds of potent NST knowledge and types of novel NST-enabled material products and NST-related decisions with which they will be confronted in their respective social roles as professionals and citizens.

Regarding 2, a comprehensive study of NST ethics must reflect the fact that NST-related ethical issues are unlikely to arise only in one particular phase or stage of the NST lifecycle for any particular nano-endeavor. On the contrary, different ethical issues may well arise in different stages of that lifecycle, from conception, funding, and basic research, to manufacturing, diffusion into society, operation and use, maintenance, monitoring, and recycling. As with the full range of relevant actor groups, this fact about diverse genesis points of ethical issues in life-cycle stages must be reflected in any inquiry into NST ethics that purports to be comprehensive.

Regarding 3, assuming that cooperation from NST principal investigators is forthcoming, it would likely be fruitful to conduct anthropological-ethical fieldwork within communities of NST practitioners, much as in recent years some bioethicists have done in communities of physicians and medical researchers. For example, ethicists could conduct interviews intended to probe NST practitioners about ethical issues in relation to their cutting-edge work, or administer surveys to them intended to elicit their core ethics-related beliefs. Technical practitioner beliefs usefully explored could include (1) how salient or frequent ethical issues are or are likely to be in their work; (2) what it is they think makes an issue an ethical (as opposed to a non-ethical) issue; and (3) their notions of key ethical concepts, such as "harm," "acceptable risk," and "social responsibility of the technical professional." Such inquiries could yield useful baseline information about perceived ethical issues in NST; contribute to identifying new possible ethical issues otherwise unperceived by ethicists; and be useful for the design of ethics tutorials and modules on ethical issues for NST-related students, practitioners, technicians, and other equipment users.

Regarding 4, there are a number of problematic beliefs held by many scientists and engineers that can impede them from recognizing or coming to terms with ethical issues that may arise in their work. One such belief is that science and technology are "neutral" or "value-free." From this contested belief many scientists and engineers mistakenly conclude that engaging in scientific and engineering activities *per se* cannot possibly raise any ethical issues. Hence, they conclude, they need not worry about ethical issues when acting as technical professionals, only possibly when acting as managers, policy officials, or citizens. Technical professionals of this persuasion sometimes believe that ethical issues arise only because and when the neutral products of scientific or engineering endeavor are utilized in problematic, non-neutral ways by non-technical parties in the larger non-technical world, and that venturing into this realm is not their moral responsibility as technical professionals. NST scientists and engineers should be induced to ponder the possibility that ethical issues that their work may raise depend on the interaction of the technical features of their work with the fact that that work unfolds within—and its fruits are or will be deployed and utilized within—a particular national and global social order, one with certain discernible characteristics. For example, the fruits of their work will likely be exploited in societies with certain reward systems, ones that reward being first to market and that make the application or non-application of their work in some areas and not in others, and the modes of and constraints on those applications, reasonably foreseeable. The full range of NST-related ethical issues should not be artificially narrowed by idealized decontextualization of the work of NST practitioners.

A related core belief that hinders many scientists and engineers from coming to grips with ethical issues raised in or by their work is the long-entrenched idea that there is an inviolable moral right to free scientific inquiry that should not be abridged by social forces offended by what they view as controversial research. The tragic events of 9/11 have resulted in this epistemological sacred cow being subjected to critical scrutiny. Scientists working in areas of biology related to weapons of mass destruction are for the first time being compelled to consider whether the potency of such weapons in the current global political context might dictate that the moral right to free scientific inquiry needs to be put on a much tighter leash than previously thought morally justifiable, as regards both the permissibility of certain experiments and the publication of certain findings. Similarly, given the mobility of NST knowledge, the minute scale of many NST outcomes, and the potency and virtual indetectibility likely to characterize many products enabled by that knowledge and those outcomes, NST practitioners may also be required to re-examine the same ethics-related belief about the inviolability of scientific research. Technical practitioners in certain areas of NST might be asked to consider whether their work, together with the context in which it is carried out, dictates that their ethical rights as investigators be selectively delimited and their ethical obligations expanded. In short, the virtual indetectibility and potency likely to characterize not a few outcomes of NST work, when set in the high-stakes conflictual context of contemporary global society, could warrant a transformed matrix of ethical rights and responsibilities for practicing NST professionals. It is worth exploring critically whether this is the case and, if it is, contributing to the elaboration of its major new elements.

In conclusion, upon reflection, it is not surprising that analysis of ethical issues related to NST is in such a rudimentary stage. After all, NST research and development work itself is still in its early phases, and it is far from clear at this juncture what kinds of practical possibilities and products will emerge from pursuit of this work in this and other societies. Nevertheless, it makes sense for interested ethicists to spend time "in the field" getting to know the assumptions, practices, plans, cultures, goals, and projections of NST professionals, as well as helping them develop the ability to recognize non-obvious ethical issues, especially ones that cannot be reduced to risks of consensual harms like injury, death, and disease. My hope is that in 5 to 10 years, ethicists, working closely with thoughtful NST practitioners, will have succeeded in prying open the Black Box of NST ethics to a reasonable degree and will have begun to shed bright light on some of its more intriguing and provocative contents. Such an achievement could in turn make it more likely that the astonishing technical virtuosity and creativity already apparent in NST work will wind up significantly enhancing societal well being in a variety of welcome ways.

References

1. B. Joy, Why the future doesn't need us, *Wired* **8**(4) (2000) (available at http://www.wired.com/wired/archive/8.04/joy.html).
2. M. Crichton, *Prey*, New York: HarperCollins (2002).
3. A. Mnyusiwalla, A. S. Daar, P. A. Singer, "Mind the gap:" Science and ethics in nanotechnology, *Nanotechnology* **14**, R9-R13 (2003).

ETHICS AND NANO: A SURVEY

Vivian Weil, Illinois Institute of Technology

While still at an early stage of nanotechnology research and product development, a good strategy for identifying ethical issues is to track what is actually happening. Turning away from the realm of speculation, I survey a complex array of activities actually going on close to the ground.

Up to now, ethics commentators have generally addressed ethical and societal implications either before the development of a new technical area or afterward, once the technology is in place. Ethics before development usually involves advocacy of a halt to some frontier technical work until we have a better understanding of likely consequences. Ethics afterward, when the technology is in place, requires dealing with risks and harms that have already been brought about.

The problem with ethics before is that it requires people to determine the risks and harms of speculative technologies. The ethics cannot make progress when predictions of what is to come are imagined and vague. The problem with ethics afterwards is that unnecessary significant risks and harms may have already been produced.

A way to avoid these problems with nanoscience and nanotechnology is to explore ethical dimensions of current activities in progress. I will highlight two developments. One is the growing recognition that in promoting and pursuing nanotechnological innovation, we face ethical and societal issues that must not be ignored. The other is the increasing effort by scientists to identify and experimentally investigate risks to animal organs (and perhaps human organs) from certain nanoparticles and nanotubes. In light of their effort, I draw suggestions for innovative approaches to further identify and investigate such risks and other ethical and societal implications of nanotechnology developments. One approach focuses on the organization of interdisciplinary research in the nanotechnology area. The other focuses on features, such as transparency of institutions involved in nanotechnology developments. Finally, I draw attention to ethical concerns that have come to the fore regarding nano developments in military institutions and nano enterprises viewed from an international perspective.

Concerns of Nano Scientists and Engineers

I turn first to the growing recognition that there are issues to address. It is interesting that within the nanotechnology area, some technical people have begun to acknowledge ethical concerns [1, 2]. Their response nourishes the expectation that ethical discussion and debate will not be left to outsiders alone, but that lively discussion can grow inside nanotechnology enterprises.

Nanotechnology specialists have recognized that it would be a mistake to forge ahead thinking they have no issues of ethics and societal implications to consider. They advocate investigation of potential impacts that nanotechnology developments may have on human health and the environment [1, 2]. By putting these concerns first, they acknowledge the ethical priority of regard for human welfare, that is, of taking care not to cause avoidable harm to people or the environment.

Some people urge that transparency and sharing of knowledge should characterize their enterprises. In this way, they underline the necessity for honesty and candor. These are moral virtues useful for bringing ethical problems to light and essential to building trust. The public's trust is a core ethical concern for nanotechnology enterprises.

Nanotechnology specialists acknowledge that a cautionary approach may initially slow investigation and implementation. They, nevertheless, have recommended a careful, deliberate approach to avoid irreparable harm and to gain public acceptance. In recent years in Europe and the United States, scholars and policymakers have articulated a Precautionary Principle for advancing innovative technologies in the face of uncertainties and unknowns [3]. That Principle may be a useful starting point for taking a careful approach in the nanotechnology area.

Some people working in the field have also recognized the need to look ahead in order to be prepared for impacts on the workforce and the workplace. For instance, the introduction of dramatically new production processes may devalue once sought-after workers' skills, and thereby cause workforce disruptions that will have to be addressed. In addition, new materials in the workplace may introduce risks requiring stringent safeguards for workers, perhaps similar to the safeguards for working around powdered metals. It is a requirement of justice that we try to avoid or at least mitigate foreseeable harms to target groups; indeed, the imperative "when possible, avoid doing positive harm" is part of the minimum conception of justice on which most philosophers would agree.

Another workforce issue that nanotechnology specialists emphasize is the need for interdisciplinary education. In this way, they take a standpoint of preventive ethics. They are looking ahead to prepare future technical people in the nanotechnology area to work across disciplinary boundaries and function responsibly in a complex environment featuring multiple perspectives. The nanotechnology area is marked by the convergence of disciplines. In that light, this preparation seems both practically and ethically essential.

Some people anticipate that issues of justice and fairness may emerge in the distribution of benefits from nanotechnology developments. They warn against developing a counterpart to the digital divide in the nanotechnology area, producing a gap between those benefiting from the information technologies and those left out. The term "nano divide" has already come into use. A primary concern of justice is to avoid increasing inequalities in society. Here the concern is not to further disadvantage those already disadvantaged. By producing benefits inaccessible to those already disadvantaged, nanotechnology developments would be unjust [4, 5].

While these alerts from within the nanotechnology community are not widespread, they give evidence that scientists and engineers in the nanotechnology domain recognize dangers from neglecting to follow up clues to problems that may arise. At this time, we really know little about what the problems will be. This is because we do not yet know what applications will be pursued, and there is a whole spectrum of very different kinds of risks deriving from the diversity of the nanotechnology area. In addition, nanotechnology specialists warn of unpredictable consequences in light of the stochastic interactions of molecular systems. A leader in the nanotechnology area observes that "emerging behavior doesn't come as clockwork [1]." The foregoing itemization of ethical concerns voiced by nanotechnology specialists provides a kind of map or program suggesting where and what to investigate.

Recent Developments

The sense that we are still in an early stage, at the brink of major nanotechnology developments, is widely shared. If that view is correct, there is still time to take steps to address ethical and societal issues in innovative ways that take advantage of lessons learned from the emergence of earlier radically new technologies, such as nuclear energy and biotechnology.

Yet nanotechnology economic development initiatives and networks, nanotechnology communication networks, and nanotechnology commercial ventures proliferate. The NanoBusiness Alliance reported in November 2003 that there are approximately 1,100 startup enterprises in the United States and venture capital investment has exceeded $1 billion dollars over the last three years [6]. Nearly every manufacturing company in the Fortune 500 is entering the nanotechnology field. National Laboratories are energetically involved, and government agencies, including NSF, are funding and promoting the nanotechnology area. Nanotechnology applications have led to a number of "improved" familiar products, such as sunscreens, catalysts, cosmetics, dental bonding materials, optical coatings, power machinery, and fabrics. Within the United States and at the international level, the atmosphere is already very competitive [7]. Commercial enterprises move forward at a fast pace while investigation of potential risks and harms proceeds at a "measured" pace [8].

Investigations of Health Effects

I turn now to some recent findings about health effects of certain nanomaterials based on research carried out in university, company, and government agency laboratories. Scientists reported toxic effects on mouse and rat lungs of certain single-walled nanotubes and certain nanoparticles. Presumably, effects on the lungs of mice and rats were investigated because of known toxic effects of other tiny particles—such as quartz dust and asbestos—on the lungs of humans.

An April 2003, report in *Science* and a November 2003, report in *The New York Times* highlighted findings of toxic effects of single-walled carbon nanotubes [9, 10]. The report in *Science* also covered findings of toxic effects of nanoparticles made from polytetrafluoroethylene, carbon-3, and manganese nanoparticles. The polytetrafluoroethylene particles were particularly lethal. When Carbon-3 and manganese nanoparticles were inhaled by rats, these particles found their way to rats' olfactory bulbs and then traveled throughout the brain.

At the March 2004 annual meeting of the American Chemical Society, an academic scientist reported a study indicating that buckyballs, a form of carbon and an important material in nanotechnology, can cause extensive brain damage in fish. This is the first study "to indicate destruction of lipid cells, the most common form of brain tissue [10, p.C5]." A DuPont company scientist who led a session on nanoparticles at the March 2004 national meeting of the Society of Toxicology stated that scientists are beginning to address many fundamental questions about the toxicity of nanoparticles. He commented that toxicity may depend more on how nanoparticles are coated and how quickly they clump together than on their size [10].

Whether new or unique features are shown in the toxic effects of nanomaterials is a question for further investigation. Also in question is whether the amounts now available should cause concern. If cheap methods are devised for moving up to large-scale production, exposure would increase and with it the need for better understanding of the effects of different levels of exposure. The ethical imperative for further investigation is obvious.

Environmental Impacts

Research on environmental impacts of nanomaterials seems not as advanced as investigation of health effects. The focus of early interest in environmental impacts was on positive effects. Until 2003, the Environmental Protection Agency was oriented toward supporting research on how nanotechnology could help clean or protect the environment [8].

However, a shift is occurring, with the focus moving toward investigation of potential negative consequences [8, 10, 11]. The need to understand better the effects of nanoparticles in the environment is underlined in questions materials engineers ask: for example, to what extent do effects of nanoparticles resemble those of powdered metals. In view of the perception that the nanotechnology area is advancing at a fast pace, we should call for more concerted scientific investigation of potentially harmful environmental impacts. As noted earlier, this is an ethical concern of the highest priority.

The Commercial Arena: Fortune 500 and Entrepreneurial Companies

Among the investigators who produced the findings on health effects are scientists at DuPont. One of them explained a cautious approach at DuPont where some remember unwelcome consequences from past innovations, for example, a lawsuit over health effects of a chemical in Teflon [8]. Other Fortune 500 companies may be similarly cautious. Whether or not that is the case, this instance brings out the fact that a large, long-established company has learned from past experience to be prudent, has resources for testing for health effects, and can justify the investment.

It seems unlikely that such an approach is as affordable in young, adventurous firms that are rapidly forming. These entrepreneurial firms present a distinctive challenge when it comes to building in attention to ethical implications at the front end, in decision making. These firms are by their nature fast-moving, and they operate in an atmosphere of stringent time constraints and the urgent need to acquire resources. Their business perspective does not easily incorporate careful attention to societal consequences. Yet it is important to determine what approaches can help entrepreneurial firms to consider ethical and social implications and to track health, environmental, and social impacts. The very agility of these firms may be an advantage. In any event, they are not exempt from ethical imperatives, and they must guard against harmful and costly mistakes.

In Academia and Research Settings: Expanding the Range of Disciplines

Academics involved in nanotechnology research also operate at a fast, competitive pace, and they express a sense of excitement about their projects, often pointing specifically to multidisciplinary aspects (such as the engineers and scientists organized by Carlo Montemagno in the NSF NNIN competition of 2003). Unusual and potentially valuable properties of new products, such as strength and electrical conductivity, show up right away. An ethical perspective is often needed to ask about and then explore properties, such as disposal features, that have social implications. Investment in scientific research to investigate such other properties and their impacts must be forthcoming.

To bridge the gap from raising questions to conducting research that responds to questions, it will be useful to expand the range of disciplines associated with nanotechnology projects. Such expansion should offer channels to nanotechnology specialists for easy interchange with toxicologists; materials, chemical, and environmental engineers; marine biologists; and other engineers and scientists whose expertise and investigation are needed to follow up on indications of harmful effects.

There is good reason to think that expanding the range of multidisciplinarity in this way can add to the excitement of working across disciplines. It becomes essential to fashion the interdisciplinary education advocated by nanotechnology specialists to facilitate this kind of ethically informed interchange.

Further Expansion of the Range of Disciplines

The expansion must be broad enough to include specialists from history, philosophy, and the social sciences. NSF, other agencies participating in the NNI, and the National Research Council's 2002 report *Small Wonders, Endless Frontiers* [12] all emphasize that nanotechnology research and development must build in attention to ethics and social implications.

So the expansion must embrace philosophers, including philosophers of science and ethics specialists, historians of science and technology, and social scientists, including sociologists, economists, anthropologists, and Science and Technology Studies (STS) scholars. Over the last 25 years, STS studies have surged, and a number of philosophers and historians have collaborated with engineers and scientists in addressing a range of issues, especially ethical issues, in engineering and science.

It is important to bring in not only specialists working in all of these areas, but also postdocs, in order to prepare a cadre of investigators immersed in the nanotechnology area. Considering the great scope and diversity of the nanotechnology area, specialists in the humanities and social sciences see a need for many new investigators from their fields who are equipped to work in nanotechnology endeavors.

Ethical and societal issues are closely interwoven. Within nanotechnology endeavors, specialists from the humanities and social sciences can conduct research that brings together their expertise and investigative approaches. At the same time, they can develop ways of working collaboratively with nanotechnological specialists. And they can extend their investigation to a meta level, to study these very processes of cooperation and collaboration.

With proximity to nanotechnology projects, social scientists can develop data on and insight into processes of innovation, diffusion, and network formation and evolution. They have an opportunity to study the emergence of interest group activity from its early stages. STS scholars can combine perspectives from inside and outside nanotechnology enterprises to advance technology assessment.

Ethics specialists can help nanotechnology specialists to bring specific concerns to the surface and to consider appropriate responses. For example, by observing and asking questions and by developing questionnaires for use in nanotechnology R&D facilities, they can prompt engineers and scientists to bring up questions that might otherwise remain unformulated—such as specific questions about impacts of buckyballs and fullerenes in marine environments.

Multiplying Zones of Interchange within Nanotechnology Enterprises and with the Public

Once articulated, these questions can be discussed in multidisciplinary nanotechnology settings where participants can evaluate the seriousness and immediacy of concerns raised. When questions seem to warrant further investigation, participants can consider who should be informed to ensure that the next steps are taken. The creation of more zones of interchange should be a priority. More opportunities for interconnection offer many advantages for advancing nanotechnology enterprises, not least the advantage of enabling nanotechnological specialists to relay ethical and societal concerns to appropriate specialists in other disciplines.

Through these channels, nanotechnology scientists and engineers may prompt toxicologists and materials, chemical, and environmental engineers, for example, to acquire funding and to do more research of the kind recently reported. We cannot put all our reliance on outsiders to ask appropriate questions. To bring concerns to the surface, we need an aggressively questioning environment within nanotechnology enterprises.

At the same time, people working in the nanotechnology area have to accept an obligation to bring outsiders into the conversation. The experience of interchange across disciplines should lay some basis for initiating conversation with outsiders, with members of the public, and with different publics. This might mean discussing, for example, likely uses of products of current developments or options for new products, both near term and further off. Actually engaging in dialogue with the public, making use of feedback (thereby giving the public a measure of power), and then continuing the conversation are essential to building and maintaining trust.

A democratic society requires openness and transparency in government so that citizens have access to knowledge about the actions of government that affect their well being. With that knowledge, citizens can participate effectively in deliberating and deciding about matters that affect their welfare and can hold government officials accountable for their actions [13, 14, 15]. In this way, citizens can play their part in a democratic process that decides how to balance benefits and harms from new technology.

Members of corporations involved with nanotechnology development should accept the responsibility to promote openness and transparency in their organizations. The reasons are that these features are essential for public trust, they support democratic decision making concerning technological innovation, and corporations receive explicit encouragement, if not funding, from government to pursue nanotechnology enterprises. Proprietary concerns, of course, present obstacles. To implement this responsibility, members of corporations will have to learn how to protect legitimate proprietary information while sharing information appropriate and necessary for public discussion. Certain marketing strategies, such as focus groups, already require balancing of this kind.

Governance Issues

To provide for the sorts of interchange and investigation that are needed, we look to institutional and organizational structures and to policy formation and decision making processes. These governance issues need attention in all the institutions involved with nanotechnology, in universities, private companies, government funding agencies, government laboratories, and government agencies involved in regulation for health and safety. These institutions play important roles in nanotechnology development and may be involved in control or regulation of nanotechnology activities and products. The ability of these institutions to learn and adapt, preserving core values, is critical to ethically responsible management of nanotechnology enterprises.

One way to adapt is to ensure that many points of view are brought to bear in considering research priorities and design issues, and in decision making generally. Inclusiveness is essential to obtaining multiple perspectives on likely impacts and—no less important—on desired products and outcomes.

As already noted, openness and knowledge sharing are extremely important. An insular environment in which a small number of like-minded people make decisions is to be avoided. That setting is not likely to generate ethically responsible and sound decision making about nanotechnology R&D. We can have more confidence in carefully constructed processes providing many viewpoints and feedback from the public. Brainstorming and testing of many models of dialogue are needed to construct effective mechanisms for bringing in the public and diverse perspectives.

With these efforts, we should advance our understanding of how relevant publics are constituted. At the same time, we should learn more about conditions that favor the public's acquiring knowledge. In this connection, the issue of hyperbole should be investigated, with attention to how it works in the marketing of science and what might be its appropriate role. Such research generated by the nanotechnology area may lead to more fundamental investigation and understanding of how publics acquire knowledge. This understanding may be useful, in turn, for maintaining ongoing interchange between nanotechnology engineers and scientists and their publics.

Government regulation will surely be called for as the nanotechnology area advances and we learn more about effects on health, the workplace, and the environment. It would be a mistake to try to keep regulation at bay. Rather, efforts should be directed to identifying situations to study more closely with an eye to the possibility of regulatory action. Some ethical imperatives require legal sanctions.

The Military Context

Reports of military nanoscience and engineering activities tend to heighten public fears, and development of radically innovative weaponry has the potential for reducing stability in the prevailing international system. To avoid or mitigate such outcomes nanotechnology developments in the military require the firmest commitment to transparency and knowledge sharing. The resolve to maintain openness can lay a basis for agreement on international limitations during development or testing, or even in the research stage, before any deployment of weapons [16]. Consider an abstract in a National Academy of Engineering catalogue that describes a volume from the Air Force Science and Technology Board, *Implications of Emerging Micro- and Nanotechnologies* [17, p.32]. According to the abstract, priority areas designated by the Air Force "include small, launch-on-demand tactical satellites; miniaturized ballistic missiles; low-cost autonomous surveillance and reconnaissance systems; and smaller, cheaper precision missiles and bombs."

The abstract notes, however, that "commercial investment…now overshadows military investment" and private sector interests drive product development." Nevertheless, according to the abstract, the book ends with the idea that "the Air Force can greatly influence the direction and focus of research" by funding and participating in fundamental research.

Judging by this abstract, nanotechnology R&D is still at an early stage in the military context. The information about priorities shows a level of openness that could be built upon to elicit feedback from the public and lay a basis for further conversation with the public.

The International Dimension

Finally, I turn to the international dimension. So far, scientific and economic competition in the nanotechnology area has been the main international focus. Another important aspect concerns relationships with developing countries. Their participation in the nanotechnology area, their contribution to public discussion and debate, and their vulnerability to harmful consequences should also be part of the international focus.

By giving proper weight to the interests of developing countries, decision-makers in nanotechnology enterprises and in policy processes not only avoid adding to injustice, but also act prudently, in keeping with the minimum conception of justice to do least harm to those worst off. Certain biotechnology techniques and products in agriculture provide an object lesson. U.S. producers of genetically modified agricultural products encounter strong resistance abroad, in developing and in developed countries, in spite of energetic efforts to promote their products.

To avoid such setbacks, policy and decision processes should, from an early stage, bring in perspectives and feedback from relevant constituencies, abroad as well as at home. Toward these goals, comparative studies that consider different styles of decision making in other countries and different cultural perspectives will be useful. Again, we need innovative approaches and testing of methods to achieve "breakthroughs" in responsibly managing international relations and communication involving nanotechnology enterprises and products.

Conclusion

Not long ago, speculation about cascading consequences and possible negative aspects of self-assembly, such as runaway self-assembly, captured attention. Now we are finding out about actual impacts to investigate. And innovation in self-assembly using biological mechanisms is going forward. A first step toward self-assembly of micro-electronic devices, using DNA as a scaffold for nanotubes, was reported in *Science* in November 2003 [18]. This self-assembly process and others in testing may soon be appropriate for study.

We need to find out whether such phenomena behave in line with existing models, and if they differ, what further scientific investigation is needed. In this way, we advance from assumptions and speculation to investigation and problem solving. We need the broadest multi-disciplinary approach in order to raise questions that will bring potential issues and problems to light before problems are manifest. And we need to include diverse perspectives in decision making and in policy formation to help determine both paths to avoid and directions to pursue.

References

1. C. Montemagno, Nanotechnology's implications for the quality of life. Presentation made at NNI Workshop on Societal Implications of Nanoscience and Nanotechnology, 3–5 December, Arlington, VA (2003) (Section 6, this volume).
2. Anonymous reviewer for V. Weil, Zeroing in on ethical issues in nanotechnology, *Proceedings of the IEEE* **91** (11), (November 2003).
3. D. Appell, The new uncertainty principle, *Scientific American* **284**(1), 18 (January 2001) (available at http://www.biotech-info.net/uncertainty.html).
4. J. Rawls, *A Theory of Justice*, Cambridge, MA: Belknap Press of Harvard University Press (1971).
5. D. Gauthier, *Morals by Agreement*, Oxford: Oxford University Press (1986).
6. P. Wehner (NanoBusiness Alliance), E-mail communication, U.S. Senate passes nanotechnology bill (19 November 2003).
7. B. Feder, Bashful vs. brash in the new field of nanotech, *The New York Times* **C1 C3**, (15 March 2004).
8. B. Feder, As uses grow, tiny materials' safety is hard to pin down, *The New York Times* **C1+**, (3 November 2003).
9. R. Service, Nanomaterials show signs of toxicity, *Science* **300**, 243 (11 April 2003).

10. B. Feder, Study Raises Concern about Carbon Particles, *The New York Times*, **C5** (29 March 2004).

11. V. Colvin, The potential environmental impact of engineered nanomaterials, *Nature Biotechnology* **21**(10), 1166-1170 (October 2003).

12. NRC (National Research Council), *Small Wonders, Endless Frontiers*, Washington, DC: National Academies Press (2002).

13. I. Kant, Appendix 2, Of the harmony which the transcendental of public right establishes between morality and politics. In *Perpetual peace and other essays on politics, history, and morals,* Indianapolis: Hackett Pub. Co. (1983).

14. A. Gutmann, D. Thompson, *Democracy and Disagreement*, Cambridge, MA: Belknap Press of Harvard University Press (1996).

15. A. Gutmann, D. Thompson, *Why Deliberative Democracy*, Princeton, NJ: Princeton University Press (2004).

16. J. Altmann, Military uses of nanotechnology: Perspectives and concerns, *Security Dialogue* **35**(1), 70-71 (March 2004).

17. Anonymous abstract of *Implications of Emerging Micro- and Nanotechnologies*, from the National Academy of Engineering printed books catalog (2004).

18. K. Chang, Smaller computer chips built using DNA as template, *The New York Times* **A20**, (21 November 2003).

LAW IN A NEW FRONTIER

Sonia E. Miller, Converging Technologies Bar Association

> History is, in large measure, the study of change.... But the changes now upon us are qualitatively different from those that have gone before... we are beginning to understand what truly revolutionary change means... Quite a new frontier. And, as in any frontier, it takes time and experience before the relevant mores, rules, and laws can be fully defined and the appropriate social and economic structures put into place...In the interim, we frontier people must function with one foot—or, should we say, one part of our minds and spirits—in the old world, and one in the new. (Roberta R. Katz, in *Justice Matters* [1, p.149])

With the Internet, the physical plane through which we identified and defined our relationships and reference points disintegrated. The non-physical terrain of cyberspace dissolved geographic boundaries and demarcations [2]. Borders became invisible and new business, economic, political, educational, social, cultural, ethical, and communication forms evolved. International space and time collapsed at a moment's click. The World Wide Web was foreign territory just a few years ago [3]. Yet, legislation was drafted and case law decisions made as if the world geography had gone unchanged. Efforts at fitting square pegs into round holes continue today as the American and international legal systems still base their jurisdiction over legal issues by circumscribed lines within a spatial environment, including local courts, state courts, federal courts, and regulatory agencies.

Enter nanotechnology—the building block of the small. This is a tool that represents the enabling ability to transcend the unthinkable—automatically dispense specific biochemicals and pharmaceuticals to particular body tissues as needed via smart nanoscale drug-delivery devices, detect and image worn or damaged body parts through smart embedded sensors, advance computational powers beyond Moore's Law, enhance material properties, advance molecular-cluster manufacturing, and develop carbon nanotube products and ceramic nanoparticles for use across industries.

Nanoscale materials already are embedded in many products such as cosmetics, sunscreens that incorporate titanium dioxide nanoparticles to prevent sunburn, clothing that is made from hydrophobic nanofibers to help resist stains, and power machinery. Beyond these uses, our current manufacturing processes, educational systems, business models, economic structures, health care, environment, defense, space, energy, and societal infrastructures will be transformed without recognition as further advancements in nanotechnology research and development, and commercialization of the applications of this scientific discovery are integrated. These global benefits also carry unintended societal consequences and legal ramifications.

While the need to address the legal and regulatory implications of nanotechnology is oftentimes mentioned in reports, workshops, and conferences, the broad complexity of the issues within the multiple clusters of legal practice areas remains overlooked and ignored. Historically, the law has lagged behind scientific and technological advancements by at least a decade. The law cannot continue to live in the past, nor should society and the scientific and engineering communities accept that as its standard mode of existence. With science and technology's relentless advancements and society's centrifugal forces, the civil justice system of the 21st century must come out of obsolescence to lead and help shape the new values, standards, and rules of play wrought by this new frontier.

What active blueprint for change will enable today's American and international legal systems to forge collaborative relationships with diverse partners to diffuse the societal and economic implications of nanotechnology?

Establish a consistent and accepted definition of terms and communication protocols within scientific disciplines globally and international legal systems.

What is nanotechnology, its worldwide market, applications, industry, and competitive value? What is the legal definition of "human" and of "person"? When does life form? What constitutes human life?

Develop an understanding that the law must be viewed from a broad perspective through the convergence and possible integration of multiple legal practice areas.

As nanotechnology and nanoscience converge with other sciences and technologies, and advanced computing and human-machine integration speeds forward, the resulting issues and impacts will cut across many legal practice areas:

- criminal law through DNA forensics, its determination of "free will," and human-trial experimentation
- family law as a result of genetic intervention capabilities, creation of artificial life forms, and stem cell research
- health law and the U.S. Department of Labor's Employee Retirement Income Security Act (ERISA) resulting from nanotechnology convergence with medicine and biotechnology
- environmental law due to unknown risks of radically new technologies, such as the effects of inhaled nanoparticles and the fact that they change shape as they move from liquid solutions to the air, the release of buckyballs into the air and water and its effect on pollution and the food chain, and the likelihood of animals and people being exposed to hazardous materials and toxicity
- energy as new possibilities and forms are developed and the effects of their use undetermined
- transportation as new developments are incorporated into the automotive and aerospace industries
- elder law resulting from an aging demographic due to improved medical processes
- torts because of the potential for personal injury due to product misuse or mishap whether intentional or negligent, and trespass of nanoparticles
- intellectual property, by the patenting of all life forms, both engineered and enhanced, intellectual property rights of multiple parties, the question of who owns innovation by machines and human-machine hybrids (artificial intelligence and genetic algorithms), the need for due diligence prior to commencement of research to determine possible infringement, revisitation of the Bayh-Dole Act and the commercialization of converging products from laboratory to market by multiple licensees, protection of trade secrets, copyrights and trademarks, maskworks, and international recognition of rights and assets
- corporate law and contracts, formation of new entities, agency and partnership relationships, mergers and acquisitions, licensing of multiple parties who hold multiple patents, technology transfer confidentiality and non-disclosure agreements, patent cluster portfolios, and other new forms of agreements;
- Constitutional law, the courts, the judiciary, protection of individual rights and equal protection as privacy rights, security, and surveillance become more invisible through advancements in computing, biometrics, e-commerce, and federal legislation
- employment and labor law and the potential for discrimination resulting from issues of equity, distribution, and access
- taxation of, and incentives for, new product entries

- real property law
- international trade laws, trade regulation, customs, and cross-border jurisdiction as the issues impact interstate commerce and antitrust concerns as more competing parties collaborate
- civil procedure, litigation, and medical malpractice

Provide for open discussion with stakeholders regarding the pros and cons of regulating the industry versus self-regulation.

Invest in training and development in the federal agencies, regulatory bodies, and judicial systems.

As monies are allocated for research and development in new sciences and technologies to improve humanity and the quality of life, so, too, must monies be allocated to the agencies and systems that must understand, oversee, and be ready to respond to the new frontiers. Otherwise, lack of expertise will create bottlenecks to commercialization of emerging technologies extending the timeline to market for a product. The U.S. Patent and Trademark Office (USPTO) lacks a specialized nanotechnology examining group, technology center, or art unit [4]. Examiners are not trained to understand the multidisciplinary legal issues that may result from nanotechnology. Applications could thereby be delayed slowing the commercialization process. The Food and Drug Administration (FDA) needs additional training in understanding the differences between a medical device, a drug, a biological or chemical entity, and when a product may look like food but act like medicine. The delays encountered in the patent process at the USPTO trickle down to the review process at the FDA. The Environmental Protection Agency (EPA) must have the capacity to work in concert with the USPTO and FDA, as well as the Consumer Product Safety Commission and the Federal Trade Commission, as the implications of new discoveries arising from nanotechnology intersect across several regulatory and oversight bodies. This begs the question: Should a central oversight agency or commission be developed? Are the currently enacted regulatory and policy models sufficient to resolve the issues of unknown and unforeseen risks?

Consider the adoption of specialized courts for converging science and technology cases, and court laboratories for training and development—The Courtrooms of the Future.

Judges, legislators, and policymakers need to understand the lexicon and the short- and long-term, cross- and multi- jurisdictional impact and effects of the complex issues expected to be raised by new applications. Maryland, with the enactment of Rule 16-205 on January 1, 2003, became the first state in the country to adopt a specialized court for business and technology cases [5]. Under Rule 16-205, each circuit court is charged with establishing a special track for business and technology cases, create a procedure for assigning cases to that track, assign specific judges to the program, who are specially trained in business and technology, and develop alternative dispute resolution proceedings conducted by individuals specially trained in the issues. Our judicial system continues to function based on the needs of the Industrial Age. As Maryland has recognized, we must take measures to bring the courts to the present, at least forward to the Information Age, and then to the new frontier—the nanotechnology revolution.

Educate the scientist, engineer, and technologist to the science of business, ethics, and jurisprudence.

As there is a call to educate the legal infrastructure, so too must the scientific, engineering and technological communities understand the issues from the laboratory to the marketplace, the ethical implications of research claims of novel innovations, the dissemination of research results, and the interplay of law with business and new inventions.

Call for an International Convention through the United Nations.

The effects of nanoscience and nanotechnology transcend global boundaries. Dialogue should commence to begin addressing international implications, commerce, foreign policy, and the possible need for treaties.

Begin to educate the consumer to the facts versus the fiction of nanotechnology.

Monies continue to be poured into studies to determine consumer attitudes and into research schools to convene open discussions with peers about the issues. It is time to convert the passive approach of studies and research to active consumer protection education and review of the laws on the books. The media, books, and movies are feeding the consumer while advocacy groups are providing unsubstantiated studies of the adverse effects of nanotechnology. Fears must be properly assuaged, now. This is critical for nanoscience and nanotechnology to realize their true potential.

Create legal shops for the small business and minority-owned entrepreneur.

Legal costs run high for a startup entrepreneur, thereby ousting him or her from the competitive nano-marketplace. Offering unbundled legal services can reduce the cost of attorney fees while teaching an entrepreneur how to individually manage certain legal processes. The attorney and patent agents thereby become legal information engineers in a consultancy capacity.

We live in a transitional era. While this is not a comprehensive list of legal concerns and recommendations, it is a small representation of the vision we must employ concerning the implications of this new frontier. Advocating that the laws, policies, regulations, and legal systems currently in place need to be completely overhauled or eradicated to meet the challenges of nanotechnology would be absurd. It does mean, however, that we must be vigilant to the possibilities and the societal impact that accompanies exponentially rapid change. Current laws, regulations, and policies must be analyzed to determine those that continue to apply and those that must be modified to meet today's challenges. Our judicial infrastructure must be evaluated. Our agencies and commissions must work in concert. Today, we have the opportunity to lay a new foundation—to design anticipatory legal and policy measures, propose new legislation, and develop novel and unique regulatory infrastructures to address the potential risks of tomorrow.

References

1. R. R. Katz, with P. Gold, *Justice Matters: Rescuing the legal system for the twenty-first century*, Washington, DC: Discovery Institute (1997).
2. W. Gibson, *Neuromancer*, New York: Penguin Putnam, Inc. (1984).
3. R. Susskind, *The future of law: Facing the challenges of information technology*, Oxford: Oxford University Press (1996).
4. R. Bawa, Nanotechnology patenting in the U.S, *Nanotechnology Law and Business* **1**(1), 36-38 (24 January 2004).
5. C. S. Fax, Specialized court for tech cases starts to operate in Maryland, *National Law Journal* (31 July 2003).

AN EXPLORATION OF PATENT MATTERS ASSOCIATED WITH NANOTECHNOLOGY

E. Jennings Taylor, Faraday Technology, Inc.

As this nanotechnology/nanoscience workshop was held during the centennial anniversary of the invention of flight, it is interesting to consider patent matters from the perspective of the inventions and innovations leading to commercial aviation. This working paper introduces the issues associated with patents with a case study of two early aviation patents. Then, various key components of the U.S. patent system are presented. Questions, in particular as related to nanotechnology, are raised throughout the paper.

On December 17, 1903, Wilbur and Orville Wright flew a "heavier than air" machine 120 feet and thereby demonstrated their "Flying Machine." A patent application for their flying machine had been filed March 23, 1903, approximately nine months prior to its "actual reduction to practice." Their patent application did not include a motor, as the Wright Brothers were afraid that they'd have to demonstrate flight with a motor![3] The Wright Brothers patent, No. 821,393 issued May 22, 1906. The main invention covered in the '393 patent was "wing warping" moving in concert with the rudder to provide "lateral balance."

On December 5, 1911, patent No. 1,011,106 for a "Flying Machine" issued to Glenn Curtiss and a group of inventors including Alexander Graham Bell. The '106 patent achieved "lateral balance or equilibrium" or restored the same using "rigid non-flexible wings" with "little wings," the predecessor of modern day ailerons.

Curtiss announced he was going to make an airplane as described in his patent. The Wright Brothers filed suit and received a preliminary injunction against

[3] Martin J. Adelman, Professor & Director of The Intellectual Property Program at George Washington University Law School, comments made during University of Dayton School of Law Symposium on Patents & Innovations, Dayton, Ohio November 13, 2003.

Curtiss. A patent infringement suit ensued ruling in favor of the Wright Brothers and infringement was affirmed on appeal. Henry Ford entered the picture and provided legal counsel and public relations assistance to Curtiss. In the press it was claimed that the Wright Brothers wanted such broad coverage for their invention that, to paraphrase, *"...if I went on the street corner and flapped my arms they would claim infringement of their '393 patent!"*[1]

The hostile activities of the inventors were halted by the United States government due to World War I. Specifically, the British needed the improved flying machine described in the '106 patent. After the war, the "hostile" activities regarding the '393 and '106 patents were not resumed. The use of little wings in the form of airlerons continues today.

A similar scene has been played out numerous times before and since the early days of aviation and will undoubtedly be at issue in the future, in particular as related to nanotechnology. Simply stated, initial inventors seek the broadest interpretation possible of their patents while subsequent improvers look to gain recognition and reward for their enabling inventions.

While there are some recent publications dealing specifically with intellectual property rights and nanotechnology, generally the focus is on introducing the patent system to allow inventors to protect their rights [2, 3]. Furthermore, the legal issues have been addressed from predominantly a regulatory perspective as related to medical applications of nanotechnology [4].

However, recently studies have clearly indicated an explosion of nanotechnology-related patents [5, 6]. In addition to the volume of patents and patent applications, recently indications of the growing complexity of patents issued today versus those of 20 years ago have been noted [7]. This paper focuses on potential patent issues associated with nanotechnology. Below are some preliminary questions.

- What breadth should be allowed for the first patent in a field?
- Should more breadth be allowed for so-called pioneer patents?
- How should the property rights associated with patents be partitioned between first patents and improvement patents?
- Are these issues best addressed by congressionally generated statute or common law judicial precedent?
- Is the current patent system ready for the challenges associated with the emerging field of nanotechnology?

The remainder of this working paper is presented to provide a basis for further focusing these questions and developing additional questions.

Purpose of the U.S. Patent System

The Constitutional basis of the U.S. patent system is derived from the founding fathers: "The Congress shall have the power... To promote the progress of science and useful arts, by securing for limited times to authors and inventors the exclusive right to their respective writings and discoveries [8]." There is no record of debate on this matter and presumably this provision was not contested by the founding fathers [9]. James Madison had previously written that "Evidently, this provision was uncontroversial as there is no record of any debate on this matter [9, p.72]." This clause of the constitution provided the basis for both our patent and copyright systems. This paper focuses on the patent system.

Because the word "right" is only used once in the Constitution, one may be inclined to think of an invention as a natural right. However, since statutory law limits the term of a patent as generally limited to 20 years from filing [10], it is difficult to rationalize the patent right as a natural right. Contrast for example, the copyright statute provides protection for the life of the author plus 75 years. Consequently, although derived from the same clause of the Constitution, patent and copyright statutes are implemented quite differently [11]. Even so, the natural issue related to patents continues to be debated [12].

While the U.S. patent laws were unique to our country, they were based on previous attempts by countries to control craft knowledge and their associated economic benefits [13, 14, 15]. The only President to receive a patent, Abraham Lincoln, said that the patent clause in the Constitution "...added the fuel of interest to the fire of genius, in the discovery and production of new and useful things [16, p.363]." Regarding the U.S. patent system, Thomas Jefferson said that the patent statute "...has given a spring to invention beyond my conception [17, p.376]." Jefferson further noted the success of the U.S. patent system as "...encouragement to men to pursue ideas which may produce utility [17, p.377]."

In today's parlance, "to promote the useful arts" may be thought of as "to promote economic development and well being through technological innovation." Consequently, the U.S. patent system is in effect a policy tool. More recently, the patent system has been described as the "[my emphasis] *primary* policy tool to encourage the development of new technologies [11, p.102]."

The explicit "contract" of the patent system is that an inventor is given a "right to exclude others" provided that the inventor fully describes his invention in the patent. The full public disclosure alerts subsequent inventors, i.e. improvers, to the original inventor's ideas that in turn lead to improvements on the original invention. As a result, technological progress proceeds in the form of both disruptive and incremental innovations [18].

The U.S. patent system consists of three components: statutory law based on congressional legislation, examination procedures and rules for the U.S. Patent and Trademark Office (USPTO), and common law based on judicial precedent. These are introduced below. We may ask

- Has the patent system effectively encouraged technological innovation?
- What modifications, if any, are required for the patent system to effectively promote technological innovation related to nanotechnology?

U.S. Patent Statute

The initial Patent Act of 1790 created a system based on a patent examination board headed by the Secretary of State (Thomas Jefferson), Secretary for the Department of War (Henry Knox), and Attorney General of the United States

(Edmund Randolph) [18]. They referred to themselves as the "Commissioners for the Promotion of the Useful Arts," a clear indication of their perceived role. During the approximately three years of the existence of this first patent act, approximately 57 patents issued.

Due to the difficulties in balancing the work between the respective departments and the increasing demands of the patent work, a simple registration system was established in the Patent Act of 1793 and five additional acts until the Patent Act of 1936, which generally created the basis of the modern patent system [19, 20].

U.S. patent law requires that patents be issued to an inventor or group of inventors [21]. This is in contrast to some other patent systems in the world where patents may be issued to corporations, without a named inventor. Under U.S. patent law, the rights in the patent may be assigned to entities, such as corporations, other than individuals [22], and in fact this is usually the case.

Patentable inventions are defined as "...any new and useful process, machine, manufacture, or composition of matter, or any new improvement thereof... [23, p.102]" To sum it up, patentable matter includes "anything under the sun made by man [24, p.519]."

The specification of the patent application must contain a description in sufficient detail to "...enable any person skilled in the art ...to make and use the same [25]." This requirement may be thought of as the full disclosure or enabling requirement. Recall, it is part of the explicit contract for granting the limited monopoly to said invention. The basis for the patent examiner to determine full disclosure is the "mythical" person of ordinary skill in the art.

In order to obtain a patent on the subject matter described above, the invention must be useful, new, not previously patented, not described in a printed publication, or not offered for sale for at least one year prior to the date of the U.S. patent application [26]. In addition to being new and useful, the invention must not be obvious in light of the prior art either alone or in combination "...to a person having ordinary skill in the art [27]." Again, note the basis for the examiner in determining non-obviousness is the "mythical" person of ordinary skill in the art. These considerations raise more questions:

- If an intelligence based on nanotechnology invents something patentable, who is the "inventor" of the patent?
- Should special statutory laws be developed to cover new subject matter associated with nanotechnology?
- Will new meanings to "new" and "useful" be required?
- Will the evolving definition of person having ordinary skill in the art keep up with developments in nanotechnology?

A Delicate Balance

From judicial precedent, a Doctrine of Equivalents has emerged in the U.S. patent system expanding the patented invention beyond the literal description in the patent claims. The Doctrine of Equivalents evolved to prevent "...the unscrupulous copyist to make unimportant and insubstantial changes...[28]" in order to claim around the protection granted to an otherwise valid patent. The Doctrine of Equivalents maintains the incentive for inventors to fully disclose their invention by obtaining a patent. A test for equivalency suggested by the court involves "...persons reasonably skilled in the art [28]."Again, the "mythical" person of ordinary skill in the art is invoked.

Providing balance to a Doctrine of Equivalents, Prosecution History Estoppel prevents inventors from narrowing their claims during its prosecution at the USPTO and then subsequently re-capturing the original breadth in order to claim infringement by a later patent [29]. Prosecution History Estoppel provides a mechanism for recognizing and rewarding significant and unanticipated improvements made by subsequent inventors.

A delicate balance must be maintained between the Doctrine of Equivalents and Prosecution History Estoppel in order to provide incentives for the first inventor as well as subsequent improvers. A new doctrine, The Foreseeability Bar, has been suggested to balance the incentive for first inventor protection vis-à-vis subsequent inventor improvements. More specifically, an "equivalent" cannot be claimed if it was not claimed in the original patent and would have to be foreseeable to a person of ordinary skill in the art. In this manner, the public notice required for both disruptive and incremental inventions [30] may be maintained. The proposed bar would work in concert with the statutory and common law precedent based on the person reasonably skilled in the art. We may then ask

- Are the Doctrine of Equivalents, Prosecution History Estoppel, and a proposed Foreseeability Bar applicable to nanotechnology?

The Evolving Patent System

Generally speaking, the patent statute does not distinguish between different technologies or industries [31]. Exceptions include the special obviousness standard for biotechnology-related art [32]. Biotechnology patents have grown very lengthy, reaching thousands of pages in some cases. Undoubtedly, special examination rules are required for patents in this technical area.

In a noteworthy case, special industry protection has been enacted by Congress, specifically the Semiconductor Protection Act. This "action" resulted from six years of debate, is extremely complicated, and has only been cited one time [33]! The efficacy of this special act is certainly in question.

In terms of common law derived from judicial precedent, the patent system has been clearly shown to be technology-specific. This is evident in the interpretation and definition of the person having ordinary skill in the art. This "mythical" person will be strongly dependent on the maturity and prior art resident within a specific field. Additional nanotechnology-related questions must be recognized:

- Will special patent application examination practices be required for emerging nanotechnology inventions?
- Should there be a special art unit for nanotechnology inventions?

- Should the Congress proactively legislate the patent statute to cover emerging nanotechnology patent issues?
- Are the courts adequately prepared to deal with emerging nanotechnology patent issues?
- Will judicial precedent keep pace with emerging developments in nanotechnology in order to promote technological innovation?

Conclusion

One hundred years ago, a flight of 120 feet was demonstrated, or in terms of the patent system, the invention of a flying machine was "actually reduced to practice." Subsequently, an improvement patent was issued on the invention of a flying machine. The key question in subsequent patent litigation was how broadly should the original patent be interpreted and what, if any recognition should be provided the "improvement" invention? In spite of significant patent battles, today, military and commercial aviation are commonplace. A question for today might be

- Could the technological advance of aviation have been promoted any more efficiently then with our current patent system?

The patent system consists of statutory law based on congressional legislation, an examination procedure, and common law derived from judicial precedent. The patent principle "person having ordinary skill in the art" finds itself in the patent statutory law, examination rules, common law, and in proposed fixes to the patent system. This principle has served the patent system very well. The key questions are

- Will this principle continue to serve the patent system during the technological advances associated with nanotechnology?
- How will we define the "nanobot" having ordinary skill in the art?

The fact that the U.S. Patent System plays a dramatic role in technological innovation can hardly be disputed. As recently reviewed [11], a number of models have been proposed defining the relationship between the Patent System and technological innovation, including (1) the prospect theory [34], (2) monopoly theory of J. Schumpeter [35], (3) competitive innovation [36], (4) cumulative innovation [37], (5) anticommons [38,39], and (6) patent thickets [40]. These models seem to work well for specific industries or for specific technologies. Given the lack of a single, comprehensive theory that covers all technologies, we are left with two final questions:

- What will be the impact of the increasing volume and complexity of nanotechnology related patents on the patent system (i.e. statutory law, examination, and common law)?
- What modifications are required for the patent system (i.e. statutory law, examination, and common law) to effectively and efficiently promote technological innovation as related to nanotechnology?

References

1. M. J. Adelman, Comments made during University of Dayton School of Law Symposium on Patents & Innovations, 13 November, Dayton, Ohio (2003).
2. S. B. Maebius, Key considerations in protecting your intellectual property in nanotechnology, *Journal of Nanoparticle Research* **4**, 373-374 (2002).
3. B. Bastani, D. Fernandez, Intellectual property rights in nanotechnology, *Thin Solid Films*, 472-477 (2002).
4. F. A. Fiedler, G. H. Reynolds, Legal problems of nanotechnology: An overview, *Southern California Interdisciplinary Law Journal* **3**, 593-629 (1994).
5. M. C. Roco, Broader societal issues of nanotechnology, *Journal of Nanoparticle Research* **5**, 181189 (2003).
6. Z. Huang, H. Chen, A. Yip, G. Ng, F. Guo, Z. K. Chen, M. Roco, Longitudinal patent analysis for nanoscale science and engineering: Country, institution and technology field, *Journal of Nanoparticle Research* **5**, 333-363 (2003).
7. J. R. Allison, M. A. Lemley, The growing complexity of the United States patent system, *Boston University Law Review* **82**(77) (2002).
8. Constitution of the United States, Article I, Section 8, Clause 8.
9. W. H. Francis, R. C. Collins, *Cases and Materials on Patent Law*, St. Paul, MN: West Publishing Co., (1995).
10. 35 U.S.C.154 (a)(2) (2000).
11. D. L. Burk, M. A. Lemley, Policy levers in patent law, *Virginia Law Review* **89**(7), 101-215 (November 2003).
12. S. Ghosh, Traditional knowledge, patents, and the new mercantilism (Part II), *Journal of Patent & Trademark Office Society* **85**(11), 885-924 (2003).
13. E. C. Walterscheid, The early evolution of the United States patent law: Antecedents (Part 1), *Journal of Patent & Trademark Office Society* **75**(9), 697-715 (1994).
14. E. C. Walterscheid, The early evolution of the United States patent law: Antecedents (Part 1), *Journal of Patent & Trademark Office Society* **76**(10), 771-857 (1995).
15. E. C. Walterscheid, The early evolution of the United States patent law: Antecedents (Part 1), *Journal of Patent & Trademark Office Society* **77**(2), 77-107 (1996).
16. R. P. Basler, *The Collected Works of Abraham Lincoln*, volume 3. The Abraham Lincoln Association (1953) (available at http://www.hti.umich.edu/l/lincoln).

17. P. J. Federico, Operation of the Patent Act of 1790, *Journal of Patent & Trademark Office Society* **71**(5), 373-385 (1990). Note: From Primary Reference Writings of Thomas Jefferson Ed. By H.A. Washington, Vol. VI, p.158 (1815).

18. C. Christensen, *The Innovator's Dilemma*, New York: Harper Business (2000).

19. E. C. Walterscheid, To promote the progress of useful arts: American patent law and administration, 1787-1836 (Part 1), *Journal of Patent & Trademark Office Society* **78**(1), 61-78 (1997).

20. E. C. Walterscheid, To promote the progress of useful arts: American patent law and administration, 1787–1836 (Part 2), *Journal of Patent & Trademark Office Society* **79**(1), 11-40 (1998).

21. 35 U.S.C. 116 (2000).

22. 35 U.S.C. 261 (2000).

23. 35 U.S.C. 101 (2000).

24. W.H. Francis and R.C. Collins. Cases and Materials on Patent Law 4th Ed. (1995). Discussion of Supreme Court Case of Diamond v. Chakrabarty (1980).

25. 35 U.S.C. 112 ¶1 (2000).

26. 35 U.S.C. 102 (a)(b) (2000).

27. 35 U.S.C. 103 (a) (2000).

28. Graver Tank v. Linde Air Products (1950).

29. Warner-Jenkinson Corp. v Hilton Davis Corp. (1997).

30. M. J. Conigliaro, A. C. Greenberg, M. A. Lemley, Foreseeability in patent law, *Berkeley Technology Law Journal* **16**, 1045 (2001).

31. D. Burk, M. A. Lemley, Is patent law technology-specific? *Berkeley Technology Law Journal* **17**, 1155 (2002).

32. 35 U.S.C 103(b) (2000).

33. M. A. Lemley (Professor and Director, Center for Law and Technology, University of California School of Law). Comments made during University of Dayton School of Law Symposium on Patents & Innovations, 13 November, Dayton, Ohio (2003).

34. E. W. Kitch, The nature and function of the patent system, *Journal of Law and Economics* **80,** 265 (1977).

35. F. S. Kief, Rights and property rules for commercializing inventions, *Minnesota Law Review* **85,** 697 (2001).

36. K. Arrow, Economic welfare and the allocation of resources for invention. In National Bureau of Economics Research, *The Rate and Direction of Inventive Activity*, Princeton, NJ: Princeton University Press (1962).

37. J. R. Green, S. Scotchmer, On the Division of Profit in Sequential Innovation, *RAND Journal of Economics* **26**, 20 (1995).

38. M. Heller, The tragedy of the anticommons: Property in the transition from Marx to markets, *Harvard Law Review* **111**, 621 (1998).

39. M. A. Heller, R. S. Eisenberg, Can patents deter innovation? The anticommons in biomedical research, *Science* **280,** 698 (1998).

40. C. Shapiro, Navigating the patent thicket: Cross licensing, patent pools, and standard setting. In *NBER Innovation Policy and the Economy, Vol 1.*, ed. A.B. Jaffe, J. Lerner, S. Stern, Cambridge, MA: MIT Press (2001).

THE ETHICS OF ETHICS

James R. Von Ehr, Zyvex

The National Nanotechnology Initiative has set aside a reasonably large amount of money to address the Social and Ethical Implications (SEI) of nanotechnology. Everyone is in favor of the "ethical development" of nanotechnology. The devil is in the details—who defines what is and is not ethical? Ethicists certify one another as ethicists, but in the absence of absolute ethical standards, how can we judge a good ethicist from a poor one? What is the difference between ethics and morals, and why do we call this research "ethical implications" and not "moral implications"?

I started pondering the ethics of ethics after reading about some ethicists making what I considered unethical mandates on issues in biotechnology and medicine. One example of this is letting terminal cancer patients die in pain instead of giving them sufficient pain medicine to relieve their suffering. How can a person deny quality of life to others, or sentence them to a painful death by neglect, and still be considered ethical? Thorny issues such as buying organs from deceased donor's families, suspending stem cell research, and debating the right to self-administered euthanasia still challenge biomedical ethicists. What unique issues will nanotechnology raise, and how should we make choices we can look back on as ethical ones?

Ethics or Morality?

The dictionary definitions of ethics and morality give little guidance in distinguishing the two terms. Modern usage seems to view morality as more of an individual choice and ethics as more of an interpersonal choice (how we treat one another). We used to look to priests and philosophers for guidance on how to be a good person, but today we are increasingly looking to ethicists in that role.

Ethics reflects politics. Characteristically, ethicists on the right wish to codify their moral beliefs, ethicists on the left wish to codify their vision of social equality, and libertarian ethicists campaign for personal liberty. The last decade of federal government has given us political ethics examples from both the left ("social justice," "digital divide," or "Kyoto protocol") and right (stem cell research ban, medical marijuana ban). Whichever side is in power sets the agenda for ethicists in government, to the dismay of the other side. Both the left and the right believe government has to "do something" because they don't trust individuals to make the "right" choices on their own.

A classic liberal, civil society begins with the supposition that people have natural rights, and can generally be trusted to "do the right thing" if properly socialized. Societies that are more traditional start with the supposition that people are naturally evil, and need religious guidance in moral behavior. The founders of our country were deeply religious, yet they established a classic liberal, civil society with minimal interference in personal choice. Can we establish broad ethical principles in line with their approach and take a steadier course than one dictated by the quivering compass of political ethics?

Starting from a Principle

Perhaps we could start our ethical analysis with a basic principle of human rights. One of the most basic shared American values has been "that all Men are created equal, that they are endowed by their Creator with certain unalienable Rights, that among these are Life, Liberty, and the Pursuit of Happiness." The writers of our Declaration of Independence believed that no one has rights superior to those of anyone else, and that we intrinsically possess the right to live our lives as we wish, provided that we respect the equal rights of others to do the same. Depending on how they are asked, most people would agree that these rights include the right to do stupid or harmful things to oneself, like smoke, drink alcohol, eat unhealthy food, or even take illicit drugs.

Philosophical ethicists are concerned with the difficult areas of where those rights impinge on the similar rights of others. Ethicists become moralists or collectivists when they proscribe behavior of an individual that affects only that individual. Philosophical ethicists grapple with societal trade-offs that have no simple answers (e.g., cloning). Moralistic or collectivist ethicists take a position based on their personal beliefs and mandate it for the rest of us (e.g., drugs are evil, so a terminally ill cancer patient should not be allowed to take so much pain medicine that he or she could become addicted).

Society today does not distinguish between the two types of ethicists and gives deference to anyone calling himself or herself an ethicist. The power of moralistic or collectivist ethicists to dictate their viewpoint increases as the power of government increases. Such ethicists exist on the right and left sides of the political spectrum, in secular or religious practitioners, so this issue is not going to fade away on its own.

So long as there are no neutral ethical principles of ethics, we have no ethical compass. With the principles of the Declaration of Independence as our guide, we can more clearly define ethics. When someone says he or she is limiting your freedom of action "for your own good," it is clearly a moralistic position (often justified by the a priori claim that you'll hurt someone else). When someone takes something you earned away from you to give to someone else, it is clearly a collectivist position (often justified by denying that you honestly earned it). We can't change people's politics but we can, and should, deny such political behavior the rubric of "ethics."

When individual freedom of action is limited only so much as it is necessary to respect the rights of others, that is an ethical position that springs from hundreds of years of classical liberalism, and reflects the founding principles of this country. Political ethicists are entitled to their views, but we don't owe them government support simply because of their avocation. We should only spend public money on ethicists who base their deliberations on the principle of our natural rights that include life, liberty, and the pursuit of happiness. We should dismiss politically based ethical pronouncements that ignore natural rights as unethical. It might be desirable to have an ethicist review board similar to appeals courts, staffed with people of widely different political viewpoints, to review important ethical pronouncements for conformance to such ethical principles.

The Ethics of Nanotechnology

Ethicists today grapple with comparatively "simple" issues like stem cell research and anti-aging medicine, but nanotechnology will lead to issues that are far more complex. Moralists will want to ban certain technologies as unethical or illegal once the possibilities for personal enhancement start to become available. Collectivists will worry about the potential for unequal results as technological advances initially go to those doing the work instead of being spread equally among everyone.

Some ethicists assert that we have a moral duty to die around age 80, but nanomedicine will greatly extend healthy lifespans. Some ethicists assert that self-administered euthanasia is unethical. Is it ethical to withhold a life-sustaining drug once a person turns 80 in order to meet the ethical principles of the first ethicist? For an even more difficult issue, certain new medicines may not be available to everyone—perhaps they are too expensive, or incompatible with the genetic makeup of the potential patient. Would there be a call for us to suppress this medicine for everyone? What would the decision be if that genetic difference was sex-related, or racially related?

Today we ban some molecules as illegal drugs, while we hail other similar molecules as miracle cures. Nanomedicine will create families of potent new drugs, some of which will be used for recreation or personal enhancement. How will we decide which arrangements of atoms to ban, and which ones to use? Most of our existing recreational drugs, such as alcohol, caffeine, and tobacco, would be banned because of their psychoactive properties if they were newly developed today. Will moralistic ethicists ban all new drugs that make us feel good or improve us?

Molecular manufacturing will change the economics of manufacturing, and the types of jobs available. If Americans are unqualified for these jobs because our children choose not to study science and engineering, will we accept the consequences of those choices, or will we react badly against those who do have the necessary skills? We could conceivably lose our economic dominance in less than a decade. Will collectivist ethicists attempt to restrain those aggressive capitalists?

Summary

Everyone is in favor of the "ethical development" of nanotechnology. Nobody wants to spend his or her career doing "unethical development." However, people's ethics reflect their politics, and without any common ethical principles, we can expect to argue incessantly over moralistic or collectivist issues engendered by the speed of development and technological power of nanotechnology.

I suggest that our goal should be to establish a framework of government involvement in nanotechnology that protects our natural rights as free humans, while fostering the technological developments that can help us achieve our full human potential. This will occasionally make religious or collectivist ethicists uncomfortable, so we need to strive to not make bad decisions for the majority based on the opinions of a few. Throughout human history, we have used technology to improve our lives—shelter, farming, transportation, medicine, communications, and electricity. Nanotechnology promises to be the greatest technology mankind has yet developed. For the benefit of mankind, we should encourage a philosophical ethics that supports personal freedom to explore the new possibilities. We should resist the moralistic or collectivist "ethics" that restricts personal choices in ways unnecessary to preserving the rights of others.

The stakes are high for us to get this issue of ethics right. Success will mean an unprecedented opportunity for improved lives for everyone in the world. Failure by timidity means we will relinquish our role as leaders of the free world to bolder nations. Failure by carelessness would result in reducing our well being at best, or destroying ourselves in the worst case. Humanity has always faced a similar choice with technology, and we are here because we have always managed these choices to our advantage. History leads us to the conclusion we will do the same with nanotechnology. Wise choices today can shorten the time until we can reap those rewards.

NEGOTIATIONS OVER QUALITY OF LIFE IN THE NANOTECHNOLOGY INITIATIVE

Rosalyn W. Berne, University of Virginia

We have been asked to consider what might be some of the ethical or societal implications of nanotechnology. There are a number of approaches to this question, all of which involve some elements of speculation, since it is not too clear as of yet what the nanotechnology future might bring. It is alleged to bring many new and novel new processes and devices to our health care, to electronics, and perhaps to the military. Proponents promise that myriad new consumer goods will be made possible through nanotechnology. In a report of the National Nanotechnology Initiative, it is written that nanoscale science and technology has the potential to "profoundly change our economy, to improve our standard of living, and to bring about the next industrial revolution [1, p.1]."

This is all, however, still a matter of societal negotiations.

As a trope of American determination and vision for improved quality of life, the word "revolution" is often used to identify the purpose and nature of the nanotechnology initiative. For example, when John Marburger, Director of the Office of Science and Technology Policy, spoke before a group of participants at an NSF workshop on the societal implications of nanotechnology, he called nanoscale science and engineering "a revolution—like the industrial revolution—rather than just another step in technological progress [2]."

The word "revolution" invokes some of our deepest ideologies, and is rooted in our oldest sense of identification as American people. It also goes to the core of the nanotechnology initiative. Large-scale changes that become evident over a short period of time are variously defined in terms of "revolutions." But all revolutions, including technological revolutions, require the allocation of tremendous resources at the burden of the society. They depend on wide-scale social participation, which is difficult to enlist. Therefore, they must be negotiated in the public domain before they can get underway. Unless there are clear indications of likely improvement to "quality of life," individuals refuse enlistment, and revolutions cannot be fully or successfully engaged. The public endorsement of the nanotechnology initiative will depend, in part, on the successful solicitation of public commitment. First, during the research and development phase that is underway at present. Then again, the process of social negotiation will occur during the appropriation of nanotechnology into consumer markets. Both phases involve the creation of meaning and the appeal to belief, loosely constructed in a negotiation process.

Language both conveys and constructs meaning. As such, it has the powerful capacity to take the otherwise indeterminate reality that nanotechnology represents and focus it towards determined visions and goals. The rhetorical strategy of appealing to nanotechnology's potential for improved quality of living is one means proponents and political leaders have of cultivating a sense of meaning and purpose about nanotechnology in the public domain, during the first phase. Its meaning is much like the notion of progress, which as Ellul points out, is a phenomenon we can neither contest nor grasp [3]. What is to be understood by the promise of nanotechnology to improve on the "quality of life"? While its meaning is vague, the notion of quality connotes a standard of something as measured against other things of a similar kind. It has to do with gradations and points to either increase or decrease in measurable terms. Quality invokes sentiments that are widely shared. The conceptualization of life, on the other hand, is much more difficult to define. Scholars who are interested in quality-of-life assessments have little agreement about the meaning of the term, life. In one review of quality-of-life literature, the dominant trend is seen as construing "life" to mean mental life, but it sometimes also means environmental conditions. Its authors write, "The term functions in a metatheoretical way to reference research aimed at policy outcomes. It can also be used rhetorically to create a favorable attitude towards a public official or his cause...But a precise and universally accepted definition of the concept has yet to be framed [4, p.132]."

For the purposes of this writing, let us consider "quality of life" to mean standards of living, as measured against either previous standards or that of others who are living in ways that appear to be more or less desirable or valuable than one's own. In terms of the industrialized world, for most people, notions of quality in living are no longer about basic needs, such as access to food and shelter, health and prosperity. Rather, the idea points toward an unspecified but limitless improvement to our material lives, over and above the qualities we currently have. Given the enormous prosperity that is enjoyed in these societies, what kind of appealing improvements might nanotechnology offer—increasing longevity of human life, fewer diseases, less sickness, more food, more money, more property, more stuff, faster stuff, smaller stuff? What is being negotiated for when we reach for improved quality of life through nanotechnology?

One traditional approach to measuring quality of life is in terms of individuals' success in accomplishing their desires. Another approach defines quality of life in terms of people carrying out their place in the social order. Gerson suggests an alternative, which is to approach what it means to conceptualize quality of life from the assumption that both social order and individuals "arise in and through a process of ongoing negotiation about who shall be whom, and what order shall pertain [5, p.796]." The approach proceeds on the supposition that selves are constructed through a process of cooperation, in specific negotiating contexts. The question of what is negotiated is answered in consideration of contributions participants make, and the consequences they face in participating. Let us consider the appropriation of the computer as an example of the reconstruction of self in a social negotiation process.

Computers rapidly shifted down in scale from room-sized monstrosities that individual citizens had little access to, to relatively small, very fast, convenient processors and purveyors of vital information. As such, these devices now make it possible for many of us to work with mobility: in an airport, on a train, in a hotel room, or at our own home. In its inception, this newfound capacity to work and communicate from nearly anywhere was spoken of in futuristic terms with promises of improvement to the quality of life. Using it promised to give more free time, to make everything more efficient. As such, it was widely held to be a social and economic good, and accepted with its associated costs to society. Every radically new technological change requires a negotiation of changes to our social and cultural norms and expectations. The miniaturization of computers has led to the rapid and perhaps irretrievable erosion of clear boundaries between work life, family life, social life, and personal life with their independent responsibilities and even their own sanctities. For many of us, a quality of living once considered to be without compromise—time at the family dinner table—has been supplanted by competing demands of busy living. Many young people now use the computer as their primary source of information, and as their preferred medium of communication with other humans. IM (Instant Messenger) replaces face-to-face communication.

One of the motivations of the computer's early proponents was the efficiency and speed that it offered to the processing of military and business systems of information. Another was the incredible profit potentials from the new markets worldwide, associated with producing and distributing computer hardware and software. The revolution in computing happened not simply because of the determination of business enterprise, or because of the government support for the development of these technologies. It happened because there was cooperative process that took place simultaneously; a negotiation in the public domain that involved beliefs about quality-of-life gains and perceptions of whom we might be.

Beliefs about quality of life are constructed from culturally shared, socially shaped perceptions. They are also formed from cognitive processes that reach well beyond the primary and instinctual thrust to survive to dreams and fantasies of freedom from want, need, and limitation. As the pursuit of new knowledge that has some of these myths at its core, nanoscaled science and technology researchers are engaged not only in the acquisition of new knowledge, but also in negotiated and competitive social/cultural construction of meaning. As Steve Fuller explains more generally,

> ...the so-called ultimate ends—such as peace, survival, happiness, and (yes) even truth—refer not to radical value choices for which no justification can be given, but rather to constraints on the manner in which other instrumentally justifiable ends are pursued. Thus, happiness in life is achieved not be reaching a certain endpoint, but by acquiring a certain attitude as one pursues other ends. [6, p.16]

Similarly, improvements to the quality of living have no end point, but only perpetual adjustments to attitudes, made possible by the allure of technological promise. Perhaps the nanotechnology initiative is one that offers not improvement on the quality of life, but maintenance of the quality we now have. Or is it promising to improve the quality of life for those who have not yet achieved the same high quality that we have here in the United States? Or perhaps collectively, we are simply dissatisfied with our current lives and feel compelled towards the acquisition of new and novel things and the changes to our lives that those things might bring. Planetary history certainly does point to the proclivity of all living entities to pursue change as a means towards self-improvement. That seems to be our human nature. What is called into question here is not the desire or even the worthiness of change, but rather the direction and intention of it.

The congruence of nanoscience and nanotechnology suggests that inevitable and radical material and economic changes are afoot. For me, those seemingly inevitable changes bring two worries. One is a normative concern about the emotional, psychological, and spiritual well being of the human family. The other is the conscientious care of our earthly home. I would like to believe the changes to come as a result of nanotechnology can be directed in humanitarian and earth-respecting ways. As we negotiate the contributions that individual citizens must make and the consequences societies must bear in the revolution we are embarking upon, there are meta-ethical considerations that ought to be, but have not yet been, given deep consideration in that negotiation process. Those include

1. How is it that we wish to live?
2. What is it that we hold to be most true and most important in living together, not just in this particular society, but also as a human community living on this planet?
3. What does it mean to us individually and collectively to pursue the knowledge that will further refine our abilities to manipulate the material world?
4. What does it mean to us to better and improve on that material existence?
5. How much more improvement do we crave, in what terms and to what ends?
6. What will be the cost for our improved living, and what parties will be asked to pay those costs?

If successful, nanotechnology could radically alter not just the physical, but the social, emotional, psychological, and spiritual dimensions of human life. All novel technologies do. Nanotechnology represents a collection of new tools and the development of new devices, which will be used in ways that have profound implications on the way we construe our lives, their meaning, and significance to us. This always happens with the development of new technologies. I earlier offered the example of the transference of computing technologies into the broader society. Computing technologies changed our immediate quality of living, and especially, the rate at which we communicate. The effect has been to increase the volume of information we are expected to exchange and process, without increasing the amount of time we devote to those exchanges. One consequence is feeling exhausted and overwhelmed with daunting tasks under the pressure of rapidly moving time. The way we perceive time has also changed, and the amount of work we expect to accomplish in that time has increased, simply because we have incorporated into our daily living a technology which works many times faster than the human being is constitutionally designed to respond.

Let me suggest an example of an altered quality of life, which could shift our fundamental sense of being and of living in community with others, and which will have to be renegotiated when nanotechnology is transferred into the broader society. That quality is the relatively stable emotional and psychological comfort we have now with being alive inside of our bodies. If we are successful in shrinking the size of transistors, say, 100-fold, then we will have the incredible capacity to place computing devices in and around our bodies in order to have access to information we currently do not have. In order for us to adjust to the flow of that information, a radical, perhaps subconscious reconstruction of cognitive processing about ourselves within our bodies will have to happen. There are many hundreds of thousands of things going on inside of and around our bodies about which we have no current conscious knowledge. Gaining access to even a few of those processes, such as particular bio-chemical changes, temperature changes, exposure to viruses and bacteria, and breathing- and heart-rate changes will require that we make meaning of that information. If we don't successfully make meaning of that information, we risk our mental health. But to do so will require renegotiation of who we believe ourselves to be and what it means to be alive in the body. Matters of spiritual orientation also will have to be renegotiated; as technologies increasingly lead us to trust that our bodies and its processes are at our command, our sense of trust and faith in an omniscient being beyond ourselves will need also to be reconciled. New meanings will have to be made of sickness, health, and reliance on a divine other.

In other words, as we allow revolutionary material changes to take place in the larger society, we must also change what is inside of us. Technological transformations are symbiotic to internal, cognitive process in the search for meaning and of the human need to establish a sense of self and purpose in life. To that end, we pursue visions of improved qualities of our lives. Gerson speaks of the constraining aspects of participation in any situation [5]. People make their contributions to that situation, and are also bound by its limitation. Yet, the situation is enriching, and offers resources and opportunities not otherwise available. The National Nanotechnology Initiative asks of us multiple billions of public dollars for its research and development. It requires that many hundreds of thousands of our research scientists, engineers, and graduate students devote their intellectual resources to the pursuit of knowledge about the nanoscale of phenomenon. Educational institutions from K-16 will soon be asked to make significant adaptations of teacher training, curriculum and pedagogy toward building a future workforce that is trained in the languages and techniques of this "technological revolution." Job losses, environmental effects, social restructuring, unanticipated financial costs, and revisions of public policy are just a few of the constraining factors which must be dealt with in the negotiation process.

Resources and opportunities otherwise not available are offered on the table of societal negotiations to assure the public support that is required if the revolution is to go forward. Thus, the FY 2004 national funding request claims that, "The discovery of the novel phenomena and material structures that appear at the nanoscale will affect the entire range of applications that the grand challenges identify." Among those grand challenge areas are

- chemical-biological-radiological explosive detection and protection (homeland defense);
- nano-electronics, -photonics, and -magnetics (next generation of information technology devices)
- health care, therapeutics, and diagnostics (Better disease detection and treatment)
- energy conversion and storage
- environmental improvements

These identified areas point to great possibilities for profound changes to the quality of living for the collective, as well as for the individual. Each reflects values generally upheld in our society regarding healthy life and longevity, physical security and freedom from aggression, rapid and free access to information, and cleanliness and safety in the environment. These values have evolved in tandem with new technological developments, without which they could not be assured. We affirm as true and good that which we believe to be possible, in the negotiation of improvements to who we are and how we wish to live. At our best, we agree to constraints of new technological developments as guided by enlightened self-interest. At our worst, we blindly accept and consume anything that promises us more, irrespective of the resources we have committed in the process.

What resources? Gerson suggests that we begin by looking at four, in making assessments about quality of life: Money, time, sentiment, and skill. Regarding the promises of quality of life that makes nanotechnology so alluring, I offer the following considerations.

Money

Whose money is being used to bring financial benefit to whom in the development of nanotechnology? In the research phase, principal investigators and their institutions are the primary recipients of financial resources. Once development moves to market appropriation of new goods and services, then the agencies of the federal government, industry, venture capitalists, principal investigators and their universities whose patents are registered and research findings are materialized in marketable products will all begin to see returns on their investments. Eventually, nanotechnology startup firms will be publicly traded and individuals who invest in them may reap financial returns as well. Individual citizens will purchase and benefit from the goods and services of nanotechnology. Whether those benefits will mean improvements to their quality of living depends greatly on the answers individuals give to the questions of meaning and purpose in their lives.

Time

Faster and smaller semiconductors will mean on-demand information access and processing. This, in turn, will mean less personal time, not more, as we have already learned from the computer revolution, as discussed above. Time and money will need to be allocated to industrial retooling and retraining, and to public education and consumer information exchanges.

Sentiment

How do we feel about privacy? There may be substantial loss of privacy, but some increase in feelings of national security when invisible surveillance mechanisms become ubiquitous. How do we feel about intimacy? As communication shifts increasingly from touch, smell, and other sensual perceptions to electronic media, the meaning and expression of intimacy will also shift under new domains of knowing and experiencing the other. How do we feel about power? Our insatiable desire for it may bring us face to face with our true and increased frailty, as a result of dangerous liaisons of power accumulation.

Skill

Most of our aging population is without the skills or resources to participate fully in the benefits of computers. How will any of us, especially the aged, acquire the skills needed to manage intellectually, physically, and emotionally in the strange, new nanotechnology environment? Who will have access to those opportunities for training, and at what costs?

We haven't much choice about the fact that the next technological revolution is upon us and will, one way or another, take its course. Potentially, it will provide for many new and wonderful opportunities for human health and well being. Just as with the Internet, the coming nanotechnology era will reflect all of what we humans have: our ignorance, selfishness, insecurities, hostilities, greed and hatreds, as well as our tremendous capacity for creativity, wisdom, compassion, generosity, and agape. We still have many choices to make in the negotiation of that future.

References

1. NSET (National Science and Technology Council, Committee on Technology, Subcommittee on Nanoscale Science, Engineering and Technology), National Nanotechnology Initiative: Research and Development Supporting the Next Industrial Revolution, Supplement to the President's FY2004 Budget. (available at http://www.nano.gov/ html/res/fy04-pdf/fy04-main.html).
2. J. Marburger, The future of nanotechnology. Presentation made at NNI Workshop on Societal Implications of Nanoscience and Nanotechnology, 3–5 December, Arlington, VA (2003).
3. J. Ellul, The Technological Bluff, Grand Rapids, MI: Wm. B. Eerdmans Publishing (1990).
4. K. A. Schuessler, G.A. Fisher, Quality of life research and sociology, *Annual Review of Sociology* **11**, 129–149, p.132 (1985).
5. Gerson, Elihu M., On "Quality of Life", *American Sociological Review* **41**(5), 793-806, p. 796 (Oct., 1976).
6. S. Fuller, *Philosophy of Science and its Discontents*, New York: Guilford Press, 16 (1993).

5. GOVERNANCE

PROBLEMS OF GOVERNANCE OF NANOTECHNOLOGY

Frank N. Laird, University of Denver

As new as much of nanotechnology is, many of the concerns about governance bear a strong resemblance to issues that scholars of the societal implications of technology have studied before. Yet again, society is confronted with a family of technologies that is rapidly spawning innovations and extending human capabilities. Yet again, social and policy institutions charged with governing those technologies cannot seem to keep up. As a result, those institutions appear inadequate to protect society from potential harm or to channel the innovative energy into forms that will further social goals. These problems of governance are not new. Government officials and scientists have voiced similar concerns about almost every major innovative technology of the last 60 years, from nuclear power to biotechnology. From those earlier cases, we can draw on a rich literature to address multiple concerns and point the way to fruitful future research.

Bounding Nanotechnology

One problem with trying to create governing institutions or policies for nanotechnology is ambiguity over what technologies that umbrella term covers. Therefore, an analysis of the problems of governing nanotechnology needs to examine the question of technological inclusiveness. Such boundary work will not be settled by simple *a priori* declarations, but will more likely be negotiated as advocates for particular technologies seek to get them included in government-funded R&D programs, corporate initiatives, and so on [1, 2]. As a result, governing institutions will need to be flexible in their understanding of which technologies are in their purview.

These flexible, negotiated boundaries imply an evolving agenda for research on the societal implications of NSET. Particular NSET innovations will raise societal issues, such as environmental protection or privacy rights, relevant to specific government agencies. Thus an important topic for research is identifying the relevant agencies within whose boundaries these issues will fall and assessing their ability to cope with newly emerging technologies.

Democratic Participation

From nuclear power to genetically modified organisms, the history of past controversies over new and powerful technologies tells us that various groups from the public will get involved in the governance of these technologies. Some scholars have argued that such participation in science and technology policy issues should be the norm [3, 4]. In any event, past experience with other new technologies tells us that public participation related to new technologies will, in

M.C. Roco and W.S. Bainbridge (eds.),
Nanotechnology: Societal Implications — Individual Perspectives, 207–240.

one form or another, be the norm. Therefore, the key question is how that participation will take place.

Many policy institutions now have considerable expertise in incorporating public participation into their decision making processes, though some do it better than others. So many government programs require public participation that government officials who run such activities now have their own professional association with about 1,000 members from numerous countries [5]. In addition, scholars have produced numerous case studies of participation in scientific and technological issues (for example, [6]), delineating the many available techniques for such participation, the opportunities and barriers arising out of those techniques, and the contextual factors that influence their success or failure.

Based on the existing literature, researchers can provide criteria for evaluating whether participation activities measure up in terms of being democratic and efficacious. There is considerable scope for interesting comparative work here, since various countries in Europe have developed extensive participation mechanisms and since those countries are also trying to get into nanotechnology. Important research would include both reviewing what we already know about participation in complex technological issues and how features of nanotechnology would pose distinctive challenges for such participation.

Learning

In order to study the societal effects of nanotechnology, we have to hazard guesses about both the future shape of those technologies and the social trends with which they will interact. In short, we seem to need to make predictions. However, much social science research, not to mention common experience, suggests that we have no ability to make such predictions with any level of confidence [7]. Making policy for the future of emerging technologies asks for predictions that we cannot possibly make.

Future technological systems defy prediction for a host of reasons. Most obviously, considerable uncertainty clouds our knowledge of the future values of the relevant variables. Complicating matters further, technological systems evolve by interacting with social systems, including governments, markets, moral norms, and religious beliefs. So we have interactions among complicated variables, all exhibiting a high degree of uncertainty. To make matters worse, experts may disagree as to which variables to analyze and which to leave out. (Sarewitz, Pielke, and Byerly [7] give a number of examples from the environmental field, but the lessons apply more generally.) All of these difficulties combine to render our predictions highly unreliable.

Historians of technology have studied how people in the past tried to predict the future of technologies. Some of the predictions may now sound quaint, but the salient point about them is that they tell us more about the circumstances and beliefs of the people making the predictions than about the future of the technologies, which they invariably got wrong [8].

However, all this deserved skepticism about prediction does not relieve us of the need to make future-oriented policy for this emerging set of technologies. There are far too many cases of technologies for which there was not such policy and the result was that later policymakers had to try to straighten out avoidable problems. Very often technologies are promoted for a set of narrow reasons but then turn out to have wide-ranging impacts. So much has this been the norm since World War II that both scholarly and popular commentators have repeatedly complained that society's technological abilities have outstripped its governing abilities.

Given these severe constraints on prediction, governments and societies can try several approaches to coping with powerful emergent technologies. First, one can try to embed within social and policy processes the values that we want to see the new technologies promote, such as social justice, equity, economic efficiency, environmental sustainability, security, or others. It is important that these processes be deliberative and democratic. Such processes give the outcome legitimacy, since many different people and groups will have the opportunity to shape just what we mean by the terms "just," "equitable," and so on. Deliberative democratic processes provide diverse sources of information and knowledge, what some have referred to as the intelligence of democracy [9].

Second, the institutions dealing with such extraordinarily innovative technologies as nanotechnologies will need to be able to adapt to rapid changes and unexpected developments. In short, those institutions need to be able to learn. Institutional learning is a subtle process, one that involves more than just having well-informed people in an organization [10]. It requires that the institution's culture and operating procedures make it open to new ideas or uncomfortable data. It also requires that the institution act on new ideas in a way that allows those ideas to get deeply ingrained in the institution, not just carried out as the idiosyncratic ideas of individuals, so that the learning outlives the individuals who started it. This ability to learn and adapt involves many subtle problems, not least of which is social capital, the relationships of trust in society that make experimentation and trial and error tolerable. As others have noted [11], those relationships of trust have eroded severely in recent decades. Again, researching this issue for nanotechnologies will require looking at the considerable history of other technologies in order to understand the challenges that nanotechnology will pose to learning organizations.

Sophisticated Use of Expertise

The literature on science and technology policy now has great depth in studies of expertise (for example, [12, 13]). There are numerous studies of the ways expertise gets labeled as valid, enters into policy debates, and influences policy from very senior levels to the workings of bureaucracies. The nature of expertise is itself a hotly contested issue, as different sides of technological controversies seek to de-legitimatize the expertise of opposing sides. Precisely because nanotechnology is so new, there will be considerable difficulty in deciding and publicly justifying declarations regarding who is considered an expert.

Embedding Values into Institutions and Processes

Governing a new and powerful technology will inevitably raise very sensitive norms or values. It is, *a priori*, hard to know just how any particular nanotechnology will affect some important social value. That said, we need to analyze and make recommendations for the sorts of values that seem to drive the ways that governing institutions frame the societal problems related to nanotechnology (for a discussion of institutionalized values, see [14]). If nanotechnology develops into as big a financial phenomenon as its backers contend, it will be particularly important for governing institutions to focus on values in addition to economic efficiency, since those other values will tend to get lost as financial investments in nanotechnology increase. Again, analyses of how this has played out in other cases, from biotechnology to energy, could inform and guide this research.

References

1. T. F. Gieryn, Boundaries of science. In *Handbook of Science and Technology studies*, eds. S. Jasanoff, G. E. Markle, J. C. Petersen, T. Pinch, Thousand Oaks, CA: Sage Publications (1995).
2. D. H. Guston, Boundary organizations in environmental policy and science: An introduction, *Science, Technology & Human Values* **26** (4), 399–409 (Autumn 2001).
3. F. N. Laird, Participatory analysis, democracy, and technological decision making, *Science, Technology and Human Values* **18**, 341–361 (Summer 1993).
4. R. E. Sclove, *Democracy and Technology*, New York: Guilford Press (1995).
5. IAP2 (International Association for Public Participation). (available at http://www.iap2.org).
6. J. C. Petersen, *Citizen Participation in Science Policy*, Amherst, MA: University of Massachusetts Press (1984).
7. D. Sarewitz, R. A. Pielke, Jr., R. Byerly, Jr., *Prediction: Science, decision making, and the future of nature,* Washington, DC: Island Press (2000).
8. J. J. Corn, ed., Imagining Tomorrow: History, technology, and the American future, Cambridge, MA: MIT Press (1986).
9. C. E. Lindblom, E. J. Woodhouse, *The Policy-making Process*, Englewood Cliffs, NJ: Prentice Hall (1993).
10. F. Laird, Rethinking learning, *Policy Currents* **9**(3-4), 3-7 (November 1999).
11. R. D. Putnam, Bowling *Alone: The collapse and revival of American community*, New York: Simon and Shuster (2000).
12. S. E. Cozzens, E. J. Woodhouse, Science, government, and the politics of knowledge. In *Handbook of Science and Technology Studies*, eds. S. Jasanoff, G. E. Markle, J. C. Petersen, T. Pinch, Thousand Oaks, CA: Sage Publications (1995).
13. S. Jasanoff, *The Fifth Branch: Science advisers as policymakers*, Cambridge, MA: Harvard University Press (1990).

14. F. Laird, *Solar Energy, Technology Policy, and Institutional Values*, New York: Cambridge University Press (2001).

SOCIETAL IMPLICATIONS OF EMERGING SCIENCE AND TECHNOLOGIES: A RESEARCH AGENDA FOR SCIENCE AND TECHNOLOGY STUDIES (STS)

Bruce E. Seely, Michigan Tech University

The idea that we should consider the possible consequences of emerging technologies did not begin with the National Nanotechnology Initiative (NNI). The most prominent efforts in this direction include the work of the Office of Technology Assessment (OTA), which existed from 1974 through 1995 [1]. OTA's efforts were wide-ranging and served to establish some basic approaches to the assessment, if not the forecasting, of the consequences of technology. Similarly, historian Joel Tarr of Carnegie Mellon University and others explored the concept of retrospective technology assessment in the mid-1970s, also seeking to develop methods for assessment and forecasting. Tarr believed that assessment of historical cases of technical change, where the origin and consequences were observable, could illuminate the challenges and opportunities for assessment of technical systems before their full societal implications were experienced [2–5]. The general pattern of assessment efforts, however, has been to ask about new processes or technologies after they are in place. This NNI workshop on nanotechnology and its predecessor demonstrated that the scientific, technological, and policy communities are moving to a different approach for considering the societal implications of science and technology. Following a pattern initiated by the human genome project and continued by other large-scale NSF research efforts, the nanotechnology research program has built into it a charge to examine the interaction of society and technologies, even as new developments are unfolding; similarly the Foresight Institute has given prominent attention to the inevitability of unanticipated consequences [6]. Research on the consequences, both predicted and unanticipated, of new science and technology can be advanced by building upon several decades of scholarship by the science and technology studies (STS) community. STS researchers are not alone in examining this topic, but their research is especially strong on the historical, social, and philosophical implications of emerging technologies. The STS community must, however, make concerted efforts to bring its insights and understandings to the audiences interested in nanotechnology.

The present STS community originated from at least three different scholarly activities during the 1950s, 1960s, and 1970s. One strand came from the *history and philosophy of science and technology*, fields of study that in the 1950s grew into full-blown academic disciplines with graduate programs, professional societies, and journals. Numerous scholars conducted research that illuminated questions about the nature, development, and societal implications of both past and contemporary developments in science and engineering [7–10]. A second root was university faculty interested explicitly in *science, technology, and society*.

Intellectually, scholars in this area (as in others) drew inspiration from the work of social commentators Lewis Mumford and Jacques Ellul, as well as from the debates over nuclear weapons and civilian power, environmental problems, and the war in Vietnam [11, 12]. University courses in this tradition reflected the teach-in tradition of the 1960s and often paired a scientist or engineer with a historian, philosopher, ethicist or social scientist in order to examine the "impact" of technology on society. These interdisciplinary partnerships often lasted only as long as the enthusiasm of the individuals involved. But a few programs, such as that established by metallurgist Rustum Roy at Penn State, proved more durable. Indeed, Roy became the leading disciple of this approach to studying science and technology and founded a journal and an organization, the National Association for Science, Technology and Society (NASTS), which continues to meet [13–18].

In the early 1970s, a third strand of STS emerged as *social studies of science,* formed by the coalescence of social scientists from sociology, anthropology, and political science interested in science and technology. Central intellectual influences included scholarship from the sociology of knowledge and various postmodern philosophers such as Derrida and Foucault, as well as theoretical conceptions such as social construction articulated by European scholars Barry Barnes, David Edge, and others. Actor network theory later added additional theoretical heft to this area of study. The result has been a body of scholarship, now labeled *science studies*, that at its core emphasizes understanding science as a social process. For such scholars, the societal implications of science and technology can be a fundamental subject for scholarly investigation [19–25].

By the 1990s, these three different approaches were very loosely affiliated under a common umbrella label of *science and technology studies* (STS). In combination, they can add much to the effort to understand the societal implications of nanotechnology. Perhaps the most important contribution may lie in the way that nearly all STS scholars have moved beyond examining the "impact of technology of society," a term that implied technology was an autonomous variable acting upon society. Langdon Winner's classic study *Autonomous Technology* used the story of Frankenstein as a vehicle for launching an examination of the relationship of technology to society [26]. Winner's conclusion that technologies, as human creations, could not be independent of society was reinforced by his subsequent study on the politics of technology, *The Whale and the Reactor* [27]. Building from this and other related insights, science and technology are products of society, not external elements acting upon them, and the best work by STS scholars now emphasizes the complex and mutual interactions of science and technology with society [28]. This is a conceptually richer approach than the older, deterministic focus upon impacts.

This basic position informs several STS research approaches and topics that profitably could be applied to considering the societal implications of nanotechnology. One of the most important concerns *the nature of science.* Scientists generally assume that the scientific method serves as a defining feature of their field, in its emphasis upon replicability and objective evidence as standards for the generation of new knowledge. STS scholars have suggested, however, ways that this approach to understanding the nature of science can be misleading. The landmark autobiography of a scientific breakthrough, James Watson's *The Double Helix,* certainly suggested that the pathways followed by scientists were littered with human motives such as the desire for fame and recognition and blindness to alternative elements [29]. Reinforcing the sense that science was much more than a rational search for the truth was perhaps the most influential work in the philosophy of science in the last 50 years, Thomas Kuhn's *The Structure of Scientific Revolutions* [30]. Kuhn's picture of the nature of change within scientific communities certainly differed from the ideal image of scientists motivated solely by objective data and a search for the truth. His study built the logical foundation for later examinations of science as a social enterprise [31, 32].

Historians and other scholars also have devoted attention to understanding *how science and scientists work.* Historical scholars have focused a significant amount of energy on changes in the institutions that have supported the growth and development of science and technology over time. The emergence of formalized efforts to seek innovation over the past 125 years has been of particular interest. The term R&D often defines this approach to science in corporate and government laboratories, and historians have paid substantial attention to the organizational and bureaucratic aspects of this type of research. Their work has emphasized the importance of the managerial tasks related to the pursuit of science in these institutional settings, and of the complexity of the process. These studies have shifted the focus of scholars from the process of invention and discovery to the often more complicated issues of development, commercialization, production, and marketing of science and technology and their outcomes [33–36].

The other central element running through these projects is the impact of scale, for most of these projects operated on a very large stage, involving large teams of researchers and vast sums of money. Scholars coined the term "Big Science" to describe the emergence of scientific activities that depended upon large and expensive tools tended by a team of researchers—a pattern first evident in astronomy and physics in the 1920s and especially the 1930s. But this tendency became a much more common characteristic of government-funded science and technology programs after 1940. World War II not only demonstrated the possibilities of this approach, thanks to the successful harnessing of scientific investigators to such projects as the development of radar, solving logistical problems, and the atomic bomb, but also accustomed a generation of scientists to the advantages of free-flowing funding from the military. The pattern became common during the Cold War, but encompassed more than the scientific endeavors of the United States and Soviet Union during the period 1950–1990. From the point of view of assessing the consequences of emerging nanotechnology, case studies of other large-scale projects, such as the atomic bomb project, high-energy physics (especially the development of particle accelerators and other expensive apparatus), the effort to explore outer space, and other large-scale research projects of the post-war era offer points of comparison [37–47].

Historians have been joined by other scholars who have examined the nature and working of science. Philosophers have undertaken studies of the nature of knowledge generation and creation, as well as the differences between individual fields of science [48–52]. Among philosophers (and others) we also find interest in the ethical issues related to the practice of science and engineering (including research ethics), as well as to the ethical challenges related to the emergence of new technologies. Such activities, along with related issues from medicine, as well as from the practice of business, have spawned a growing professional field and several journals [53-58]. Science studies scholars have added substantially to our understanding of the actual process of doing science through ethnographic research inside laboratories that emphasizes science is an inherently social process. These detailed accounts of the day-to-day work of scientists have added insights that confirm that the technical and social dimensions of practice and research cannot be separated. Science studies scholars have highlighted "the social shaping of technology" to show how the innovation process is subject to many pressures from outside the laboratory. Superb case studies by sociologists of science and technology demonstrate that science is highly contingent upon particular circumstances, and not driven solely by technical and scientific imperatives. Thus Sharon Traweek produced a marvelous account of the work of particle physicists, while anthropologist Lucy Suchman has described the ways that such insights were developed and applied by researchers at the Xerox Palo Alto Research Center facility. A superb example of this insight emerges from Claude Fischer's history of the telephone, which traces how users of this communication technology—especially women—turned the system into a tool for mass communication rather than the business tool that its inventors and development engineers had envisioned [59–66]. This body of fine-grained case studies offers a way of approaching the societal implications of nanotechnology via collaborative projects with the nanoscientists themselves, tracking their efforts in their laboratories.

Another pivotal STS concept that has emerged from studies of emerging sciences and technologies includes attention to *large technical systems*. Historian Thomas Hughes pioneered this topic with studies of electric power systems in several nations, showing that the history of attention to machinery and hardware could not be separated from the integral social, political, economic, and cultural elements that introduce enormous complexity into power systems. Hughes and others exploiting these insights drew attention to several aspects that offer points of departure for research into nanotechnology. These topics include the momentum such systems can achieve, in part because of their size, and the manner in which early decisions constrain later choices. Hughes' comparative study also demonstrated that national "styles" of technology could appear as different business, social, economic, regulatory, and cultural considerations led to variations in similar technical systems—in this case, electric power. Other scholars have applied Hughes' categories of analysis to postwar technical systems as diverse as computers and information technology and highways. There seems little doubt that this line of thinking has application to the development of nanotechnology [67–71].

A final strand of STS scholarship that contributes to understanding the development of new sciences and technologies themselves, as well as their consequences, builds from the recognition that predictions are fraught with peril and uncertainty. Engineers, not to mention savvy managers, know that changing one element within a large system alters the dynamics of the entire system, and predicting the consequences of even apparently simple substitutions is very difficult. In her classic study of women and household technology, Ruth Schwartz Cowan found that the introduction of these "labor-saving" technologies affected women in many unanticipated ways, with the most surprising outcome being little reduction in the amount of time women spent on household chores [72, 73]. STS scholars have shown that assessing the societal implications of new technologies is an even more difficult task, for new systems not only displace old ways of doing things, but also allow entirely new possibilities to unfold. Changes almost always stress existing systems, and some prove harder to cope with than others. STS scholars have long probed the unanticipated consequences of such changes, which represent one of the most important societal implications of new technology. Without necessarily being anti-technology, some scholars have been highly critical of the implementation and application of large technical systems and suggested that the failure to consider possible consequences in advance has contributed to problems and even disasters. The best of these case studies detail how decision making structures and the preliminary assumptions of managers of large systems can lead to disastrous outcomes, as in Kim Fortun's study of the Bhopal chemical plant disaster in India and Diane Vaughan's review of the circumstances behind the Space Shuttle Challenger explosion [74, 75]. Other scholars have focused on the general circumstances surrounding complex systems, formulating variants on Murphy's Law—the assertion that things that can go wrong will, and that complicated systems are more likely to malfunction than simple ones. Among the more recent studies along this line are Edward Tenner's extended essay *Why Things Bite Back*, which considers changes in medicine, the environment, sports, and the computerized office and urges greater caution, alertness, and understanding of the technological domain within which we live [76–80].

Yet the call for caution from Tenner and others has rarely been adopted within the American science and technology policy process. In part, concerns about unanticipated effects have been overridden by the tendency and desire in some quarters to view new technological developments as revolutionary in their implications. Beginning with aviation and radio and moving on to nuclear power "too cheap to meter" and the promised efficiency of the paperless office, advocates and enthusiasts have promoted the promises of one "next big thing" after another. We have lived through the Jet Age, the Space Age, a materials revolution, and the age of the digital computer and the Internet revolution [81–91]. The traditional optimism of American citizens and occasional tendency to utopian expectations are at work here, but this celebratory rhetoric is also driven by a desire to attract government funding and venture capital [92, 93]. George Whiteside, Mallinckrodt Professor of Chemistry at Harvard, observed during the history and ethics working session at this workshop that the adversarial politics so characteristic of the American system encourages hyperbolic language for staking out positions, while the necessity of compromise ensures more moderate outcomes.

Whiteside added, however, that scientific choices should be shaped by a cooler, more dispassionate process. Yet hyperbole clearly characterizes current public discussions of nanotechnology, as perhaps best witnessed by the wide dissemination of images of nanobots. A gallery of such drawings can be found at the Web site of the Foresight Institute, a leading advocate of nanotechnology (http://www.foresight.org/Nanomedicine/Gallery/index.html). Nano-probes, nano-dentists, or nano-vaccine injectors certainly capture the imagination, without regard to an assessment of the capability of constructing such devices. Significantly, a heated debate has developed between K. Eric Drexler, the early leader of the field and a visionary advocating nanobots and the technical conception underlying them (molecular assembly) and other scientists, notably Richard Smalley, who consider that the optimistic visions of Drexler are not only unrealistic, but also potentially damaging to the future of nanotechnology [94].

The intended purpose of such images, apart from an exercise in imagination, seems to be the creation of a sense of inevitability and optimistic purpose for nanotechnology. Such optimistic rhetoric dominated several presentations at the workshop, just as it appears in much of the published literature. Importantly, nanoscience advocates present not possible or potential impacts, but much more certain-sounding and highly optimistic statements of what *will* happen. They discuss the impact of nanotechnology and extrapolate its consequences, promising this will be the biggest thing since computers, since printing, even since agriculture [95–103]. All of this breathless enthusiasm suggests an important question for research by STS scholars, namely, the role of enthusiasm and hyperbole in big science projects, especially in the development of nanotechnology.

Indeed, STS research might help introduce some realism into discussions about the societal implications of nanotechnology. Our lives may indeed be very different because of the unfolding ability to manipulate nature at the nanoscale. But STS studies of societal implications could perhaps temper the easy revolutionary assumptions and analogies so easily invoked to celebrate the promise of nanotechnology. Even the idea that nanotechnology is somehow qualitatively different from other technological developments needs to be examined critically. The bases of this difference need to be defined and reviewed. Comparative research certainly would suggest that early advocates of other new technologies rarely identified the most important eventual uses of their inventions and discoveries.

This short list of possible research topics related to nanotechnology does not exhaust the opportunities for STS contributions to the study of the societal implications of nanotechnology. But to bring their historical, social, and ethical insights to bear on the development of nanotechnology, STS scholars must broaden their horizons and their approaches to research. Most STS researchers work as individuals on investigator-defined projects. Rarely are they responsive to RFPs, agency announcements, and the desires of funding sponsors in the manner of researchers in the physical science and engineering. Few STS scholars possess the training to work in teams with other STS scholars, much less with scientists and engineers. Fewer still commit to the painstaking effort of developing the personal and professional connections that allow such collaborations. The STS community will contribute more fully to the discussion of the implications of nanotechnology by adding such approaches to the existing pattern of individual studies. Such efforts need to be encouraged. Indeed, two examples of such efforts devoted to nanotechnology are already underway with funding from the National Science Foundation: one at the University of South Carolina under the direction of David Baird (http://www.cla.sc.edu/cpecs/nirt/index.html), and the other at the University of Virginia under the direction of Michael Gorman. Both researchers participated in this workshop.

STS scholarship has much to contribute to the important debates about the implications of nanotechnology, and the scientific and engineering audience should encourage their participation. An expansion of research styles will not only better prepare STS scholars to conduct research on the societal implications of nanotechnology, but also should broaden the audience for the findings that will result. Above all, STS scholars can bring a comparative perspective to the discussion of the societal implications of nanotechnology. They possess no crystal balls, and past examples are not a roadmap for the future. But reflection on the experiences of technical innovations can provide a useful reality check when contemplating the future implication of the emerging science and technologies of the nanoscale.

References

1. The OTA legacy (available at http://www.wws.princeton.edu/~ota/).
2. J. A. Tarr, *Retrospective Technology Assessment—1976: Conference on Retrospective Technology Assessment*, San Francisco: San Francisco Press (1977).
3. V. T. Coates, A. Finn, *A Retrospective Technology Assessment: Submarine telegraphy: The transatlantic cable of 1866*, San Francisco: San Francisco Press (1979).
4. I. de Sola Pool, *Forecasting the Telephone: A retrospective technology assessment*, Norwood, NJ: ABLEX Publications (1983).
5. J. A. Tarr, *The Search for the Ultimate Sink: Urban pollution in historical perspective*, Akron, OH: University of Akron Press (1996).
6. Foresight Institute. Preparing for nanotechnology (available at http:// www.foresight.org/).
7. B. E. Seely, SHOT, the history of technology and engineering education, *Technology and Culture* **36**, 739–772 (October 1995).
8. V. L. Hilts, History of science at the University of Wisconsin, *Isis* **75**, 63-94 (1984).
9. M. Claggett, Teaching the history of science, *Isis* **43**, 51-53 (April 1952).
10. A. Thackray, R. K. Merton, On discipline building: The paradoxes of George Sarton, *Isis* **63**, 473-495 (1972).

11. L. Mumford, *The Myth of the Machine: Technics and human development*, New York: Harcourt, Brace Jovanovich (1967).

12. J. Ellul, *The Technological Society*, New York: Knopf (1964).

13. S. H. Cutcliffe, C. Mitcham, *Visions of STS: Counterpoints in science, technology, and society studies*, Albany: State University of New York Press (2001).

14. S. H. Cutcliffe, *Ideas, Machines, and Values: An introduction to science, technology, and society studies*, Lanham, MD: Rowman & Littlefield Publishers (2000).

15. R. Roy, J. Lerner, The status of STS activities at U.S. universities, *Bulletin of Science, Technology & Society* **3**, 417-432 (1983).

16. R. Roy, STS: Seeing large things that are large, *Bulletin of Science, Technology & Society* **13**(1), 1-2 (1993).

17. IASTS (International National Association for Science, Technology and Society), formerly National Association for Science, Technology, and Society (NASTS) (available at http://www.nasts.org/index.htm).

18. J. Sanford, Science, technology, and society program survives, thrives, *Stanford Report* (August 2001) (available at http://news service.stanford .edu / news/2001/august22/sts-822.html).

19. B. Barnes, D. O. Edge, *Detailed Science in Context: Readings in the sociology of science*, Cambridge, MA: MIT Press (1982).

20. W. E. Bijker, J. Law, *Shaping Technology/Building Society: Studies in sociotechnical change*, Cambridge, MA: MIT Press (1982).

21. M. Callon, J. Law, A. Rip, *Mapping the Dynamics of Science and Technology: Sociology of science in the real world*, Basingstoke, UK: Macmillan (1986).

22. J. Law, J. Hassard, *Actor Network Theory and After*, Oxford; Malden, MA: Blackwell/Sociological Review (1999).

23. S. W. Fuller, *Philosophy, Rhetoric, and the End of Knowledge: The coming of science and technology studies*, Madison: University of Wisconsin Press (1993).

24. S. Jasanoff, G. E. Markle, J. C. Petersen, T. Pinch, eds., *Handbook of Science and Technology Studies*, Thousand Oaks, CA: Sage Publications (1995).

25. M. Biagioli, ed., *The Science Studies Reader,* New York: Routledge (1999).

26. L. Winner, *Autonomous Technology: Technics-out-of-control as a theme in political thought*, Cambridge, MA: MIT Press (1977).

27. L. Winner, *The Whale and the Reactor: A search for limits in an age of high technology*, Chicago: University of Chicago Press (1986).

28. M. R. Smith, L. Marx, eds., *Does Technology Drive History? The dilemma of technological determinism*, Cambridge, MA: MIT Press (1994).

29. J. D. Watson, *The Double Helix; a personal account of the discovery of the structure of DNA*, New York: Atheneum (1968).

30. T. S. Kuhn, *The Structure of Scientific Revolutions,* Chicago: University of Chicago Press (1962).

31. P. Horwich, ed., *World Changes: Thomas Kuhn and the nature of science*, Cambridge, MA: MIT Press (1993).

32. S. Fuller, *Thomas Kuhn: A philosophical history for our times*, Chicago: University of Chicago Press (2000).

33. G. Wise, Willis R. Whitney, *General Electric, and the origins of U.S. industrial research*, New York: Columbia University Press (1985).

34. L. S. Reich, *The Making of American Industrial Research: Science and business at GE and Bell*, 1876-1926, Cambridge, UK; New York: Cambridge University Press (1985).

35. D. A. Hounshell, J. K. Smith, *Science and Corporate Strategy: Du Pont R&D, 1902–1980*, Cambridge, UK; New York: Cambridge University Press (1988).

36. P. L. Galison, *How Experiments End*, Chicago: University of Chicago Press (1987).

37. D. J. Kevles, *The Physicists: The history of a scientific community in modern America*, New York: Vintage Books (1979).

38. P. L. Galison, B. W. Hevly, *Big Science: The growth of large-scale research,* Palo Alto, CA: Stanford University Press (1992).

39. J. Krige, *Choosing Big Technologies,* New York: Harwood Academic Publishers (1992).

40. R. G. Hewlett, *The New World, 1939/46*, University Park, PA: State University Press (1962).

41. R. W. Seidel, *Los Alamos and the Development of the Atomic Bomb*, Los Alamos, NM: Otowi Crossing Press (1995).

42. L. Hoddeson, G. Baym, *Critical Assembly: A technical history of Los Alamos during the Oppenheimer years, 1943-1945*, Cambridge, UK; New York: Cambridge University Press (1993).

43. H. Gusterson, *Nuclear Rites: A weapons laboratory at the end of the Cold War*, Berkeley: University of California Press (1996).

44. R. D. Launius, *NASA: A history of the U.S. civil space program*, Malabar, FL: Krieger Publishing Co. (1994).

45. R. W. Smith, *The SpaceTtelescope: A study of NASA, science, technology, and politics*, Cambridge, UK; New York: Cambridge University Press (1989).

46. J. Krige, A. Russo, L. Sebesta, *A History of the European Space Agency 1958-1987,* Noordwijk, Netherlands: European Space Agency (2000).

47. J. Krige, *History of CERN*, Amsterdam; Oxford: North-Holland (1996).

48. S. Fuller, *The Cognitive Turn: Sociological and psychological perspectives on science*, Dordrecht, Netherlands; Boston: Kluwer (1989).

49. N. J. Nersessian, *The Process of Science: Contemporary philosophical approaches to understanding scientific practice,* Dordrecht, Netherlands; Boston; Hingham, MA: M. Nijhoff (1987).

50. V. G. Hardcastle, ed., *Where Biology Meets Psychology: Philosophical essays*, Cambridge, MA: MIT Press (1999).

51. M. Fortun, H. J. Bernstein, *Muddling through: Pursuing science and truths in the twenty-first century*, Washington, DC: Counterpoint Press (1998).

52. S. Mitchell, Dimensions of scientific law, *Philosophy of Science* **67**(2), 242-266 (2000).

53. M. Davis, G. P, Cartwright, The conscience of computer science: Organizations that target the ethical & social implications of technology, *Change*, 61–63 (January-February 1998).

54. K. Shrader-Frechette, V. Weil, Ethics of scientific research, *Ethics* **106**(4), 879-881 (1996).

55. R. D. Hollander, N. H. Steneck, Science- and engineering-related ethics and values studies: Characteristics of an emerging field of research, *Science, Technology & Human Values* **15**(1), 84-104 (Winter 1990).

56. D. G. Mayo, R. D. Hollander, eds., *Acceptable Evidence: Science and values in risk management,* New York: Oxford University Press (1991).

57. M. Davis, Continuing and emerging issues in engineering ethics education, *The Bridge* **32**(3), 8-13 (2002).

58. J. R. Herkert, Engineering ethics in the USA: Content, pedagogy and curriculum, *European Journal of Engineering Education* **25**(Part 4), 303-314 (2000).

59. B. Latour, S. Woolgar, *Laboratory Life: The social construction of scientific facts*, Beverly Hills: Sage Publications (1979).

60. B. Latour, Science in Action: How to follow scientists and engineers through society, Cambridge, MA: Harvard University Press (1987).

61. S. Traweek, *Beamtimes and Lifetimes: The world of high energy physicists*, Cambridge, MA: Harvard University Press (1988).

62. L. Suchman, L. Bishop, Problematizing "innovation" as a critical project, *Technology Analysis and Strategic Management* **12**(3), 327-333 (2000).

63. M. E. Gorman, W. B. Carlson, Interpreting invention as a cognitive process: The case of Alexander Graham Bell, Thomas Edison, and the telephone, *Science, Technology & Human Values* **15**(2), 131-164 (Spring 1990).

64. D. A. MacKenzie, *Inventing Accuracy: A historical sociology of nuclear missile guidance*, Cambridge, MA: MIT Press (1990).

65. D. A. MacKenzie, J. Wajcman, ed., *The Social Shaping of Technology,* 2nd ed., Buckingham, UK: Open University Press (1999).

66. C. Fischer, *America Calling: A social history of the telephone to 1940*, Berkeley: University of California Press (1992).

67. T. P. Hughes, *Networks of Power: Electrification in western society, 1880-1930*, Baltimore: Johns Hopkins University Press (1983).

68. T. P. Hughes, *Rescuing Prometheus*, New York: Pantheon Books (1998).

69. R. Mayntz, T. P. Hughes, ed., *The Development of Large Technical Systems*, Frankfurt am Main: Campus Verlag; Boulder: Westview Press (1988).

70. J. Summerton, *Changing Large Technical Systems*, Boulder: Westview Press (1994).

71. A. C. Hughes, ed., *Systems, Experts, and Computers: The systems approach in management and engineering, World War II and after*, Cambridge, MA: MIT Press (2000).

72. R. S. Cowan, *More Work for Mother: The ironies of household technology from the open hearth to the microwave*, New York: Basic Books (1983).

73. G. I. Rochlin, *Trapped in the Net: The unanticipated consequences of computerization*, Princeton: Princeton University Press (1997).

74. K. Fortun, *Advocacy after Bhopal: Environmentalism, disaster, new global orders*, Chicago: University of Chicago Press (2001).

75. D. Vaughan, *Challenger Launch Decision: Risky technology, culture, and deviance at NASA*, Chicago: University of Chicago Press (1996).

76. E. Tenner, *Why things bite back: Technology and the revenge of unintended consequences*, New York: Knopf (1996).

77. E. Tenner, *Our Own Devices: The past and future of body technology*, New York: Alfred A. Knopf (2003).

78. C. Norman, *The God that Limps: Science and technology in the eighties*, New York: Norton (1981).

79. W. Rybczynski, *Taming the Tiger: The struggle to control technology*, New York: Viking Press (1983).

80. H. P. Segal, *Future Imperfect: The mixed blessings of technology in America*, Amherst: University of Massachusetts Press (1994).

81. S. Hilgartner, R. Bell, R. O'Connor, *Nukespeak: Nuclear language, visions, and mindset,* San Francisco: Sierra Club Books (1982).

82. J. L. Wright, *Possible dreams: Enthusiasm for technology in America*, Dearborn, MI: Henry Ford Museum & Greenfield Village (1992).

83. J. J. Corn, *The Winged Gospel: America's romance with aviation*, 1900–1950, New York: Oxford University Press (1983).

84. J. J. Corn, B. Horrigan, K. Chambers, *Yesterday's Tomorrows: Past visions of the American future,* New York: Summit Books; Smithsonian Institution Traveling Exhibition Service (1984).

85. C. Stoll, *Silicon SnakeOoil: Second thoughts on the information highway*, New York: Doubleday (1995).

86. W. E. Burrows, *This New Ocean: The story of the first space age*, New York: Random House (1998).

87. P. A. Winters, *The Information Revolution: Opposing viewpoints*, San Diego: Greenhaven Press (1998).

88. M. L. Dertouzos, *What Will Be: How the new world of information will change our lives*, San Francisco: HarperEdge (1997).

89. F. Williams, *The Communications Revolution*, Beverly Hills: Sage Publications (1982).

90. J. Zygmont, *Microchip: An idea, its genesis, and the revolution it created*, Cambridge, MA: Perseus (2003).

91. D. E. Sichel, *The Computer Revolution: An economic perspective*, Washington, DC: Brookings Institution Press (1997).

92. H. P. Segal, *Technological utopianism in American culture,* Chicago: University of Chicago Press (1985).

93. D. Boorstin, *The Americans: The ational experience*, New York: Random House (1965).

94. Nanotechnology: point-counterpoint, *Chemical and Engineering News* **81**(31 December 2003) (available at http://pubs.acs.org/cen/coverstory/8148/8148counterpoint.html).

95. K. E. Drexler, C. Peterson, G. Pergamit, *Unbounding the future: The nanotechnology revolution,* New York: Morrow (1991).

96. D. Mulhall, *Our molecular future: How nanotechnology, robotics, genetics, and artificial intelligence will transform our world*, Amherst, NY: Prometheus Books (2002).

97. Interagency Working Group on Nanoscience, Engineering and Technology, *National Nanotechnology Initiative: Leading to the next industrial revolution: A report*, Washington, DC: National Science and Technology Council (2000).

98. D. Fatz, *The microengineering & nanotechnology revolution*, Richmond, Surrey, UK: Clinica Reports (2000).

99. M. Schulenburg, *Nanotechnologie: Die letzte industrielle revolution?,* Frankfurt am Main: Insel, (1995).

100. T. Lowe, The revolution in nanometals, *Advanced Materials & Processes* **160**(1), 63-65 (January 2002).

101. R. Fleeter, The coming nanospace revolution, *Ad Astra* **11**(6), 22-24 (November/December 1999).

102. D. Voss, Nanotechnology: The hope and the hype, *MIT's Technology Review* **102**(2), 55-59 (1999).

103. G. Stix, Little big science, *Scientific American* **285**(3), 32-37 (2001).

INSTITUTIONAL IMPACTS OF GOVERNMENT SCIENCE INITIATIVES

Tobin L. Smith, AAU

The institutional impact of large-scale governmental science initiatives, such as the National Nanotechnology Initiative (NNI), is a topic in which research universities have a significant and growing interest. Universities are interested in these initiatives for a few reasons. As major partners with the government in carrying out federally funded research, universities are trying to find new and better ways to address cutting-edge scientific challenges. Certainly, large science initiatives offer one means by which to tackle major societal and scientific challenges. Universities are also interested in educating students to take on these new scientific challenges and in providing them with the technical skills and knowledge that will be required by both academia and industry in the future.

Universities also have a vested interest in ensuring that adequate federal support exists to pursue exciting new research opportunities. To achieve this goal, universities recognize the need to capture the imagination of the general populace and key policymakers concerning the potential that science has to grow our economy, ensure our national security, improve our health, and enhance our overall quality of life. As will be discussed later, large-scale government science initiatives can go a long way in helping to focus both the attention of the general public and policymakers on how science can address specific societal needs.

Universities are paying attention to new large, multi-agency, and multidisciplinary science initiatives for yet another reason: initiatives such as the NNI promise to profoundly transform not only how science is conducted, but also how universities, university researchers, and faculty conduct their research and educational activities.

This paper attempts to address the following four questions: (1) Do large-scale science initiatives increase government and public support for research? (2) What are the positive aspects of such initiatives both from a federal science policy perspective and from a university perspective? (3) What institutional challenges/barriers exist in trying to deal with these new initiatives and do they have any negative implications? And finally, (4) What steps might be taken by universities and the government to assist the research community in responding to new large-scale and interdisciplinary science initiatives and to maximize the effectiveness of these initiatives?

1. Do large-scale science initiatives increase government and public support for science and university-based research?

Certainly, the answer to this question is yes. Duncan Moore, former Associate Director of the Office of Science and Technology Policy, said in a May 2001 talk:

> As we look at the R&D portfolio, there seems to be two ways we can increase the level of funding. One way to do this is to say that science is good, so fund more of it. Scientists believe this is a compelling argument that everyone should accept, but, in fact, it is not a very compelling argument. It does not go very far outside the scientific community, or with Congress. It is good enough to get an increase of the inflation rate plus 1 or 2 percent, but at that rate, it will take 35 years to double R&D funding.
>
> Another way of getting increased funding is to use the initiative-based argument for increasing R&D. With this method, you make such a compelling argument that a certain area is so important for some reason (you have to define the reason), that we should put huge amounts of money into this area. This is how the National Institutes of Health can get such huge budget increases. [1, p.273]

Indeed, past large-scale science initiatives have helped to establish government priorities for where investments in science should be made and have served as the primary impetus for significant federal funding increases for scientific research. By focusing attention on a specific societal need or scientific challenge, government science initiatives have helped to rally both public and governmental support for specific areas of research. They also focused the attention of the scientific community on a particular scientific or technical challenge.

Such efforts have tended to highlight mission-oriented research supported by an individual federal agency. For example, in the 1960s, tremendous amounts of money were invested in space R&D through the National Aeronautics and Space Administration (NASA) in order to beat the Soviet Union to the moon. As oil prices rose in the 1970s and gas lines grew, we saw significant increases in national investments in energy R&D and growth in the research conducted by the Department of Energy (DOE). During the 1980s significant new funding was provided for basic and applied research at the Department of Defense (DoD) in the name of protecting national security. More recently in the 1990s we have seen significant funding increases in health research, with Congress doubling the budget of the National Institutes of Health (NIH), as shown in Figure 5.1 [2]. And, with the events of September 11, 2001, and the subsequent anthrax attacks on Capitol Hill, we have seen a significant investment into science and technology research aimed at homeland security. This new R&D funding has flowed primarily to the Department of Homeland Security (DHS) and to the National Institute of Allergy and Infectious Diseases (NIAID) for bioterrorism research.

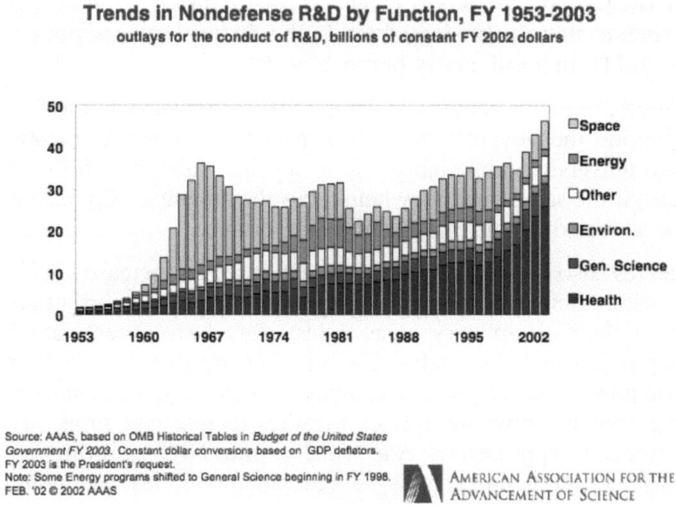

Figure 5.1. Nondefense R&D trends over 50 years.

When government has undertaken these major research efforts in the past, not only have they often been targeted to meeting the missions of specific federal research agencies, but also they have resulted in increased resources for universities and university researchers. However, because the earlier initiatives have emphasized and relied heavily upon specific scientific disciplines, they caused certain departments, centers, laboratories, and schools at our universities to experience significant infusions of funding and growth, while others received little if any benefit. Such initiatives have also resulted in significant influxes of students into certain research fields. A good example of this is the large jump in the number of students who entered nuclear engineering in the late 1970s and early 1980s when the federal government was investing heavily in nuclear energy research. This increased interest among students was in part likely due to the public awareness of the need for alternative energy sources as well as engineering schools and colleges being well funded to conduct nuclear research by the federal government.

But the Information Technology (IT) and National Nanotechnology Initiatives (NNI) are quite different. These efforts are attracting significant new funding for science. Yet unlike past initiatives, funding for both the IT initiative and NNI is not coming primarily from one research agency, but from multiple federal agencies. Moreover, these initiatives are truly multidisciplinary, involving faculty and researchers from several scientific disciplines and campus departments. And finally, while certainly there will be societal benefits derived from these initiatives, their initial focus is on the need to achieve breakthroughs in a particular scientific field as opposed to achieving a larger societal goal (e.g., a nanotechnology initiative out of which one of the potential societal benefits might be increased energy efficiency as opposed to an energy efficiency initiative). This new research model raises new issues for both the government and universities that will inevitably challenge both government and university institutions.

2. What are the positive aspects of large, multidisciplinary, multi-agency initiatives, such as nanotechnology from both a federal science policy perspective and from a university perspective?

Certainly, there are many benefits in large-scale initiatives such as NNI. These initiatives provide the government with a means to focus the attention of the research and university community on a specific (and hopefully important) research area. At the same time, they help focus the attention of policymakers and the public on an area of significant scientific importance.

These initiatives also provide financial incentives that increase the interaction among and within scientific disciplines. Individual investigators often will not venture out of their disciplinary areas unless they have incentives. The NNI, however, is proving to be a catalyst for bringing together people from various scientific disciplines and all parts of campus. The promise of nanotechnology is also bringing people within disciplines together to talk and think of new and innovative research approaches. And finally, the NNI is bringing together researchers from different universities, the federal laboratories, and industry.

The ability of these multidisciplinary initiatives to bring researchers together is something that I witnessed firsthand when working for the University of Michigan. There, I helped organize several workshops to discuss with faculty from across campus how they might pull together a campus team that could develop a university-wide strategy to respond to new government solicitations for not only NNI, but also Information Technology and the Biocomplexity in the Environment initiatives. In conducting these workshops, it was always interesting to watch as two researchers from two different departments who did not know each other speak about their research and learn that they were both involved in similar and related work. History has proven that such cross-pollination of research efforts can have tremendous benefits and, in fact, has resulted in some of the most significant scientific advances.

Initiatives like the NNI also are changing how our *government* works to address important scientific issues. One of the NNI's greatest, yet often overlooked, benefits is that it has prompted government agencies to work together and to focus attention on their *shared* S&T objectives.

Certainly, NNI represents an excellent example of the coordinating mission that was envisioned with the creation of the National Science and Technology Council (NSTC). The NSTC was the multi-agency council established under the auspices of the Office of Science and Technology Policy (OSTP) to coordinate multi-agency S&T needs.

NNI is a valuable model for interagency coordination in other areas. One current example where such coordination could be helpful is in homeland security science and technology (S&T). It is a significant accomplishment that there exists a National Nanotechnology Coordination Office. The existence of this office is tremendously beneficial. Because of these coordination efforts on behalf of nanotechnology, the research community and the broader public can go to one Web site (www.nano.gov) and find out what any government agency is doing with regards to nanotechnology—conferences, agency solicitations, training opportunities, and even a page for kids. Certainly, such interagency coordination would be beneficial were it to be provided in other major areas of national need, such as homeland security, where science and technology can play a major transformational role.

Coordination among federal research agencies offers them the opportunity to periodically check to see if they are on course and to benchmark themselves against other agencies. It is also an efficient way of ensuring that research agencies create complementary, not duplicative or competing, research programs.

Finally, by focusing attention and resources, the government can help to capture the public's imagination and to spark real excitement about this emerging scientific field. This public interest can, in turn, help increase public and government support for key areas of science and increase the number of U.S. students interested in pursuing those scientific fields.

3. What institutional challenges/barriers exist for the government and universities in trying to deal with these new initiatives and are there any negative implications?

Government Challenges

While large-scale research initiatives do force agencies to work together, those efforts can be impeded by the structural and cultural differences that exist across agencies. Thus, even when initiatives work to promote cooperation, the level of coordination and interaction between agencies might not be all that good.

Also complicating efforts to plan and fund multi-agency initiatives is the decentralized nature of Congress. Multiple House and Senate committees are responsible for various agencies responsible for R&D. For example, in the Senate, responsibility for oversight of the Departments of Defense and Energy and the National Science Foundation (NSF) fall to different committees. Unlike the House, there is no overall Senate committee with responsibility for science. Even the House Science Committee does not have jurisdiction over all of the key science agencies involved in nanotechnology. The House Science Committee has no jurisdiction over NIH or over the Department of Defense, both important players in nanotechnology. These agencies are overseen instead by the House Energy and Commerce Committee and the House Armed Services Committee.

This decentralization of oversight responsibility in Congress can make it difficult to reach agreement on the goals and focus of multi-agency initiatives. One way to avoid the problem is through special authorization bills specifically aimed at guiding multi-agency initiatives, as opposed to individual agencies. The obvious example is the 21st Century Nanotechnology Research and Development Act, P.L. 108-153, approved by Congress and signed into law on December 3, 2003. While useful, this initiative-oriented legislation has the potential to come into conflict with legislation approved specifically to authorize entire federal agencies, such as the NIH and NSF. Oversight and authorization by Congress may become increasingly tricky as more and more initiatives are multi-agency based. And it will be interesting to see how both chambers of Congress deal with this fact.

Another organizational challenge posed by Congress's decentralized nature is that funding also is provided by multiple appropriations subcommittees. This has led to situations in which the Administration has proposed funding in several agencies for a particular research initiative, only to have Congress choose not to provide funding in one or more of the agencies responsible for carrying out specific parts of the initiative. Indeed, this happened during the first year of the Information Technology Initiative. As a part of the IT initiative, the Administration proposed $70 million for a new Scientific Simulation Initiative in the Department of Energy, but Congress failed to provide any funding for the DOE portion of the IT effort [3]. Indeed, some members of Congress, failing to recognize the complementary and very different nature of the DOE's information technology and computing efforts, even suggested that the National Science Foundation could do a much better job overseeing these programs. Therefore, they felt, there was no need for DOE to be involved in the IT initiative in the first place.

Another funding-related challenge posed by such initiatives is the perception that they divert funds from traditional and core scientific disciplines. Yet it can be reasonably argued that if such initiatives did not exist, additional funding would not otherwise flow into these disciplines. Perhaps the initiatives themselves help to drive up overall funding for research.

A final challenge for the government is how to measure the effectiveness of these multi-agency and multidisciplinary initiatives. The Government Results and Performance Act of 1993 (GPRA), P.L. 103-62, and, more recently, implementation of the President's Management Agenda announced by President Bush in the summer of 2001, place more emphasis on developing effective metrics to assess the effectiveness of specific government programs, including government-sponsored research programs. To do this, the Office of Management and Budget (OMB) has developed the Program Assessment Rating Tool, also known as the PART. It will, however, be particularly difficult to develop and apply a set of uniform metrics to evaluate large-science initiatives, because they involve multiple agencies with very different missions and objectives.

University Challenges

As with federal agencies, universities have many cultural barriers to overcome in order to respond effectively to multidisciplinary initiatives such as the NNI. Despite growing interest on university campuses in fostering interdisciplinarity and the proliferation of new research centers on campuses to address multidisciplinary research efforts, universities are still primarily organized around traditional disciplines and departments, as are university funding and reward structures. The degree to which universities are able to respond to the NNI will be determined by their ability to overcome traditional disciplinary silos.

For some faculty, multi-agency, multidisciplinary initiatives such as the NNI are not the most attractive research opportunities. Such large initiatives can indeed address problems of scope, size, cost, and complexity that cannot be done in single-investigator projects. But large projects also tend to be more developmental and highly structured and may be perceived as squeezing out the innovation of the individual investigator. Scientists value highly the flexibility to control the direction of their research program. That can be lost in a major research initiative such as the NNI. To help foster a multidisciplinary atmosphere and to be successful in large-scale competitions, universities will need to devote significant amounts of their own time and resources. This required commitment of time and resources might serve as a disincentive for some universities to participate, while others' unwillingness to commit institutional resources might result in their failure to be competitive.

Another barrier to participating in such initiatives might be the scope, size, and complexity of the proposals. This may be particularly true in competitions for nanotechnology centers, where universities are likely to be expected to form significant consortia of both other universities and industry. Moreover, many of these center proposals require K-12 and public education outreach components, activities which faculty often view as having little or nothing to do with their research. Because faculty members either have little interest in these aspects of the proposal or little knowledge about how to build in these components, a significant institutional commitment is often needed to support these activities. Those universities with experience in forming such collaborations and in developing such outreach components are likely to have a great advantage in competing for major awards.

Another university issue is the risk of investing significant institutional resources in an effort that fails to secure a grant. Initiatives often draw significant interest and as a result, success rates for grant proposals are often lower in large-scale initiatives than in other research programs. Faculty and institutions that invest significant amounts of time and resources in preparing unsuccessful proposals can become discouraged and may choose not to reapply.

Finally, and perhaps most significantly, initiatives such at the NNI will ultimately change the focus and fundamental nature of the research conducted at university campuses. They will also impact education and training at universities. Institutions that are successful in competing for large research awards or centers are likely to find that such successes also bring changes in how people work. When funding ends, institutions face the significant challenge of how to sustain such programs beyond the life cycle of the award, which may last only 3–5 years. This is particularly true in instances where large university centers receive their initial support from a center grant. It is important that institutions have a well-formulated exit strategy that ramps down such activities in an orderly fashion, just as they should ramp up in an orderly way.

4. What steps might be taken by universities and the government to help the research community respond effectively to new large-scale and multidisciplinary science initiatives and to maximize the effectiveness of these initiatives?

Although there are multiple challenges for universities in responding to the NNI and other similar large multidisciplinary science initiatives, some steps can help reduce these barriers and encourage more institutions and faculty to participate.

These steps include

- *Providing government and institutional incentives to ensure broad-based participation by researchers from multiple disciplines*—For the NNI to be truly successful, it will have to ensure that there are ample opportunities for funding across academic disciplines and work so that faculty from all disciplines who want to participate are able to do so. Some areas or disciplines might be unintentionally locked out, not because they are not important, but because it is unclear to grantees what they can offer to the initiative. To ensure participation in the NNI from a broad array of disciplines, special outreach efforts might be needed to explain what researchers from specific disciplines—for example, mathematics—can contribute to NNI. Specific funding or other special incentives aimed at attracting individuals from many academic disciplines might also be useful. For example, the NNI's specific emphasis on ethical, legal, social, and economic issues surrounding nanotechnology and funding for research into these areas illustrates the type of efforts that might be undertaken. To encourage faculty participation, universities might provide additional incentives of their own such as release time from teaching or service requirements, shared faculty positions and joint appointments, and additional administrative support or lecture series/symposium grants.
- *Ensuring support for the work of individual investigators and small research teams as well as large nanotechnology centers*—Within the university community, much attention is focused on opportunities for obtaining and competing for large new universities' centers. It is important, however, to ensure that large multidisciplinary initiatives also build in opportunities for individual investigators and small research teams. The agencies involved also need to make sure that faculty located across university campuses and working in many different departments and disciplines are aware of these smaller funding opportunities.
- *Provide planning and seed grant funding*—Because of the cultural barriers at universities that make creating true multidisciplinary efforts difficult, it is often useful to use seed grants or planning grants to pull people together and to focus attention on multidisciplinary scientific initiatives. Although such grants are often provided by institutions, the government should consider providing funding in areas such as NNI specifically for institutional planning activities. Other useful incentives might include small equipment grants or administrative support for faculty willing to plan activities aimed at making the university competitive for grants in a new and emerging scientific fields such as nanotechnology.
- *Encourage partnerships at all levels*—Universities should look not only at their own resources, but also at what industrial and other partners might be able to provide. For example, an industrial partner might help establish a focus for proposals, which might be submitted jointly. Other partnerships might involve other universities, liberal arts and community colleges, and/or federal laboratories.

- *Linking new initiatives to the academic and educational missions of universities*—It is critical that new government science initiatives encourage not only cutting-edge research, but also creation of new courses and curriculum. Certainly this applies to nanotechnology. A university might develop and offer a new undergraduate course reaching non-science majors that focuses on nanotechnology or develop specific new degree programs with a focus on nanotechnology- related fields.

Conclusion

Because of the cultural and institutional barriers at universities, responding to the NNI will not be easy for many institutions. Perhaps that is a good thing, since the very intent of the NNI is to promote investigation into a critical field of science that up until this point had not been viewed in a coordinated fashion, either on university campuses or within the federal government. Initiatives such as NNI are by their nature intended to challenge traditional campus cultures and ways of conducting science.

Thus, the government is exerting its power to focus the efforts of major research agencies and of the scientific community on an area where the potential scientific payoffs are immense. Were it not for the government's work, the scientific community would have tremendous difficulty in organizing such an effort and in overcoming the cultural barriers that characterize the institutional structures of academic institutions.

Certainly, nanotechnology is an area where major scientific advancements are more likely to be realized because of the major initiative that has been undertaken by the government. Among other things, nanotechnology research has the potential to speed computation, increase engine efficiency, improve human health, and result in new lightweight and stronger materials. The strong emphasis placed by the government on nanotechnology through the NNI will help these applications become a reality sooner rather than later.

The importance of the government's role in promoting large research initiatives, such as the NNI, to help to foster new scientific collaborations and advance science was perhaps best summarized by French mathematician, Pierre Louis Moreau De Maupertuis, who stated that

> There are sciences over which the will of kings has no immediate influence; it can procure advancement there only in so far as the advantages which it attaches to their study can multiply the number and the efforts of those who apply themselves to them. But there are other sciences for which for their progress urgently need the power of sovereigns; they are all those which require greater expenditure than individuals can make or experiments which would not ordinarily be practicable. [4, p.276]

References

1. D. T. Moore, Establishing federal priorities in science and technology. In *AAAS science and technology policy yearbook, 2002*, ed. A. H. Teich, S. D. Nelson, S. J. Lita, Washington, DC: American Association for the Advancement of Science (2002).
2. G. J. Knezo, Federal research and development: Budgeting and priority setting issues, 107th Congress, *CRS Issue Brief for Congress,* Washington, DC: Congressional Research Service, The Library of Congress (24 September 2001).
3. M. E. Davey, Research and development funding: Fiscal year 2000, *CRS Brief for Congress,* Washington, DC: Congressional Research Service, The Library of Congress (15 October 1999).
4. P. L. M. de Maupertuis, quoted in C. C. Gaither, A. E. Cavazos-Gaither, *Scientifically Speaking: A Dictionary of Quotations*, Bristol, UK; Philadelphia: Institute of Physics Publishing Ltd. (2000).

NANOTECHNOLOGY FOR NATIONAL SECURITY

Judith Reppy, Cornell University

The reports from previous meetings in this series have amply documented the Department of Defense's interest in nanotechnology. W.M. Tolles writes [1, p.175]

> Contributions of nanotechnology to traditional defense systems will be many. It takes little imagination to elucidate developments that will lead to advanced materials, sensing and signal processing, information technology, battle management, casualty care, or medical procedures and medicines.

The Department of Defense (DoD) has regularly ranked first or second in its share of funding in the National Nanotechnology Initiative (NNI) since the NNI was established in 2000. In FY 2005 it is slated to receive $276 million [2], or 28 percent of the total. Thus, when we think about societal impacts of nanotechnology, our thoughts should turn to military applications. There are three important points to be made concerning the promise and perils of military nanotechnology.

Time Is on Our Side

Because nanotechnology encompasses so many fields, from novel materials to computer chips to biomedical devices, it is difficult and probably unwise to generalize about its likely impacts. Altmann [3] lists many potential military uses. Because most of the R&D in nanotechnology is concentrated on electronics and computer applications and the development of new materials, it is likely that—at least in the first instance—military applications will be embedded in components of existing types of weapons. This suggests that, despite the language celebrating nanotechnology as revolutionary, its impact on military capability is likely to be evolutionary, if only because it takes many years for new equipment to make its way through the development and procurement cycle in sufficient numbers to have a real impact on military capability. While the cumulative effect of improvements in computers and materials across a wide range of military equipment could eventually produce a genuine revolution similar to that brought about by the introduction of the internal combustion engine around the time of World War I, one should not underestimate the difficulty of realizing expected benefits in the military context.

For example, reducing the average load carried by the soldier has been a long-standing Army goal. But despite the availability of lighter-weight materials, loads are still unacceptably high: up to 100 lbs. for some specialties. As the weight of some items has decreased, more items (and the batteries to power them) have simply been added to the pack [4]. Even though advances in nanotechnology, such as those envisaged in the MIT Institute for Soldier Nanotechnologies, may produce more energy-efficient equipment and better batteries, the tendency to add more and more items to the soldier's pack may limit the impact of the new technology on military effectiveness. One notes with some dismay that the first prize in the Institute's 2004 soldier design competition was won by a rocket-launched aerial photography system, an item not currently included in the soldier's load [5]. This is a mundane example of a more general tendency for the promise of new technology to be compromised by the organizational incentives to add more and more advanced features until the original goal is lost from view.

Most Military Applications Will Be Dual Use

A second general observation is that most of the military applications imagined so far are dual use, with civilian uses leading the way in a number of instances (for example, medical applications). While dual-use technologies offer many advantages to the military, they also pose problems. Buying commercial products off the shelf can offer substantial savings, but military requirements often preclude simple adoption of a commercial product. Even in cases where the military has led the way in the development of a technology, a growing civilian market may marginalize the military user. This happened in the semiconductor industry. Originally the industry's principal customer, by the late 1970s DoD's share of the market had shrunk to 7 percent, and it was increasingly forced to pay for custom designs because its requirements diverged from those demanded by commercial customers [6]. Given the multitude of civilian applications driving research and development in nanotechnology, the military may face a similar situation in the market for nanotechnology products.

More important for international security and stability, it is generally not feasible to control the spread of dual-use technology to other users. In the post–Cold War era, controls on dual-use technology are less robust than they were when the Soviet Union and Warsaw Treaty Organization were clearly identified as the enemy. The Wassanaar Arrangement and other multilateral export control regimes on which we currently depend are voluntary, non-binding arrangements that do not include all the countries of interest [7]. In the field of nanotechnology, we must expect that other states—and, indeed, non-state actors—will develop or otherwise acquire military-related nanotechnologies, eventually with only small time lags behind U.S. developments. Although some of these technologies may come under existing export controls, others may not. Absent a strengthened international control regime for these technologies, an action-reaction dynamic is probably inevitable, with knowledge of what is possible fueling competing developments in military applications.

Intersections with International Law

There are some projected military applications of nanotechnology that raise alarm because they may violate accepted standards of international law. I leave for others to speculate on the potential impact of working molecular assemblers or nanorobots on defense or deterrence strategies, should such technologies prove feasible. There are, however, more immediate applications that may cause problems in the near future. The Biological and Toxin Weapons Convention (BTWC) and Chemical Weapons Convention both contain language that prohibits the development of means of delivery of chemical and biological agents for hostile purposes or armed conflict. The BTWC, for example, states

> Each State Party to this Convention undertakes never in any circumstances to develop, produce, stockpile or otherwise acquire or retain:
>
> Microbial or other biological agents, or toxins whatever their origin or method of production, of types and in quantities that have no justification for prophylactic, protective or other peaceful purposes;
>
> Weapons, equipment or means of delivery designed to use such agents or toxins for hostile purposes or in armed conflict. [8]

While the medical uses imagined for nanotechnology clearly are designed for prophylactic and peaceful purposes, a number of the technologies under development to deliver drugs to humans via "smart fabrics" or other nano-scale delivery systems, would appear to be equally capable of delivering harmful agents. (The recent case of the Army patent for a "rifle-launched non-lethal cargo dispenser," in which the cargo is described as projectiles loaded with non-lethal aerosols including biological agents, shows how fine the line dividing permitted from non-permitted use may be. Many experts consider this development to be in violation of the BTWC [9]). In addition, any of the developments that promise very small scale, autonomous weapons systems run the risk of violating the international law of war that requires combatants to be distinguished from non-combatants. Traditionally the distinction has been made clear by requiring combatants to wear uniforms, something that becomes problematic when the weapon is both autonomous and very small. Autonomous devices are in general problematic for international law because they remove human agency from exactly those situations where judgment is most needed. There may be still other developments, such as the projected use of nanotechnologies to enhance the mental or physical performance of soldiers, which could be abused in violation of human rights law, which guarantees the right to freedom of thought, presumably to include freedom from interference with thought processes.

The call to incorporate social and ethical concerns into the NNI deserves a response that goes beyond vague generalities to deal with specific issues and applications. In the case of military applications, there should be mechanisms for legal review to ensure compliance of military developments in nanotechnology with our existing international legal obligations. To the greatest extent possible, this review should be in the public domain, both to provide for critical review by independent experts and to reassure the public that the most sensational fears, as propagated in the popular press, are not realistic. In addition, we should initiate early negotiations to construct an international regulatory regime that would prevent the most dangerous/destabilizing uses of nanotechnology, while protecting as much as possible beneficial civilian uses.

The three considerations listed above contain the seeds of a response to the problems they identify. Taken together, they provide a space for the development of a control regime. The fact that the more troublesome military applications—including the possibility of nanobots—lie in the future provides time; the close connections between the military and civilian technology base provides a degree of transparency; and existing international treaties and regulatory regimes provide a template that can be used to guide policy in this emerging area. Given the pace with which nanotechnology is advancing, however, we cannot afford to be complacent.

References

1. W. M. Tolles, National security aspects of nanotechnology. In *Societal Implications of Nanoscience and Nanotechnology*, eds. M. H. Roco, W. S. Bainbridge, Dordrecht, Netherlands: Kluwer (2001).
2. Office of Science and Technology Policy (news release). National Nanotechnology Initiative, Research and Development Funding in the President's 2004 Budget (2003) (available at http://www.ostp.gov/html/budget/ 2004/2004.html).
3. J. Altmann, Military uses of nanotechnology: Perspectives and concerns, *Security Dialogue* **35**(1), (2004).
4. S. I. Erwin, Lighter load is cyber-soldier dream, *National Defense* (November 1999).
5. MIT, Institute for Soldier Nanotechnologies, Rocket-launched photography system wins first place, (2004) (available at http://www.mit.edu/afs/athena/org/i/isn/newsandevents/designcomp/pressrelease2.html).
6. J. A. Alic, L. M. Branscomb, H. Brooks, A. B. Carter, G. L. Epstein, *Beyond Spinoff*, Cambridge, MA: Harvard University Press (1992).
7. GAO (General Accounting Office), Nonproliferation: Improvements needed for controls on exports of cruise missile and unmanned aerial vehicle technology. GAO-04-493T (9 March 2004).

8. BTWC (Biological and Toxic Weapons Convention) (available at http:// www.opbw.org/).

9. K. Boyd, U.S. Army obtains patent for bio-weapon launcher, *Arms Control Today* (June 2003).

IN DEFENSE OF NANOTECHNOLOGY IN DEFENSE

William Tolles, Consultant

The emergence of nanotechnology as a "field of study" was, in part, stimulated by the "dual use" emphasis within the Department of Defense over the last decade. The term "dual use" was used to emphasize the importance of pursuing research within DoD that was also of importance to the civilian sector. The term was even used for specific programs labeled as "dual use" [1]. Funding levels dropped with time as the program became overtaken with other initiatives (see Table 5.1).

Table 5.1
Funding for the Dual Use Program

Year	Funding Level
1997	$65 M
1998	$68 M
1999	$30 M
2000	$30 M
2001	$30 M

Yet, programs appear to be continuing today (see, for example, http:// www.afrl.af.mil/dualuse/). Variations on this idea emerged a decade ago under the rubric Technology Reinvestment Project (TRP). An analysis of TRP reveals a key sentiment associated with viable DoD research programs:

> Programs such as TRP will attract support from a broader coalition of political forces only if they are embedded in a broader set of public purposes, such as widely perceived threats to health, safety, or the environment, and linked to a set of more rapidly demonstrable results. [2]

Yet, the emphasis for "dual use" stimulated considerable introspection among researchers in DoD, and this was combined with the concepts emerging from nanotechnology to produce powerful arguments for the interdisciplinary field as we know it today. An additional benefit of the emergence of nanotechnology is to challenge academia to set goals broader than a localized subset of problems within a single discipline, and which are related to the larger issues of interest to society as a whole. Priorities for researchers in nanotechnology are to develop appropriate products from the R&D, to develop accurate and credible visions for future thrusts, and to combine all into a package that demonstrates the viability of the program for future support (as long as the critical ingredients are maintained).

The applications of nanotechnology are many. In fact, the term "nanotechnology" is becoming a euphemism for research in the condensed phase, since so many of the phenomena we observe are related to the behavior at nanometer dimensions within the condensed phase of matter. A short list of some of the areas of application appears in Table 5.2.

Table 5.2
A Few Areas of Application for Nanotechnology

Adjustable camouflage	More reliable explosives
Artificial tissues	Mosquito-sized missiles
Cheaper, flexible displays	Nanoassemblers
Chem/bio decontamination	New medicines
Chem/Bio sensors	Robotic ants
Corrosion protection	Self-healing materials
Downloadable memories	Semipermeable fabrics (chem/bio protection)
Drug delivery vehicles	Sensitive reliable IR sensors
Electronics-nerve interface	Smaller, lower power, higher capacity memories
Greater energy from explosives	"Smart" tags
Greater energy from propellants	Stealth materials
High temperature coatings	Stronger coatings
Higher density batteries, fuel cells	Stronger composites
Improved obscurants	Stronger fabrics
Low threshold lasers	Stronger lightweight armor
Magnetic sensors/probes	Swarms
Medical diagnostics (lab on a chip)	Thermal conduction materials
Molecular computers	Thermal protection materials
Monitor life processes (sensors)	Water purification
More efficient solar converters	Wear reduction

A challenge for Defense researchers will be to assess which of these areas is most likely to provide an impact for the agency. Simultaneously weighing the probability of progress with the importance to DoD should provide some guidance for this challenge.

Nanobots

Good arguments can be made for the importance of understanding the behavior of matter through advancing knowledge at the nanometer scale. Just because an application can be envisioned, however, does not mean it may be achievable. In fact, history is replete with attempts to sell and accomplish the impossible. Alchemy represents one of the early well-established attempts to achieve the impossible. The development of thermodynamics gave rise to a firm understanding of the futile efforts to develop perpetual motion as a source of energy. Further, thermodynamics and kinetics represent the basis for material transformations today. Although nearly impossible to make predictions from first principles except for reasonably simple transformations, the tools are becoming increasingly powerful. With these tools, it should be possible to set limits on what may be achievable with such concepts as nanobots. Although not addressed directly by existing works (to the knowledge of this author), some efforts to advance knowledge at our disposal to set such limits should be undertaken. The conclusions would then serve as ammunition to dispel irrational discussion about the subject.

Self-Replicating Species

Considering the visions that have developed and are popular for nanotechnology, there are controversial elements that seem to resemble the alleged promise of perpetual motion a century ago. Science advances by building on a firm foundation, and by advancing the frontiers of knowledge through carefully crafted experiments and theoretical developments. However, some visions emerging that are related to nanotechnology are extrapolations of trends to extravagant limits, or are based on analogies that are most likely not applicable.

The complexities of life are being unraveled slowly, revealing an exceptionally complex set of phenomena at the nanometer scale. The functions of an array of large molecules (or nanostructures) appear to carry out a self-assembly process of extraordinary complexity with a large number of complex chemical reactions taking place in sequence, with material transport that involves specialized structures such as molecular motors, and passage of appropriate materials across cell walls. The fact that such a process can assemble large self-replicating entities, the process we refer to as life, is amazing. Yet, to our knowledge, each step follows the dictates of thermodynamics and the "rules of material transformation" as we understand them today. To pose a program that would produce a material considered thermodynamically infeasible is comparable to making a statement about the incorrect or inapplicable nature of thermodynamics today. As Richard Hamming (once Vice President of Research at Bell Laboratories) would say, "Stand on the shoulders of other researchers, and not on their toes." Although it is difficult to prove, assuming certain elements containing carbon can be mixed together to form diamonds under room temperature and room pressure conditions appears to be counter to what is thermodynamically possible.

Assembling species that are self-replicating represents an exceptionally complex sequence of chemical reactions and nanoscale interactions. The subject is one of considerable interest today [3]. One must assume that nature has almost exhausted the combinations of elements that emerge as life on our planet under today's conditions. That is not to say that variations of life forms that are harmful are not possible. Simple viruses represent self-replicable nanostructures that continue to plague mankind with misery. Any self-replicable species must have sufficient matter in which to build additional members of the species. It seems that carbon-based life on earth represents the only kind of nutrient that is sufficiently abundant to be a source of raw material for a self-replicating species. Further, through millions of years of evolution, with billions of variations in genetic code every day over those millions of years, a very large number of alternatives have been tried. The type of self-replicating species that may be of most concern to us is one that is similar to those we have already encountered, but represents a variation of today's troublesome species (bacteria, viruses, etc.). As mentioned in an earlier discussion of this topic [4], self-replicating species that attack a form of life are self-limiting. If they are too virulent, they eliminate the source of nourishment and become extinct.

Limits to Computational Power

Another vision that appears to have stimulated considerable concern is that of molecular computers that are many orders of magnitude more powerful than silicon computers today. Recognizing the power of the brain to recognize images and to take action in a time far less than that of today's digital computers, a "leap of faith" has generated this vision of super-powerful miniature computational devices. An ant has some degree of logic and memory and represents a very small, low-power device, with fairly crude logical operations. Hence, the alarmist's vision of super-powerful computers attached to self-replicating species, the problem enunciated by Bill Joy [5]. The semiconductor industry has amassed an R&D "army" of major proportions, highly informed of the nature of carrying information among highly specialized devices. Some of the proposed nanodevices may fly in the face of known limits to device technology. Proponents for a vision of highly capable computational devices would do well to arm themselves with the knowledge of the semiconductor industry prior to making pronouncements of a vision built on merely extrapolating trends of an industry.

Elements of a Viable R&D Program

For an R&D program to be viable, it must (1) build on a foundation of advancements having significant impact for a sponsoring agency, (2) develop new ideas making use of emerging knowledge, and (3) develop a credible vision in which it is plausible that emerging knowledge will have a positive impact. Claims that are not credible detract from a program, and could serve to derail the critical need for a stable, well-conceived program. We must assemble what we know about material transformations and attempt to set limits on what is possible, especially in the category of self-replicating species and computational capabilities of computers. With our current knowledge of the materials and information-technology disciplines, it may be possible to assess what is possible, rather than chasing the rainbow and finding no pot of gold.

References

1. Dual Use Science and Technology Program (available at http://www.dtic.mil/dust/).

2. J. Stowsky, The dual-use dilemma, *Issues in Science and Technology Online* (Winter 1996) (available at http://www.nap.edu/issues/13.2/stowsk. htm).

3. S. Rasmussen, L. Chen, M. Nilsson, S. Abe, Bridging nonliving and living matter, *Artificial Life* **9**, 269-316 (2003).

4. W. M. Tolles, National security aspects of nanotechnology. In *Societal Implications of Nanoscience and Nanotechnology*, eds. M. H. Roco, W. Bainbridge, Dordrecht, Netherlands: Kluwer (2001).

5. B. Joy, Why the future doesn't need us, *Wired* (April 2000) (available at http://www.wired.com/wired/archive/8.04/joy.html).

6. PUBLIC PERCEPTIONS

SOCIETAL IMPLICATIONS OF NANOSCIENCE: AN AGENDA FOR PUBLIC INTERACTION RESEARCH

Jane Macoubrie, North Carolina State University

In this position paper, my perspective is that of a communication theorist and social scientist. My current research program applies and builds communication theory relevant to the problem of citizen participation in technology policy making. First, I view increased knowledge of critical factors in group processes' design (communication process design) as a critical research area that can support practical and more satisfactory methods for involving the public in policy deliberation. In addition, I view citizen deliberation, defined as cognitive weighing and evaluation of facts and different perspectives [1, 2], as the best objective for processes for citizen involvement in nanotechnology policy making. Deliberation as an objective is not a new idea; what have been lacking, however, are clear operational definitions and links between deliberation and efforts (such as my own) to test strategic participation designs.

Strategically developed process designs are important, I believe, and may be particularly important for an emergent science area such as nanotechnology. Far too little is known in general about participation processes involving citizens, and certainly—for citizen participation in an emergent science area where knowledge is inherently speculative—carefully constructed and rigorously supported process designs seem critical.

Process designs are also especially relevant to technology policy because it is national-level policy with potentially enormous public impacts. Yet 100 million citizens cannot reason together simultaneously [3], and public deliberation aimed at informed, reasoned consensus-building requires reasoning together, i.e, repeated turns at responsive interaction. These principles mean that group interaction is inherently at the center of citizen deliberation. By applying existing communication theory and conducting new research, for example, my research team has documented that well-designed participation processes can achieve the following [4]:

- citizen consensus versus increased polarization [5]
- significant content-learning by citizens, which in deliberation on GMFs (genetically modified foods) lessened risk concerns
- interesting changes in reasoning as a result of informed participation and dialogue
- effective Internet-mediated processes taking one-third the time of conventional face-to-face processes, with equal satisfaction, learning, and consensus

M.C. Roco and W.S. Bainbridge (eds.),
Nanotechnology: Societal Implications — Individual Perspectives, 241–295.

As Mendelburg [6] warned, all group processes and certainly citizen dialogue have a potential downside. For example, motivated reasoning (a bias towards a solution that biases information intake) may influence citizens; polarization can be worse after group activities; and groups can be influenced by the negative effects of directive (solution-oriented) leadership, as opposed to participation-oriented (evidence-oriented) leadership. Conversely, some groups converge on a false consensus, driven by social pressure, power, lack of motivation and interest, lack of time or information, and other factors.

These points are well-taken. That group outcomes *can* vary is the operative problem, and yet, that concerns such as increased polarization and false consensus may be directly linked to citizen deliberation process is evident, based on existing research. Structured processes are known to be particularly helpful to groups making complex decisions, where any structured process is better than none [7–10]. Keeping social forces from driving poor decisions can be managed, with consent of the group, and given proper management and/or facilitation.

But structured processes are not equal in their ability to accomplish reasoned deliberation and to avert the potential downsides of group behavior, rather than foster a basis for consensual public policy. Three studies investigating polarization in *deliberative political* communication, i.e., using political tasks, found greater *integration* of ideas, not greater polarization of views [11–13]. These results contrast with studies that find greater polarization, including in several public processes using Fishkin's "deliberative polling" method of capturing public opinion [14–16]. Gastil and Dillard considered the possible reasons for the different studies' results, and concluded that the *type of deliberative process in use*, i.e., the prescribed strategy for conducting each study's deliberative forums, may explain the variations in results. In further support of this conclusion, note that in all six citizen-participation processes so far conducted by my research group, all six have not polarized [17]. Clearly, avoiding greater civic polarization over complex issues like nanotechnology would be helpful, and process designs appear to affect that.

Support for a healthy research program investigating critical factors in participation process design and consensus development is thus called for. I also believe that some of the principles I am personally investigating can be applied not only to group processes, but to improved information design and to development of interactive, technology-based solutions for engaging larger citizen groups in active democratic deliberation, perhaps through technology and science centers.

In addition to research on participation processes, if policy makers and scientists wish to work productively with the public on nanotechnology's potential for both positive and other outcomes, research should support development of better public information systems. In our recent research at North Carolina State University, in which we engaged real citizens in deliberations on genetically modified foods, our citizen panels pressed for development of much better public information systems. They viewed current information systems as nearly invisible to the public, and thus inaccessible, and as designed around experts' views of the problem. These systems did not allow the public to get answers to *their* questions.

Public information must be designed to address citizen concerns, and interactive systems that allow citizens to ask and answer their own information questions should be developed. A typology of questions such as that existing in rhetorical theory could be a good framework for designing such a system.

We must undertake research that attempts to see citizen concerns about nanotechnology as combined sets of premises: Research finds that the public sees technology and scientific controversies in general as embedded in social contexts and sets of social implications [17]. Leggett and Finlay concluded that "for our participants, science, technology, society, and environment are all inextricably linked…" and that "to overlay the technical without understanding the substrate is a gamble, and is likely to be unproductive." [17, p.169] Priest's study [18] on public views of technology found that people perceived the risks of biotechnology as higher when seen *within its social context.* To the technology alone, participants in Priest's study attributed lower risks.

Expert arguments from only the expert point of view will not, in other words, meet the public's needs. This also means that we need research that clarifies the way citizens *combine* specific concerns about technological and social issues. Alleviating fears over such radically new technologies as nanotechnology might be addressed by seeing the issues as citizens see them, which then can become the basis for problem solving about actions resolving their concerns, as well as the basis for improved public information. I also note that at the NNI/NSF Workshop on Societal Implications of Nanoscience and Nanotechnology (December 2–5, 2003), Vivian Weil reported a study showing that nanoscientists themselves see the technology in context of its social implications. The possibility that citizens and scientists might actually agree on the general societal implications of nanotechnology deserves much more attention.

To illustrate how citizens might envision societal implications as a result of a deliberative, informed process, consider our 2001 citizen panels on GMFs. These citizens ultimately emphasized three areas of GMFs requiring policy solutions: (1) the potential of *corporate control* of the food supply, (2) the desirability of *genetic* solutions to food problems rather than other alternatives, and (3) the nature of *oversight and tracking* mechanisms for knowing adverse effects. The public interest, they contended, is not served by corporate control of the food supply through control of patented seed and limits on other seed availability. They saw this as a U.S. problem, and as a Third World problem. In the case of GMFs, they also questioned using genetic modification rather than other means of solving the given problem, since its risks are poorly understood. The nature of regulatory processes for scarcely understood technological advances such as genetically modified organisms, and the panel's growing awareness of the state of regulatory control—the apparent lack of central oversight mechanisms and clear lack of systems for monitoring potential adverse effects—concerned the citizens. Without both of these, they reasoned, the risks are both unknown and unknowable, and unmanageable, hence inherently undesirable. They did not reject genetically modified foods, notably, but called for significant advances in management of a technology with such important potential impact.

In summary, understanding the issues around nanotechnology as citizens see them is a critical step that will be ongoing, so that controls and limits may be set that satisfy fears of unintended consequences. Toward this end, additional research must be undertaken on improved public information tailored to audience needs and research-based, tested methods for promoting public deliberative input.

With advances in these areas, public management of emergent technologies may be much enhanced, citizen concerns clearly understood (and hopefully, addressed through creative problem solving), and eroding trust between the public and its government restored.

References

1. S. Burkhalter, J. Gastil, T. Kershaw, A conceptual definition and theoretical model of public deliberation in small face-to-face groups, *Communication Theory* **12** (2002).
2. D. Mathews, *Politics for People: Finding a responsible public voice*, Chicago: University of Illinois Press (1994).
3. M. Walzer, Deliberation….and what else? In *Deliberative Politics: Essays on democracy and difference*, ed. S. Macedo. New York: Oxford University Press (1999).
4. J. Macoubrie, Understanding the public: Citizen logics and technology policy; Internet-mediated deomocratic deliberation (available at http://www.ncsu.edu/chass/communication/www/faculty/faculty_profiles/macoubrie/).
5. R. C. Luskin, J. S. Fishkin, R. Jowell, Considered opinions: Deliberative polling in Britain. *British Journal of Political Science* **32**, 455–488 (2002).
6. T. Mendelberg, The deliberative citizen: Theory and evidence. In *Political decision making, deliberation, and participation: Research in micropolitics 6*, ed. M. X. Della Carpini, L. Huddy, R. Y. Shapiro. Greenwich, CT: JAI Press (2003).
7. R. Y. Hirokawa, Group communication and problem-solving effectiveness II: An investigation of procedural functions, *Western Journal of Speech Communication* **47**, 59-74 (1983).
8. R. Y. Hirokawa, Discussion procedures and decision making performance: A test of a functional perspective, *Human Communication Research* **12**, 203-224 (1985).
9. B. J. Broome, L. Fulbright, A multi-stage influence model of barriers to group problem solving: A participant-generated agenda for small group research, *Small Group Research* **26**, 5-55 (1995).

10. M. S. Poole, Decision development in small groups III: A multiple sequence model of group decision development, *Communication Monographs* **50**, 321-341 (1983).

11. J. Barabas, Uncertainty and ambivalence in deliberative deliberation models: Citizens in the Americans discuss social security forum. Paper presented at the annual meeting of the Midwest Political Science Association, Chicago, IL (2000).

12. J. Gastil, J. P. Dillard, Increasing political sophistication through public deliberation, *Political Communication* **16**, 3-23 (1999).

13. V. Price, J. N. Cappela, L. Nir, Does disagreement contribute to more deliberative opinions? *Political Communication* **19**, 95-112 (2002).

14. D. Denver, G. Hands, B. Jones, Fishkin and the deliberative opinion poll: Lessons from a study of the Granada 500 television program, *Political Communication* **12**, 147-156 (1995).

15. J. Gastil, Democratic education and the National Issue Forums. Unpublished doctoral dissertation, University of Wisconsin-Madison (1994).

16. R. C. Luskin, J. S. Fishkin, R. Jowell, Considered opinions: Deliberative polling in Britain, *British Journal of Political Science* **32**, 455-488 (2002).

17. M. Leggett, M. Finlay, Science, story, and image: A new approach to crossing the communication barrier posed by scientific jargon, *Public Understanding of Science* **10**, 157-171 (2001).

18. S. H. Priest, Information equity, public understanding of science, and the biotechnology debate, *Journal of Communication* **45**, 39-54.

COMMUNICATING NANOTECHNOLOGICAL RISKS

David M. Berube, University of South Carolina

Implications involve consequences. Consequences have probabilities—all of which involve risks. The National Nanotechnology Infrastructure Network and the Center for American Technological Preparedness funded within the 21st Century Nanotechnology Research and Development Act of 2003 will need to dedicate a portion of its mandate for communicating nanotechnological risks. *nano*STS at the University of South Carolina (USC) has begun to do just that.

Determining Acceptable Risks

Risk is incredibly difficult to define because it is ubiquitous. Our world is simply a risky place in which to live. We decide whether or not to act based on our willingness to take the chances associated with action or inaction. While it would be foolish, if not deadly, to engage in an activity that was certain to threaten life, it might not be foolish to engage in the activity if it carried a lower probability of being life-threatening.

There are many theories on risk but the one that has recurred in debates over environmental matters has been the precautionary principle. While much more popular in European circles, it has appeared in veiled forms in American legislation and regulation.

Generally, the precautionary principle calls for inaction if there is an alternative that carries a lower likelihood of occurrence at similar benefits. A lower probability of cost given similar estimates of benefits would justify relinquishing a threatening choice for another that would be much less threatening.

In general, arguments against this principle involve a laundry list of concerns. Risk estimates are context-specific and there might be a plethora of reasons why acceptable risk at a time and in a place might be higher than it would be in a more utopian setting. Risk estimates tend to reflect the ideology of the estimators. If those willing to take the risky behavior are not the same as those exposed to the consequences, then estimates might be highly divergent. Risk can be counterbalanced by other variables. Individuals may be willing to expose themselves to a high level of risk if sufficiently compensated. However, given the tragic disparities of wealth in our society, what might constitute fair and equitable recompense might be unfair and inequitable to another of means.

Risk Assessment

Calculating risks associated with events seems relatively easy given a sufficient database. Risk has been defined as the product of probability and loss (see Fig. 6.1). Determining probability requires a tabulation of similar or identical events over a sufficient amount of time. Loss is actually the sum of benefits and costs associated with the event [R = Risk; Pr = Probability; P = Preparedness (loss mitigation)].

$$R = \frac{\Pr(\text{loss})}{\text{Preparedness}}$$

$$R = \frac{\Pr_1 L_1 + \Pr_2 L_2 + \ldots \Pr_n L_n}{\text{Preparedness}}$$

Figure 6.1. Risk Assessment.

Assessing risks for common events tends to be easier given the large database. Assessing risks for uncommon events, like accidents, is more problematic given the accidental nature of such an event. Its unpredictability vitiates probability estimates.

Data from natural hazards tend to be more reliable than data from technological hazards because the history of natural events is longer. Technological hazards have another problem: few are like others. Even an oil spill from a tanker (a fairly predictable event) is affected by an array of contextual variables.

Assessing technological hazards associated with hypothetical technologies is highly problematic since there is no database. While extrapolated values can be interpolated, they remain suspect.

Risk Communication

> One aspect of risk communication is figuring out how to scare people.... The other component of risk communication is figuring out how to calm people down again.... The risks that kill people and the risks that upset them are completely different. [1, p.7]

So, what is risk communication? "Risk communication encompasses the transmission of information among individuals and organizations in all segments of society, including government, industry, academia, and the general public" [2, p.52]. Also, "risk communication includes *all* messages and interaction that bear on risk decisions" [1, pp.1-2].

Risk = Outrage + Hazard

$$R = f(H,O)$$

Figure 6.2. Risk Communication.

Risk communication is not risk assessment. "The public cares too little about the hazard, and the experts care too little about the outrage.... They therefore tend to overestimate the risk when the hazard is high and the outrage is low, and underestimate the risk when the hazard is low and outrage is high" [1, pp.1,2].

How the message is crafted (who writes it and how it is written), how the message is communicated (face-to-face, public forums, newspapers, and television), and how the message is decoded (highly objective assessment, highly emotional fear-driven assessment) will impact the communication and its overall effect.

The determination of risks is challenging because assessment calculi are complex, sometimes baffling:

- Assessment is affected by the tolerance of reactions, meaning that some catastrophic events can be absorbed by the victims and the environment more than other events. For example, a release of nuclear radiation in a remote setting is less consequential than in a highly populated area.
- Levels of volition aggravate exceptions, meaning voluntary risks are self-imposed and have less value than involuntary risks. For example, if smokers want to kill themselves that is their decision, but when second-hand smoke affects others, that decision is not theirs to make.
- Indirect effects appear less concrete than direct effects, meaning that the immediate consequences of a disaster receive the most notice while longer-

term secondary consequences are often ignored altogether. For example, an accidental release of oil into an ecosystem carries long-term effects that may not show up until they have been metabolized along the food chain.

- Gains following apparent negative events are discounted; disasters may have advantages. For example, a major fire may clear away entire neighborhoods that allow development that may produce more benefits for more people.
- Population endurance is undervalued, meaning some populations are more resilient than others. For example, someone with a reduced immunity might be more highly vulnerable to a biotoxin than another with a more aggressive immunity.
- Acceptability variables are terribly indeterminable, meaning some events are acceptable when compared to alternative events. For example, do the ozone depletion consequences of fossil fuel energy production trump the effects of nuclear fission energy generation?
- Extreme event analyses often have insufficient databases, meaning catastrophic occurrences are so infrequent that any reasonable database of occurrence is non-existent. For example, the probability of the explosion of a small nuclear weapon in an urban environment has never occurred and while we may extrapolate data from Hiroshima or Nagasaki, those data are contextually vulnerable.
- The durability of assessments is exaggerated, meaning we cannot assume future events will be addressed by current solutions. Rather, they will be addressed by future, often undetermined solutions. For example, the inhalation of nanotubes in a production facility will be addressed by technologies yet to be discovered or invented.
- Bounded rationality overemphasizes quantitative variables, meaning it is often difficult to include qualitative values into an assessment calculus. As such they are either discounted or ignored altogether. Qualitative variables, such as dread and outrage, make each calculus problematic; e.g., school shootings convinced parents children were especially vulnerable when schools are statistically very safe places.
- Expressed preferences from opinion data are continually suspect, meaning that decisions on the probability or impact of events drawn from surveys and polls are unreliable because intellectualizing about a disaster produces less reliable data than responses generated from an event. For example, asking people (in a survey) how they would react to storing nuclear waste in their town will produce different results than asking people how they feel about the waste stored in their town.
- Hazard response options are ideologically ignored, meaning some responses are determined unacceptable due to the current ideological direction of the assessment teams. For example, evacuating a city in the case of a chemical attack is only impossible given our reluctance to declare martial law except under extreme circumstance.
- Sabotage and terrorism as agencies are discounted, meaning that in a post-911 world a new variable must be included in disaster assessments and in risk

perception calculations. For example, once we learned that terrorists would be willing to sacrifice their own lives to complete a mission, the textbooks on responding to a hijacking situation changed entirely.

Risk assessment is a computation; it does not translate into risk communication once the assessment is shared. The audience, often the general public, does not translate objective data into risk perception. As such, they tend to exaggerate events that are improbable and fear phenomena that are highly improbable, especially if the event has a high impact. Many people are becoming more and more concerned about smaller and smaller risks… People pay more attention to the size of the consequences and ignore both the magnitude and the uncertainty of very low probability estimates. The result would be a much-increased concern about catastrophic risks and a corresponding increase in opposition to technologies that pose them [3]. Indeed, low-probability, high-consequence events, called mini-max, are perceived as far more probable than they should be. Nonetheless, if we are concerned with communicating risk, we must factor in this fallacious reasoning.

In general, risk communication has the following problems:

- Perception and assessment should not be confused—meaning dread and outrage, while not necessarily logical reactions, remain variables in any risk communication strategy.
- Message credibility is affected by a plethora of variables:
  - real or perceived advocacy of the source
  - reputation for deceit, misrepresentation, or coercion
  - inconsistent advocacy
  - self-serving framing of the message
  - contradictory messages and refutation from other sources
  - actual or perceived professional incompetence or impropriety
- Perceived legitimacy of the process is affected by additional variables:
  - legal standing of the source to the risks addressed
  - justification provided for the communication program
  - access afforded to affected parties in the decision making process
  - degree to which conflicting claims are given fair and balanced review
- Improbable events tend to be exaggerated when the following circumstances occur:
  - The event impact is monumental. This explains why travelers are more concerned about air safety than road safety.
  - The event has special proximity, meaning an event is more important if its setting is nearer in time or space.
  - The event impacts especially vulnerable populations, such as the very young and the very old or those already suffering from other impacts.

- The event involves an involuntary risk, meaning there is more risk associated with events about which we have no input or choice.
- The event is exotic, meaning disastrous occurrences the audience has never experienced might appear more risky.
- The event is associated with governmental action, meaning people have lost much confidence in government as a product of a series of failures and media coverage of them.
- The event is heavily covered by the media, meaning the more an event is newsworthy, the more likely it will be determined to be a risk.

Technological hazards have another set of challenges. For example, public knowledge is limited regarding many scientific fields, especially a subject as exotic as nanotechnology. "Opinion surveys and tests of U.S. students' knowledge show that public understanding of science and technology is weak. Even Americans with advanced training in non-scientific fields often know little about the revolution in biology or the amazing new materials being produced in laboratories" [1, p.7]. Indeed, if persons actually attempt to read or learn about science and the risks associated with it, they often have "limited access to expert opinion leaders to help interpret scientific and technical information" [4, p.768].

Technological risks have additional problems:

- Their accidental nature, meaning technology is presumed safe if it is marketed and the public assumes industry will avoid liability by only marketing safe applications. But these disasters are unpredictable because they are the product of a set of highly unlikely variables, e.g., a meltdown of a nuclear reactor requires the occurrence of a series of unlikely events, including the failure of multiple precautionary measures.
- Varied duration of effects, meaning there is a series of effects from the initial occurrence to long-term implications. But the long-term implications would be addressed by responses yet to be determined, e.g., how a release of polychlorinated biphenyls will affect future generations will be affected by advances in gene therapy.
- Unclear and unproven response strategies, meaning how we respond to technological disasters may involve a set of highly technical solutions that may only have been considered hypothetically. For example, during the Apollo 13 conundrum, the ground crews in Houston found themselves designing means to reduce energy consumption and reduce carbon dioxide build-up by creative problem solving.
- Their inherent inexactness, meaning that there are no data to construct the true dynamics of a technological disaster, which often means we defer to outlandish and exaggerated impacts. For example, the likelihood that nanobots will consume the ecosystem and turn our blue and green planet into a grey sphere does not encapsulate the true consequences of inadvertent release into the environment [5].

Clearly, emerging technologies like nanotechnology will have many implications, both favorable and unfavorable. Determining what they may be might be a very interesting academic exercise, but reporting the findings, and communicating them to all parties, both technical and public, involves communicating risks, something we have not done well in the past (drug abuse, alcoholism, tobacco addiction, safe sex, and communicable disease). This paper has highlighted some of the hypotheses and theories that *nano*STS at USC has begun to study and evaluate.

References

1. P. Sandman, Responding to community outrage: Strategies for effective risk communication, Fairfax, VA: American Industrial Hygiene Association (1993).
2. A. J. Liston, Risk communication and health protection. In *Prospects and problems in risk communication*, ed. W. Leiss. Ontario: University of Waterloo Press (1989).
3. National Research Council. In *Improving Risk Communication*, Washington, D.C: National Academy Press (1989) (available at http://www.nap.edu/catalog/1189.html).
4. S. Hornig, Science stories: risk, power, and perceived emphasis, *Journalism Quarterly* **67** (Winter), 768 (1990).
5. K. Smith. In Environmental hazards: Assessing risk and reducing disaster, London: Rutledge (2001).

A PROPOSAL TO ADVANCE UNDERSTANDING OF NANOTECHNOLOGY'S SOCIAL IMPACTS

D. S. Hope and P. E. Petersen, Rochester Institute of Technology

In order to advance the understanding of the ethical and social impact of "nanotechnology," we believe it necessary to first assess the status of the term in public usage. *Nanotechnology* is a common term in scientific and engineering communities and in popular and mass communication discourse. Most technical and scientific disciplines sponsor topical nanotechnology conferences [1–6]; there are a growing number of research centers for the study of nanoscience and nanotechnologies [7–9]; and nanobusiness opportunities abound [10–12]. Additionally, press coverage of nanotechnology stories has multiplied in number and narrative over the past 10 years [13, 14] and visualizations of nanoimagery adorn the cover of magazines and permeate the Internet [15, 16]. Science fiction and fantasy based on nanoscience, (e.g., M. Crichton's *Prey* and Kathleen Ann Goonan's four-book series) are increasingly popular [17, 18], and nanotechnology advertisements appear on TV as in the Hewlett-Packard "Everything is Possible" campaign ad titled Nanotechnology [19]. William Safire points out that "nano" has entered the language as the common prefix for denoting that which is extremely small or extremely short in duration, as in *nanosecond*: "With its prefix rooted in the Greek *nanos,* 'dwarf,' that word zipped past millimicrosecond in 1958 as the metric system became dominant, to become the word that filled the desperate need for 'billionth of a second'" [20, p.18]. As the term gains wide usage in scientific, technical, and popular discourse, there is a corresponding need to evaluate its meanings.

Many definitions exist in both the popular press and the technical literature. The National Nanotechnology Initiative (NNI) defines nanotechnology as having these constituent parts [21]:

1. Research and technology development at the atomic, molecular, or macromolecular levels, in the length scale of approximately 1–100 nanometer range.
2. Creating and using structures, devices, and systems that have novel properties and functions because of their small and/or intermediate size.
3. Ability to control or manipulate on the atomic scale.

Yet, a widely accepted general definition does not exist. The reason a precise and generally accepted definition has not emerged is that "nanotechnology" as a concept is extremely broad—too broad for a single person to fully comprehend and be fully conversant in, too encompassing for a single discipline to totally cover, and too ambiguous for the general population to grasp with any hope of entering genuine public debate. As James Surowiecki wrote for the *New Yorker's* financial page, "The very fact that nanotech is so bewilderingly broad in its potential applications—from window-cleaning to chemotherapy—is what explains the hype, of course" [22, p.68].

Researchers in chemistry, biology, physics, and engineering are developing new knowledge and understanding of diverse "nanotechnologies" for applications in medicine, robotics, and environmental control, weaponry, communication, and consumer products. Even if we segment nanotechnology into its constituent parts, there are but a few persons who are truly knowledgeable in any one segment. The problem of meaning goes beyond research needs. For example, New York State Attorney General Eliot Spitzer was asked to investigate Merrill Lynch for misuse of the term and possible fraud:

> What exactly is nanotechnology? The definition is no longer academic as more investors become attracted to anything that carries a nanotech label. On Thursday, Asensio & Company, an investment firm, faxed a letter to Eliot Spitzer, the New York attorney general, charging that misuse of the nanotechnology label has become a favorite tactic for fraudulent stock promotion. [23, p.C2]

Despite the plethora of information that floats about via the Internet, popular press, and technical literature, there is no easy way, if there is one at all, to assess the credibility of the available information, cautions, or forecasts or to evaluate the meaning of "nanotechnology" for multiple publics. The confusion and misuse of the term *nanotechnology* can and will negatively impact education, funding, development, and public debate. For example, there is a clear need for the development of standardized research criteria to evaluate the potential benefits or risks of nanotechnology. Yet the use of the term to describe technological advances in such disparate fields as biological systems, electronics, materials, quantum phenomena, fluidics, sensor systems, and the like prevents rigorous development of standardized methods of risk assessment or potential benefits. Even if we segment nanotechnology into its constituent "nanotechnologies," the necessity for interdisciplinary research and development remains crucial. Social scientists and citizens approach nanotechnologies from differing ethical concerns than do scientists, technologists, or profit-seekers. The accessibility of information in terms of the realities and demands of *diverse nanotechnologies* at the same time that interdisciplinary projects go forward is necessary. To address this situation we need to consider (1) the best direction for research practices relevant to the social impacts of nanotechnologies, and (2) the need to make accessible to multiple audiences information required for diverse purposes, including the promotion of public discussion.

Relevant Research

To understand well the potential social impacts of nanotechnologies, it is necessary to conduct research directed toward that end. Such research requires multidisciplinary teams of scientists, social scientists, ethicists, technologists, and citizens who are not vested in the process of research or product development. The technologies have developed to the extent that it is now desirable to perform research in specific applications in the context of ethical and societal implications. For example, working to determine the potential environmental impact of all of nanotechnology would be difficult at best and probably meaningless at worst. More relevant and useful are research studies designed to investigate the impact on the environment of a specific subset of nanomaterials, their dispersal, their waste, and the resulting social impacts as determined from multiple viewpoints. Examples of such research include, but are not limited to, the following:

- the impact on community relationship to natural environments
- the impact on democratic processes
- the impact on diverse cultures and populations
- the impact on groups of varying affluence
- the impact on future workforce requirements

Such an interdisciplinary approach necessitates the maintenance of genuine working relationships between research groups and a continual feedback loop between and among researchers.

Funding for interdisciplinary projects rests on the productivity of interdisciplinary teams of researchers. Pragmatically, a problem exists in the ill-preparedness of researchers in the various disciplines to work together. Sometimes this problem is due to different paradigms; in other instances, scholars and researchers are narrowly trained and struggle to overcome rigid disciplinary barriers. Still other problems exist within the hierarchies of scholarly systems, promoting an unequal distribution of resources and a consequent lack of cooperative initiatives. These problems indicate the pressing need for reform in the educational process, especially as education impacts the social and ethical considerations for future sciences and technologies, research and development of applications and products, and citizen concerns, e.g., all associative issues surrounding nanotechnologies.

The *Nanonews Now Monthly Report* states, "the people charged with making policy in this area [nanotechnology] often lack a complete understanding of the most fundamental concepts of nanotechnology" [24].

The point is that the leaders and decision makers in industry, government, and academia need to be well-versed in the technologies, their potential applications, and resulting potential benefit and harm. Too often these leaders receive ad hoc briefings and committee reports that may have a built-in, but unintended, bias. It would be very useful to have special balanced modules readily available for these leaders. Likewise, scientists and technologists need more training opportunities in the methodologies and foundational principles of the social sciences and humanities. When scientific researchers are uninformed as to the effect of sociological theory and data on cultural attitudes towards technological change, productive investigation of the social impacts of specific applications of nanotechnologies is impeded. Philosophical, rhetorical, and political presumptions apparent in historic reactions to technologies need to be considered as research is presented in public discourse and circulated in the mass media. While restructuring the entire educational system is utopian, interdisciplinary modules can be created and distributed for all levels of education and research. Such modules might be created by teams of scholars funded by NSF in cooperation with centers of academic learning and should be accessible to multiple seekers of nanotechnology information.

System of Accessibility

Both the promises and the perils of nanotechnologies are oftentimes overstated or underestimated. In the wealth of published reports, opinions, and technical developments, facts merge with fictions in press reports as well as in visualizations. Consequently, users of such information are not fully informed, and many are misinformed. To address this situation we need to develop a systematic approach to make information available to the many segments of the population that seek information about various nanotechnologies. We believe that the field of nanotechnology is now sufficiently developed such that it is necessary and possible to address in a systematic way the specific subsets of nanotechnology one by one. Each of these subsets presents its own set of potential and realized benefits with an attendant set of risks, known and unknown.

A search of the Web reveals the problem. Using a popular browser (Google), a search on "nanotechnology" resulted in 1,340,000 Web pages; another (MSN), 244,565 pages; and another (Excite), 106 sites. Restricting the search to "social change and nanotechnology" resulted in 83,700 pages; "nanotechnology and ethics" 47,200; "nanoelectronics" 64,600; and "nanobiotechnology" 21,500. Obviously the interested novice and perhaps even the expert can be easily overwhelmed. In many instances the credentials of the site's author and the date at which the information was posted are not easily accessed.

Significant efforts have been developed to chronicle nanotechnology information. Perhaps the most noteworthy is the database site by Loyola College in Maryland [25] that was supported by NSF from 1999 to 2002. The University of South Carolina under NSF sponsorship is developing a database of abstracted materials in nanotechnology [26]. However, even with these excellent resources it would be difficult for an interested novice to navigate. We propose for consideration the development of a system-like approach that can be interactively queried by a range of interested persons.

This system-like approach would begin with a set of questions for the person querying the database. The first set of questions would query about the type of information that is desired. For example, is information desired about a specific technology like nanoelectronics, nanobio, nanomaterials? Or is information desired about the applications of specific aspects of nanotechnology as computation, medicine, environmental impact, etc.? Or is information required about economic aspects, public policy, and the like? The second set of questions then would seek information about the viewer, such as education level (K–4, 5–8, 9–12, undergraduate, advanced degree, discipline); knowledge base (novice, generalist, expert); and finally, years and type of experience in the field. For each of these categories it is necessary to understand what kind of information is sought and the general uses that the consumer desires for that information.

To develop such a system it would first be necessary to form a multidisciplinary and eclectic team to develop the appropriate set of questions. The system, of course, should also allow ready access to the extensive set of knowledge workers and a catalog of research projects and ongoing investigations. It is desirable to have all information clearly dated and the authors and their credentials clearly established. Finally, although increasingly difficult, it would be desirable to rank each site as to content and usability relative to the purpose of the person making the query. This could well be a Web-based system. Such a system is, of course, difficult to build and would need to evolve over the years. We believe that NSF could be a potential candidate to support the development of this system.

There are, of course, excellent programs working to develop effective user-responsive databases. The work by Professor Shneiderman and colleagues at the Human–Computer Interaction Laboratory [27] is an outstanding example. Such knowledge and expertise would need to be utilized prior to moving forward with the proposed approach.

In advancing this system of accessibility, it is very important to chronicle the information relative to possible social impact, both positive and negative. The effort required would be overwhelming for a single center. Rather it would be more effective to establish a national network of interactive regional centers for the study of the ethical and societal implications of nanotechnology. These centers would investigate the topic areas discussed above, but with specific emphasis on the social impact on their specific region. A non-bureaucratic superstructure would be useful to help ensure that needless duplication does not occur, yet that there is sufficient overlap to ensure effective communication. It is, of course, important for these centers to have a mechanism for interaction on a regular basis.

Combining our approach to relevant research and a system of access would enable bringing an understanding of specific aspects of nanotechnology to the diverse groups of persons desiring information and knowledge about nanotechnology, i.e., K–4, 5–8, 9–12, undergraduate, graduate, working professionals, the public at large, and finally leaders in government, academia, and industry. Each group requires a different set of data, each group requires a different vocabulary, and each group needs a way to consistently redefine itself as knowledge is evaluated. Again the list of questions should be developed by a group of knowledgeable experts in educational systems, communication, social systems, and the technology.

With this information, appropriate learning modules could be developed. It is well-known that for each learning category there are multiple learning styles. Consequently, no one module will work for all; several modules for each category, taking into consideration the different learning styles, will need to be developed. There are excellent modules available today:

- NanoKids™ developed at Rice University [28]
- the special K–12 programs that have been developed by several research groups, for example the "Exploring the Nanoworld Program" developed at the University of Wisconsin–Madison [29]
- the CD developed by the Institute of Nanotechnology [30] that could be used, incorporated and possibly expanded

Finally, the new undergraduate initiatives being developed under NSF leadership as the Nanotechnology Undergraduate Education (NUE), Nanotechnology Instructional Material Development (NIMD), and National Institute for Science Education (NISE) programs provide an excellent resource.

Above we have described what we believe is a useful concept to advance the understanding of the ethical and social impact of the constituent elements of nanotechnology. The next step in implementing such a concept would be to form a multidisciplinary planning and evaluation team to develop the concept more fully and develop a road map for future work.

References

1. Fourth IEEE Conference on Nanotechnology, 17-18 August, Munich, Germany (2004) (available at http://www.nano.ei.tum.de/ieeenano2004/).
2. NSTI (Nano Science and Technology Institute) Nanotechnology Conference and Trade Show, 7-11 March, Boston, MA (2004) (available at http://www.nanotech2004.com).
3. ASME 3rd Integrated Nanosystems Conference, 22-24 September, Pasadena, CA 2004) (available at http://www.nanotechnologyinstitute.org).

4. International Conference on Nanospintronics Design and Realization, 24-28 May, Kyoto, Japan (2004) (available at http://www.dyn.ap.eng.osaka-u.ac.jp/ICNDR2004/).
5. NIH Nanomedicine Initiative: Project Launch Meeting, 4 May, Bethesda, MD (2004) (available at http://www.capconcorp.com/roadmap04/).
6. Imaging and Imagining Nanoscience and Engineering Conference, 3-7 March, USC (2004) (available at http://www.cla.sc.edu/cpecs/nirt/conferences.html).
7. Center for Biological and Environmental Nanotechnology, Rice University, Houston, TX (available at http://www.ruf.rice.edu/~cben/).
8. Nanoscale Science and Engineering Center: Northwestern University, Evanston, IL (available at http://www.nanotechnology.northwestern.edu).
9. Center for the Directed Assembly of Nanostructures: Rensselaer University, Troy, NY (available at http://www.rpi.edu/dept/nsec/).
10. 3rd Annual NanoBusiness 17-19 May, New York, NY (2004).
11. Nano Business Partnering Forum, 17 March, Tokyo, Japan (2004) (available at http://www.prweb.com/releases/2004/3/prweb110751.php).
12. NanoBusiness Alliance, New York, NY (available at http://www.nanobusiness.org).
13. L. F. Stephens, News narratives about nano: How journalists and the news media are framing nanoscience and nanotechnology initiatives and issues. Paper presented at Imaging and Imagining Nanoscience and Engineering Conference, USC, 3-7 March (2004) (available at http://129.252.169.87/rstephens/stephensnanomsUSC Mar04.pdf).
14. B. V. Lewenstein, J. Radin, Media images of nanotechnology. Paper presented at Imaging and Imagining Nanoscience and Engineering Conference USC, 3-7 March (2004).
15. A. Anderson, S. Allan, and A. Petersen, Pictorial and textual representations of nanotechnology in the UK press. Paper presented at Imaging and Imagining Nanoscience and Engineering Conference USC, 3-7 March (2004).
16. D. S. Hope, Seeing the invisible: The compelling rhetoric of nanoscience imagery. Paper presented at Imaging and Imagining Nanoscience and Engineering Conference USC, 3-7 March (2004).
17. M. Crichton, *Prey*, New York: HarperCollins Publishers (2002).
18. K. A. Goonan, *Queen City Jazz,* New York: TOR Trade (1994); *Mississippi Blues,* New York: Avon (1997); *Crescent City Rhapsody,* New York: Avon (2000); and *Light Music,* New York: HarperCollins Publishers (2002).
19. Hewlett-Packard, "Everything is possible" television campaign (available at http://www.hp.com/hpinfo/newsroom/hpads/everything/tv.html).

20. W. Safire, Attaboy, attosecond! Prefixating on the large and small, *New York Times Magazine*, 18 (7 March 2004).

21. NNI definition of nanotechnology (available at http://nano.gov).

22. J. Surowiecki, Bring on the nanobubble. *The New Yorker*, 68 (15 March 2004).

23. B. J. Feder. Concerns that nanotech label is overused, *The New York Times*, C2 (12 April 2004).

24. NanoNews Now Monthly Report, April (2004) (available at http://www.nanotech-now.com/products/nanonewsnow/issues/010/010.htm#Quotes).

25. Nanotechnology Database, Loyola College in Maryland (available at http://www.wtec.org/loyola/nanobase/alt/frmwelcome.htm).

26. Nanoscience Abstracts, University of South Carolina Database (available at http://www.cla.sc.edu/cpecs/nirt/nanodb.html).

27. Human–Computer Interaction Laboratory, University of Maryland (available at http://www.cs.umd.edu/hcil/).

28. NanoKids™ (available at http://nanokids.rice.edu/).

29. MRSEC Interdisciplinary Education Group, Exploring the nanoworld, Univ. Wisconsin–Madison (available at http://mrsec.wisc.edu/edetc/).

30. The Institute of Nanotechnology: What is nanotechnology? (available at http://www.nano.org.uk/cd.htm).

NANOTECHNOLOGY IN THE MEDIA: A PRELIMINARY ANALYSIS

Bruce V. Lewenstein, Joanna Radin, and Janie Diels, Cornell University

One component of modern society is the constant attention to "emerging technologies." From computers to biotechnology to genomics, excited rhetoric about "revolutionary changes" tied to emerging technologies seems to fill the discourse space—or so it seems to those who are already paying attention to those technologies. The classic description of the media's effect on public debate suggests that the media "may not be successful much of the time in telling people what to think, but it is stunningly successful in telling [people] what to think *about*" [1, p.13]. This "agenda-setting" role of the media has been demonstrated in a range of emerging technologies, most recently biotechnology [2].

One of the newest "emerging technologies" is nanotechnology [3]. The very definition of what constitutes nanotechnology is still in contest, but it generally covers research and development of materials and processes that are best understood at the nanometer scale—a billionth of a meter; a few atoms across; 100,000 times smaller than the thickness of a human hair. The canonical history of the field traces itself to a 1959 speech by physicist Richard Feynman entitled "There's Plenty of Room at the Bottom" [4], but much of the current popularity of nanotechnology research dates to K. Eric Drexler's 1986 book, *Engines of Creation* [5]. In that book, Drexler outlined a vision of molecular machines that would revolutionize manufacturing, materials, health care, information processing, and virtually all aspects of the material world. Research in the area accelerated in the 1990s, a federal government-wide coordinating committee was created in 1998, and in January 2000, President Bill Clinton announced the National Nanotechnology Initiative (www.nano.gov). On 3 December 2003, President George W. Bush signed into law the 21st Century Nanotechnology Research and Development Act, authorizing $3.7 billion in funding for 2005–2008.

Among elite audiences with an interest in science, nanotechnology is well-known (see, for example, the special issue of *Scientific American* that appeared in September 2001 [3]). Much of the excitement within that community is for the technological possibilities of nanotechnology. But even in his 1986 book, Drexler highlighted social and ethical issues associated with nanotechnology, and attention to those issues has been amplified by environmental and toxicological concerns as research progresses. In a *Wired* article in 2000 that is widely cited within the community of experts paying attention to nanotechnology, Sun Microsystems chief scientist Bill Joy questioned whether the benefits of nanotechnology outweighed its risks [6]. In early 2003, a Canadian non-governmental organization that had been active in opposition to genetically modified foods issued a report questioning the safety of nanotechnology [7]. By early 2004, research questioning the safety of some forms of nanotechnology was beginning to appear [8].

But despite these debates in the expert community, and despite the need for political support to generate appropriations, wide public attention to nanotechnology appears to have been slim. The label "nanotechnology" has been used in some pop culture books and movies (most notably, perhaps, the "Terminator" series starring Arnold Schwarzenegger, now the governor of California). And in November 2002, writer Michael Crichton, author of *Andromeda Strain*, *Jurassic Park*, and other science-based thrillers, released *Prey* [9] in which rogue nanobots threaten to take over the world. 20th Century Fox has purchased rights to the film; no release date has been set.

The elite community has responded to emerging concerns about nanotechnology with fear. Leaders of the nanotechnology community do not want their work to be "derailed" by what they see as "unfounded" public concerns—and they alternately cite and deny any comparison with the case of genetically modified foods. To forestall some of the concerns, the nanotechnology community has long supported work on "social and ethical issues" in nanotechnology [10, 11].

But at this time, it is simply not clear whether the discussions about emerging technologies that take place among elite participants are in any way connected with discussions taking place in the wider public world. This study is a first attempt to explore those connections. To assess the place of nanotechnology in the media, we conducted a preliminary content analysis of three elite media outlets (*New York Times, Washington Post, Wall Street Journal*) and one general media outlet (the Associated Press) for the period 1 January 1986 to 30 June 2003.

Methods

This study presents preliminary analysis of media coverage of nanotechnology. The methods and analysis are modeled on earlier studies of media coverage of other emerging technologies, such as biotechnology. Articles for the analysis were identified in the Lexis-Nexis electronic database and the *Wall Street Journal*'s online index, searching for all articles with "nano*" in the title or text for the period 1 January 1986 to 30 June 2003. The start date was chosen to include everything since the publication of Drexler's *Engines of Creation*; the end date was purely practical, as the original research was conducted in August 2003.

We used an analytical approach developed by a European team that has been looking at public interactions with biotechnology since the mid-1990s [12–14], which itself drew on sociological analysis of public debate about nuclear power [15]. The approach involves identifying key "themes" (topics) in the media coverage, as well as "frames" (perspectives) in the articles. The analysis is based on about 350 articles over a 17-year period, drawn from a pool of about 700 articles that contained the words "nanotechnology" or "nanoscience."

In addition to basic descriptive information, articles were coded for up to three themes and three frames. The presence or prominence of positive and negative assessments of nanotechnology was also coded; an article could have both positive and negative assessments present. Two coders trained on an initial set of 25 articles; after training, intercoder reliability was assessed on a new set of 25 articles. All variables had intercoder reliabilities above 80 percent.

Results

Total Coverage

The basic trajectory of coverage shows that elite media attention to "nano" was essentially nonexistent for the first 12 years after Drexler's book. It began to go up in 1998, about the time that the U.S. government created a task force to coordinate nanotechnology research and development among different agencies. The number of articles rose quickly from just a few articles a year to more than 100 in 2002 (Fig. 6.3; the apparent dip in 2003 is an artifact of having data for only half the year; the total will clearly far surpass the 2002 total, with a count of approximate 140–150). During the period with active coverage, from 1998 onward, the *New York Times* had the most frequent coverage (Fig. 6.4).

Themes

As with coverage of biotechnology, most stories could be classified into a relatively small set of themes: safety, financial issues (investment), politics, policy, and applications (Fig. 6.5). The "other" category included science fiction scenarios (especially in 2002, with coverage of Michael Crichton's *Prey*). Although one might expect the financial newspaper, the *Wall Street Journal*, to focus more on financial or business issues, that did not appear to be the case—the balance of different themes appeared relatively even across publications (Fig. 6.6).

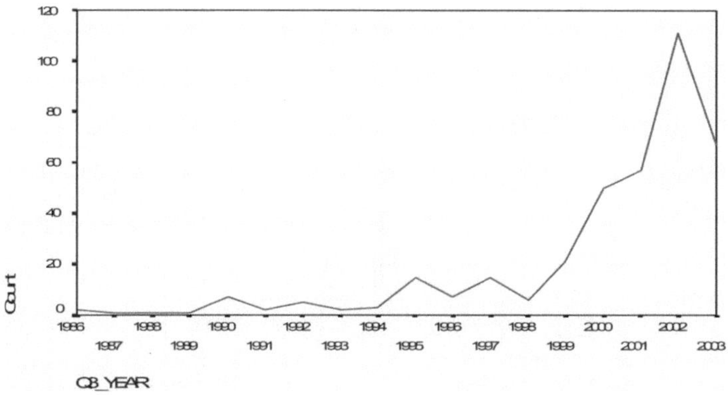

Figure 6.3. Media stories about nanoscience and nanotechnology in the New York Times, Wall Street Journal, Washington Post, and Associated Press, 1986–2003.

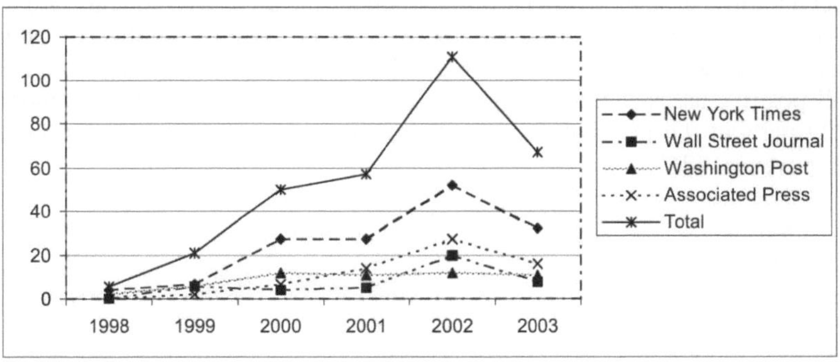

Figure 6.4. Media stories about nanoscience and nanotechnology in the New York Times, Wall Street Journal, Washington Post, and Associated Press, 1998–2003.

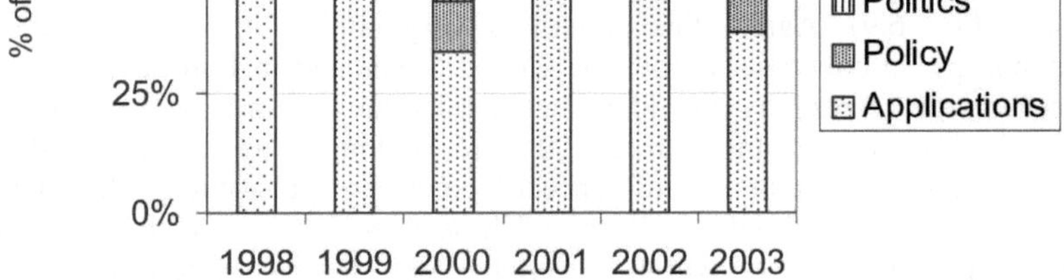

Figure 6.5. Themes in nanotechnology coverage, 1998–2003.

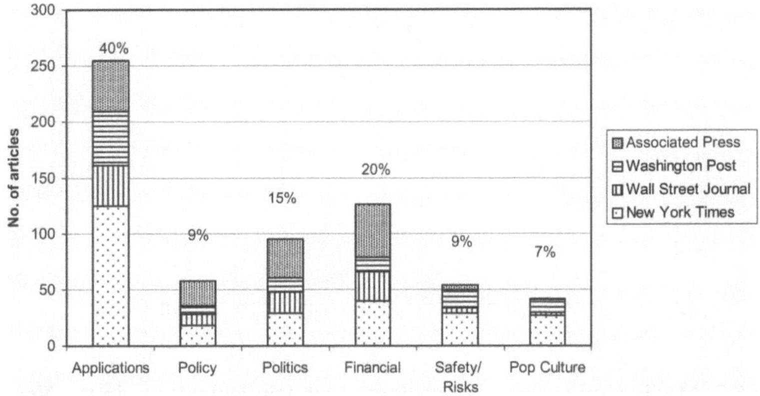

Figure 6.6. Themes in nanotechnology coverage, by publication.

Assessing the tone of the articles for positive or negative assessments to nanotechnology, articles overwhelmingly contain positive assessments (Fig. 6.7). Only in articles exploring risk are there any appreciable number of negative assessments; however, among all such articles, there are about the same number of overall positive assessments as negative assessments (Fig. 6.8). These tendencies roughly parallel the positive and negative assessments of biotechnology seen in earlier coverage [2].

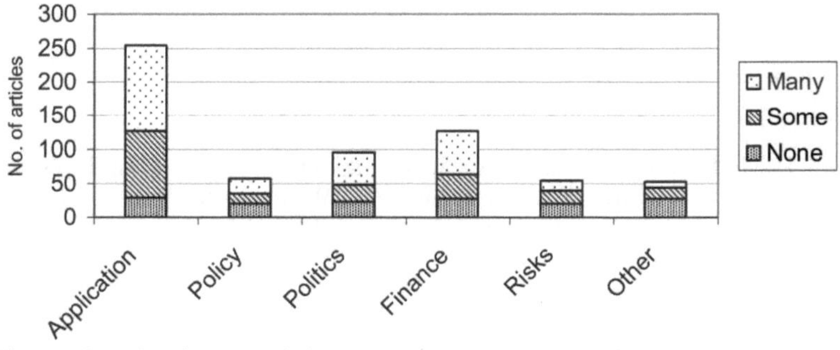

Figure 6.7. Stories containing many, some, or no positive assessments of nanotechnology.

Frames

Irrespective of year, most articles tend to frame nanotechnology in terms of progress (Fig. 6.9). Again, this pattern is very similar to the one seen for biotechnology, in coverage in both the United States and in Europe. A few new themes emerge, especially the idea that applications of nanoscience will not appear for years into the future ("they're a long way away") and that nanotechnology is part of a confluence of emerging technologies that include biotechnology and artificial intelligence. These new themes accounted for only about 20 percent of the overall coverage (Fig. 6.10), suggesting a media perception of nanotechnology as a distinctive new source of progress.

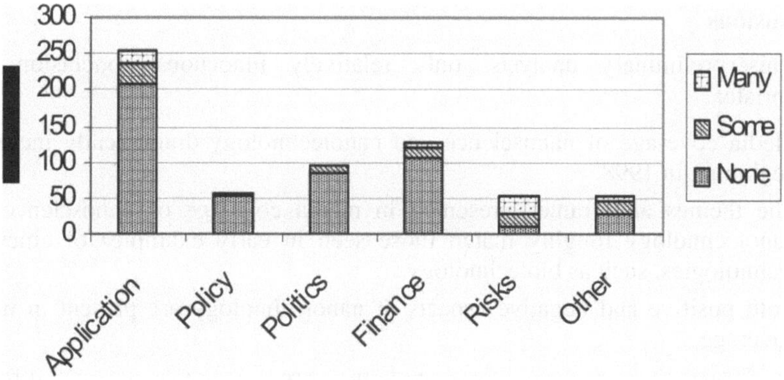

Figure 6.8. Stories containing many, some, or no negative assessments of nanotechnology.

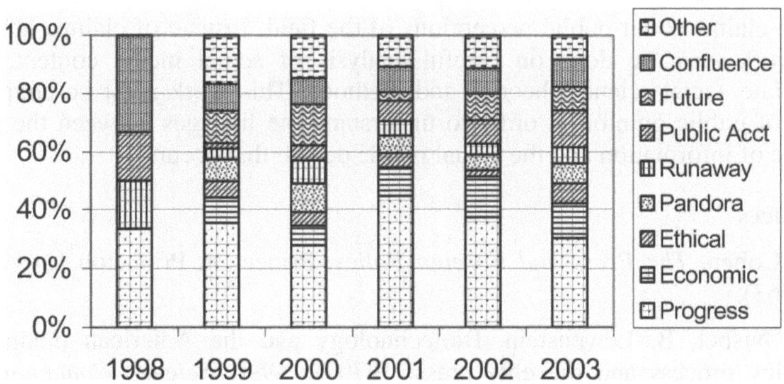

Figure 6.9. Media frames on nanotechnology by year.

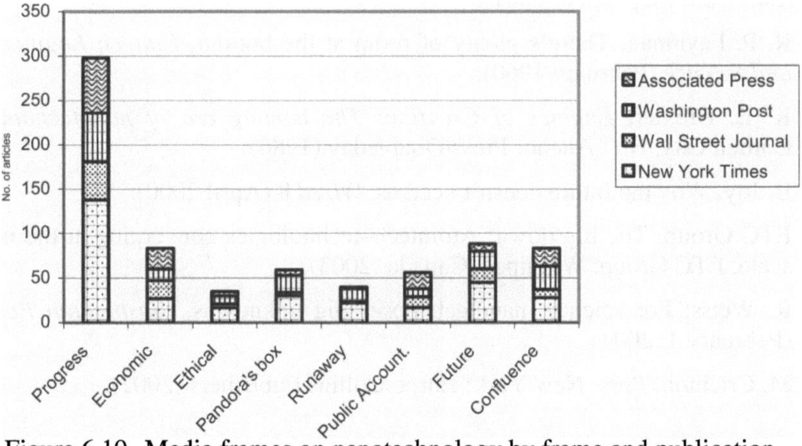

Figure 6.10. Media frames on nanotechnology by frame and publication.

Conclusions

For this preliminary analysis, only relatively innocuous conclusions are appropriate:

- Media coverage of nanoscience and nanotechnology dramatically increased beginning in 1998.
- The themes and frames presented in media coverage of nanoscience and nanotechnology roughly match those seen in early examples of emerging technologies, such as biotechnology.
- Both positive and negative aspects of nanotechnology are present in media coverage.

In future analyses, we expect to explore the differences between the different publications and assess any patterns in the positive and negative tones of coverage.

It is easy for both promoters and skeptics about nanoscience and nanotechnology to make claims about public perceptions of the field. Instead of claims, however, more work must be done on careful analysis of actual media content, using appropriate social science theories and methods. This work must be coupled to studies of public opinion in order to understand the linkages between the public presence of information and the actual public debate that occurs.

References

1. B. Cohen, *The Press and Foreign Policy*, Princeton: Princeton Univ. Press (1963).
2. M. Nisbet, B. Lewenstein, Biotechnology and the American media: The policy process and the elite press, 1970 to 1999. *Science Communication* **23**(4), 359-391 (2002).
3. Scientific American editors, Nanotechnology (special issue), *Scientific American*, (September 2001).
4. R. P. Feynman, There's plenty of room at the bottom, *Caltech Engineering and Science* (February 1960).
5. K. E. Drexler, *Engines of Creation: The coming era of nanotechnology*, Garden City, NY: Anchor Press/Doubleday (1986).
6. B. Joy, Why the future doesn't need us, *Wired* **8** (April 2000).
7. ETC Group. The big down: Atomtech–technologies converging at the nano-scale, ETC Group: Winnipeg, Canada (2003).
8. R. Weiss, For science, nanotech poses big unknowns, *Washington Post*, 1 (February 1, 2004).
9. M. Crichton, *Prey*, New York: HarperCollins Publishers (2002).

10. J. Radin, Scientists in government: Framing the environmental and societal implications of nanotechnology. Unpublished master's thesis, Cornell University, Ithaca, NY (2003).

11. M. C. Roco, W. S. Bainbridge, National Science and Technology Council (U.S.) Subcommittee on Nanoscale Science Engineering and Technology, *Societal Implications of Nanoscience and Nanotechnology*, Dordrecht, Netherlands: Kluwer (2001).

12. M. W. Bauer, G. Gaskell, eds., *Biotechnology: The making of a global controversy*. Cambridge, U.K.: Cambridge University Press (2002).

13. J. Durant, M. W. Bauer, G. Gaskell, eds., *Biotechnology in the Public Sphere*. London: Science Museum (1998).

14. G. Gaskell, M. W. Bauer, eds., *Biotechnology 1996–2000: The years of controversy*. London: Science Museum (2001).

15. W. A. Gamson, A. Modigliani, Media discourse and public opinion on nuclear power: A constructionist approach, *American Journal of Sociology* **95**, 1-37 (1989).

PUBLIC ENGAGEMENT WITH NANOSCALE SCIENCE AND ENGINEERING

Carol Lynn Alpert, Museum of Science, Boston

> "Any sufficiently advanced technology is indistinguishable from magic." (Arthur C. Clarke [1, p.5])
>
> "Any technology distinguishable from magic is insufficiently advanced." (Gregory Benford [1, p.5])

Clarke's oft-quoted observation and Benford's more pointed corollary together express the paradox at the center of our public communication gap with respect to nanotech. The leading edge of innovation stretches beyond even the researcher's capacity to fully explain ... or contain. We are tinkerers in a strange new world, poking around with atoms and electrons, making new molecules, testing how they will respond, finding plenty of odd behaviors to keep the quantum theoreticians burning the midnight oil. With new theory and new tools, a vast new frontier is now open for exploration and development, and eager homesteaders and gold-diggers are rushing in. This frontier exists on such a rarified level that few can view it in its entirety; others catch glimpses of this or that region; most rely on remote sensing; some go on trust alone. For the uninitiated, this territory is magical: access to it is far removed from the solid Newtonian world we know and trust. Magicians awe and entertain, but they also conjure up fear and *dis*trust. To whom does this magician answer? Will his amazing new powers be put to work for good or evil? Will he trick us with his hyped-up claims? Is a rekindled Frankenstein already stirring in the grave?

Like magic, nanotechnology thrives on hyperbole, of both the positive and negative kind. The need to market its potential to win support and funding fosters dramatic claims that spark equally dramatic counterclaims. Insiders describe an impending upheaval on the scale of the industrial revolution, but have trouble characterizing the anticipated watershed and pinning it down with timelines and predictable results. The message reads something like this: "We know we're on to something really big here, but we're not quite sure exactly what it will turn out to be, nor how or when it will occur." This monumental uncertainty makes everyone nervous, from prospective investors to potential consumers.

Now there's also the predictable Hollywood-style "nanophobia" to contend with, providing a kind of smoke screen hovering over more sober issues of concern: unintended impacts on health, personal privacy, environment, and movement of capital and labor.

Nanophobia as Sideshow

The new nanotechnology research enterprise, along with its government, industry, and venture capital boosters, voices serious concern over the potential impact of popular nightmarish fantasies of evil scientists and self-replicating technology run amuck: Could such a backlash have the power to derail public support for rapid, competitive development of this economically and militarily critical set of technologies? [2] Negative public reaction to nuclear power generation, to recombinant DNA, to genetically modified foods (especially in Europe), and to therapeutic cloning (especially in the United States) have all been cited as Luddite-breeding precedents that bear close scrutiny. Perhaps the chief similarity among these otherwise quite disparate historical models is the perceived susceptibility of the public to hyped-up claims and fear-mongering promulgated by special interest groups, as well as the media's well-known tendency to pit extreme against extreme, aggravating the polarization of opinion—no doubt because careful and sober analysis does not win audience share. However, it is important to acknowledge the more significant common characteristic among these previous cases: in each one, the real fears centered on the ownership, control, and regulation of such advanced technologies, and whether government institutions could be counted on to uphold the public's long-term interests over and above short-term commercial considerations.

One can plausibly argue that the nanophobic Drexler-Joy-Crichton apocalyptic scenarios are simply a sideshow—*not* the major threat to public support for nanotechnology R&D that some fear. The public is just not that naive. Most Americans know how to take the high-adrenalin Hollywood treatment in stride and to even use it to stimulate their further interest and curiosity in the science itself. Public attitude surveys show that interest in science fiction and interest in science and technology are highly correlated. (See, for example, the National Science Board's *Science and Engineering Indicators—2002*, Section 7–35. [3]) Indeed, a recent visitor research survey conducted at Chicago's Museum of Science and Industry concluded that fear was overrated:

> In most cases visitors [who were familiar with nanotechnology] said they were not troubled by any of the claims of nanotechnology. Most viewed it as a straightforward science with immediate positive benefits. Some visitors compared nanoscience to more controversial innovations, and said that in contrast to subjects like genetics, nanotechnology had no apparent dangers and seemed relatively neutral … .Very few visitors talked about science-fiction-related fears. Some made reference to science fiction as a way of joking about ungrounded fears. [4, pp.11-12]

On the other hand, public constituencies have repeatedly voiced legitimate concerns that appropriate safeguards be taken in the development and ownership of new materials and technologies, and have sought assurances that adequate regulatory procedures are provided and applied appropriately. It would be counterproductive to lump these reasonable concerns in with the science fiction fantasy fears and dismiss them all out of hand, falsely concluding that the public cannot meaningfully be engaged in reasonable discussion regarding appropriate safeguards for future research, development, and integration of nanotechnologies into our society.

Public Engagement

As has been pointed out many times, public engagement is a critical factor in the sustained development of new technologies and their successful integration into the lives of our communities, particularly if potentially negative health, safety, environmental, social, and ethical issues are involved. Public engagement also has the benefit of leading to faster uptake of commercial applications, broader investment, and increased involvement of young people in educational pathways that lead to further development of the new sector.

The term *public engagement* updates previous conceptualizations of the "public understanding of science" as a one-way street: white-coated scientists patiently lecturing on the brilliant products of their research to admiring but underwhelmed public audiences. The contemporary model of public engagement connotes interactivity and truly meaningful multidirectional discussion over the implementation of new technologies—discussions in which scientists, industry, investors and government regulatory agencies work together with citizen representatives of the diverse communities that are most likely to experience the impact of the new technologies and will need to deal with whatever unintentional fallout may occur. This notion assumes the presence of an educated and literate public essential to any functional democracy, and more particularly, the presence of a *scientifically literate* public essential to a 21st century techno-democracy. It also requires scientists, engineers, and CEOs to develop a broader perspective and a dose of humility. As *Public Understanding of Science* journal editor Bruce Lewenstein commented,

> It's really critical that scientists recognize that their assessments of what's "important" are not the only valid positions. It's also

> important for scientists to hear, truly hear, that the taxpayers who fund their work have a legitimate right to have some say in what questions they address and what levels of safety and uncertainty they accept as reasonable. Without that kind of MUTUAL respect and MUTUAL learning, any hope for true engagement is just a pipe dream. [5]

In Europe, the mishandling of science, technology, and society issues, most notably the Mad Cow disease debacle and the GM foods debate, led to a marked decline of public trust in science, industry, and government over the last decade [6]. In comparison, for better or worse, Americans tend to think optimistically of transformative technologies and tend to trust that risks will be handled reasonably well: more than 75 percent of Americans believe the benefits of technology outweigh the risks. This level of confidence is much higher than in Europe [7]. Americans also trust the institutions associated with research, expressing greater confidence in the leadership of medical and scientific institutions than in that of the Supreme Court, business, educational, financial and religious institutions, the press, and media [3]. Nevertheless, one has only to look as far as the recent debate in Congress over stem cell research and therapeutic cloning to know that as soon as technology gets personal, and begins to stretch the edges of the fabric of our social and ethical consensus, mechanisms for reasoned public engagement in dialogue and debate on the cost/benefits calculus and social and ethical implications of scientific or technical issues are, indeed, in short supply.

The idea of engaging the public in discussion of nanotechnology goes well beyond an interest in calming nascent fears of catastrophic consequences. This is a question of how we as a society move forward on behalf of *all* of us, with *all* of our short- and long-term interests on the table, as consumers, taxpayers, regulators, researchers, educators, politicians, investors, and CEOs. As Lewenstein commented "...questions of social justice, equitable distribution of risks and benefits, ethical concerns about privacy and about introduction of new materials (and even capabilities) into human bodies are questions that people can and should be addressing" [5].

We have much going for us in this endeavor. As already noted, we have little to fear from sensationalized science fiction fantasies. We have a public that thinks positively about the benefits of R&D, and is receptive to new technologies. We have investors willing to back R&D pioneers. We have an initial government commitment to deal forthrightly with the broader social implications and with potentially harmful health, environmental, and economic concerns. We have a regulatory system and a free press.

There are two key things we're short on, however: that previously mentioned "*scientifically literate* public essential to a 21st Century techno-democracy," and those "mechanisms for reasoned public engagement in dialogue and debate on the cost/benefits calculus and social and ethical implications of scientific or technical issues." Clearly, we cannot at this point rely on our formal education system to supply these necessities, although we ought to be working very hard in that direction. Neither are our commercial media up to the task.

Nanoscale science, with its convergence of fundamentals in physics, chemistry, biology, computing, and engineering, may well inspire, as well as require, a radical rewrite of the nation's K–12 science curricula, but this can occur only at a slow and uneven pace across the patchwork of independent school districts that make up the nation's formal education "system." In the meantime, science museums, and other informal science education institutions, may be able to make a substantial contribution in coming years to the quest for greater citizen engagement with this leading edge of current research.

Science Museums: Facilitators for Public Engagement

America's science centers and museums have the potential to reach significant populations with interactive exhibits and programming. A whopping 66 percent of American adults surveyed in 2001 reported that they had visited a science or technology museum at least once during the past year, the highest level of museum attendance ever recorded by the NSF survey. The figure has been rising since 1983. While traditionally regarded as destinations for school field trips and family weekend entertainment, science museums have also begun to emerge in recent years as venues well-suited for continuing adult engagement with science and technology, and as ideal educational outreach partners for university and institute-based researchers.

Science museums could also potentially morph into becoming those missing public spaces where researchers, policymakers, representatives of interest groups, and citizens can engage in forums, discussions, and facilitated consensus-building activities of the type advocated by Jane Macoubrie and others. Such activities have been prototyped at La Cite des Sciences et de l'Industrie in Paris, the Science Museum in London, and the Museum of Science, Boston, addressing issues of heightened public concern like genetics testing, GM food technology, and stem cell research. Granted, it is easier to find audiences willing to discuss these more accessible biotech issues, with their obvious personal and social implications, than it will be to involve audiences in the intricacies of non-biologically-oriented nanoscale science and engineering. We will need to work very hard at bridging that nanotechnology communication gap.

What the Heck is Nanotech?

Engaging public and school audiences in nanotechnology is challenging: even the most basic explanation seems to require a parenthetical statement (to explain the scale indicated by the prefix *nano,* for example, or that *atoms* are the building blocks of *matter*, whatever that is). Nanotechnology is hard to pin down in a brief non-technical description. It is everything, it seems—an umbrella term—but it is also nothing—nothing one can see, hear, or feel. The scale is incomprehensible, the language inaccessible. Effective communication of nanoscale processes—even with the aid of metaphors, analogies, and rich graphics—seems to require the assumption of a certain set of shared *a priori* experiences as well as extraordinary conceptual abilities. Yet these apparent cognitive barriers to nanotechnology communication mask an even more formidable threat, a widespread, but little recognized, phenomenon in our culture: physics phobia.

Most people believe that physics is beyond their reach: at best, a foreign country where only geniuses dare tread; at worst, an irrelevant and wasteful mental exercise. Popular culture tends to relieve the tension by elevating the very human Einstein to a pantheon of superhuman icons, revered as God-like savants, thus allowing the rest of us to stick to what we mere humans do best (i.e., *not* physics). As for the irrelevant and wasteful mental exercise strand of thought, here's former Senate Majority Leader Trent Lott, addressing an audience of high school and college students on C-Span in January 1997, and clearly winning their approval:

> When I was in high school, if you were in the so-called pre-college curriculum, you had to take four years of science and four years of math: a waste of *my* time, a waste of the teacher's time, and a waste of space. You know, I took *physics* ... for *what?* (Cheers, laughter, applause). [8]

Indeed, the National Science Board's *Science and Engineering Indicators* show that few American adults know what an atom or a molecule is, nor which is composed of the other. Only 13 percent were able to provide a correct explanation of a molecule. Jon Miller, the principal investigator, commented:

> This result is both surprising and troublesome. The term "molecule" has become a part of journalistic discourse on television and is often used in newspaper articles without additional explanation. An analysis of the open-ended responses indicated that many adults knew that molecules are very small but did not know whether atoms were composed of molecules or molecules are composed of atoms. Some individuals knew that a molecule is a basic building block and is very small, but could not say anything else about it. [9, p.279]

In other words, forget physics, forget chemistry, and forget molecular biology. Miller concluded, "Minimally, it is essential that science communicators recognize the limited nature of public understanding of the structure of matter..." The Chicago Museum of Science and Industry survey also corroborated the finding of general public unease with terms like atom, molecule, and the term matter itself.

As a result, mentioning a fundamental nanotechnology notion like "building a transistor atom by atom" may result in a massive audience attention loss. What's an atom? What's a transistor? In this climate, one is quickly dissuaded from venturing on to interpret other key areas of nanotechnology research involving, say, quantum dots or scanning tunneling microscopy.

Dealing with Physics Phobia

Clearly, the challenge with nanotechnology is to find multiple pathways to penetrate physics phobia, provide entry points to this rarified world beyond the senses, and empower public and school audiences with the experience of constructing and testing their own inquiry-based conceptual models. The increased confidence this learning process may engender could go a long way toward making public dialogue on societal concerns a real possibility, and may also open the door to further individual engagement and learning.

Cognitive penetration of the nanotechnology world may require something beyond the normal textbook or classroom lecture experience approach. Because that world is so anti-intuitive, so contrary to practical experience and so inaccessible to the senses, new multidimensional approaches should be explored, possibly involving large-scale interactive models enhanced by audio-visual media, and kinesthetic, sensory, and motor experiences. We also need expert communicators, skillful at creating mental and physical analogs for atomic-scale processes and making them centrally relevant to diverse audiences.

Science centers and museums are beginning to serve as laboratories for testing innovative methods of teaching and learning nanoscale science and exploring cognitive connections. Most program and exhibit designers take visitor research very seriously. Front-end studies help determine what potential audiences already understand about any given subject and the associations those understandings hold for them. Typically, exhibit designers move slowly, in frequent communication with the target audience, carefully prototyping and making iterative adjustments of exhibit concepts and activities as they proceed through the development process.

Here is a partial list of recent U.S. science museum efforts to interpret nanoscale science and engineering for school and public audiences, all developed in close cooperation with university-based researchers:

- It's a Nano World: Traveling exhibit developed by the Ithaca Science Center to introduce young children to the concept of scale; also debuted successfully at the Epcot Center
- NanoZone: Exhibit and multimedia project at the Lawrence Hall of Science at Berkeley, targeted at 8–14 year olds
- Nano: Art and science installation pieces offering experiential conceptualizations of nanoscale science and engineering at the Los Angeles County Museum of Art
- Nanotechnology: Extensive exhibit at Tokyo's Museum of Emerging Science and Innovation

Current Science & Technology

At the Museum of Science, in Boston, we launched the Current Science & Technology Center in 2001 as an experimental model for providing in-depth programming on recent research. The Center offers daily live presentations and exhibits, cablecasts, current science theater/forum performances, and multimedia on a broad spectrum of science and technology topics. Often, staff can seize on a topic that's currently getting a lot of media attention, and use it as a hook to bring our audiences into a more in-depth understanding of the science and technology involved, as well as the research process. Nanotechnology subjects are not often the stuff of front page news, and so we have improvised several other approaches.

Nanotech-related presentations developed by staff member Joel Rosenberg have titles such as The Wonderful (and Not So Wonderful) World of Carbon

Nanotubes, The Incredible Shrinking Transistor, Hooked on Photonics, and Quantum Computing. They begin with compelling ideas that link the subjects immediately to personal experience and include a little history, a core of science, a dose of personality, and a glimpse of future directions. Joel attracts a good teen and adult audience for these lively 20-minute, multimedia- and prop-rich stage events, which are often followed by more intimate audience Q&A and discussion.

The Museum's nanotechnology programming is produced in collaboration with our Nanoscale Science and Engineering Center partners, Harvard, MIT, and UCSB. The NSF-funded NSEC supports a full-time education associate in the Current Science & Technology Center. Joel has an engineering background and is well-informed on research across the entire nanotechnology field, and he develops and delivers presentations, cablecasts, and multimedia. He also curates a guest researcher speaker series, working closely with the researchers to adapt their more typical formal presentations to a style that works with our diverse audiences. We've had considerable audience interest in these encounters, and the researchers who have participated, including Eric Mazur, Eric Heller, Howard Stone, and Charlie Marcus, have been more than generous with their time. Some of the presentations have been videotaped and edited and posted on the Web site, linkable from mos.org/cst/nano. (Researchers who have the knack for engaging lay audiences with their excitement, accomplishments, and motivations in accessible language are everywhere to be sought out and emulated, as they provide a key link in the process. Graduate students ought to be encouraged to develop their communication skills—not only will it help us bridge this two-cultures gap, but it will also help them write grants, attract venture capital, and feel more at ease at cocktail parties.)

We are engaged in a three-year evaluation protocol of these various approaches with the Institute for Learning Innovation. Initial results from the formative studies show that the staff and guest presentations are on target with their approach, style, and content.

Future Applied Research: Public Engagement with Nanotechnology

We are currently exploring ways to network formal and informal science education institutions, pooling together their research, resources, and prior experience in nanotechnology education and forum activities and jointly developing and sharing new work. With education and outreach funding so scarce, none of us can afford to repeat failed experiments. Just as published research alerts scientists around the world to new findings that either discourage or encourage new avenues of investigation, so should education and outreach professionals, museum exhibit and program developers, and public engagement specialists develop and share a robust body of applied research as they further their efforts to move these fields forward.

Networking research and information about best practices among interdisciplinary collaborators can help us all take these practices to new levels of effectiveness and to new audiences, further stimulating innovation. Nanotechnology research is notorious for demanding interdisciplinary expertise. Developing effective practices for public engagement with nanotechnology may demand even broader collaboration. The current disciplinary divide between formal and informal educators robs each sector of valuable knowledge, tools, and resources. A conference bringing together museum exhibit designers, instructional material developers, K–12 education researchers, research institution outreach coordinators, and multimedia developers focused on potential synergies in nanotechnology research and societal implications for communication strategies might help break down the barriers and seed more creative interdisciplinary approaches. Such a conference might also facilitate broad dissemination of high-quality materials and catalyze greater market uptake.

Overall, it is important that government agencies, foundations, and R&D institutions deepen their commitment to the education, engagement, and dialogue that are integral to their funding for nanotechnology research. While this may include support for academic research on public opinion, ethical frameworks, and historic precedents, it may prove of greater and lasting value if a significant portion of the funding is invested in "applied research," devising and testing a variety of forms for engaging all citizens in the aspirations, substance, methods, risks, and benefits of this remarkable new world of nanoscale science and engineering. It doesn't have to be magic.

Conclusion

In this paper, I have argued that we needn't fear a hysterical public response to alarmist portrayals of nanotechnology dystopias disseminated through science fiction, feature films, or speculative science commentators. Our real concern should be addressing the public's fundamental interest in health, safety, environmental protection, and fair distribution of costs, benefits, and risk. Experience shows that new technologies integrate better into democratic societies when potential hazards are clearly and openly addressed and citizens can trust that adequate safeguards and regulations are in play. Communities need to be well-informed, and they also need to be listened to: it would be unwise and ultimately counterproductive to leave the regulation of such a potentially powerful new technology in the hands of a research and industry techno-elite or simply to market forces. Social consensus consistent with forward progress is better achieved in the presence of a scientifically and technologically literate citizenry. This is a key reason why we continually advocate for better K–16 teaching in STEM subjects, multiple public engagement strategies through science museums and media, and the development of "honest broker" forum spaces, for learning, listening, and coming to consensus. Nanotechnology education is particularly challenging due to its highly abstract nature and a culture of physics phobia in this country; it makes sense to fund innovative applied research in this area and stimulate more synergies between formal and informal educators. As a global community, we should support and vigorously fund the development of forum-style infrastructures for facilitating information sharing among the public and the various stakeholders, including joint assessment of risks and benefits and integration of societal values with science and technology research, with the goal of anticipating and resolving future conflicts.

References

1. T. Pratchett, I. Stewart, J. Cohen, *The Science of Discworld,* New York: Random House/Ebury Press (1999, revised and expanded edition 2002, ISBN 009 1886570). Gregory Benford is a physicist, astronomer, and science fiction writer at UC Davis.
2. As witnessed at the Social Implications of Nanotechnology Workshop, 3-5 December, Arlington, VA, (2003), and from source material provided thereat.
3. National Science Board, *Science and Engineering Indicators—2002* (available at http://nsf.gov/sbe/srs/seind02/start.htm).
4. J. Gilmartin, *Nanotechnology: Front-End Evaluation Report*, prepared for the Museum of Science and Industry (MSI), Chicago and the Northwestern University Nanoscale Science and Engineering Center (NU-NSEC). Unpublished; obtained from MSI Director of Science and Technology Barry Aprison (2002).
5. B. Lewenstein, email communication, 10 December 2003.
6. Public Opinion Analysis Sector of the European Commission, *Eurobarometer surveys* (1997, 1999, 2001, etc.) (available at europa.eu.int/comm./ public_opinion/index_en.htm).
7. J. D. Miller, R. Pardo, F. Niwa, *Public perceptions of science and technology: A comparative study of the European Union, the United States, Japan, and Canada,* Madrid: BBV Foundation (1997), reproduced in *Public understanding of and attitudes toward scientific research: What we know and what we need to know: A report to the National Science Foundation*, J. D. Miller, 31 May 2002.
8. C-Span, 18 January 1997. I am indebted to Eric Mazur for alerting me to this cablecast.
9. J. D. Miller, Public understanding of, and attitudes toward, scientific research: What we know, and what we need to know, *Public Understanding of Science* **13**(3), 273-294 (2004).

NANOTECHNOLOGY: MOVING BEYOND RISK

Julia A. Moore, National Science Foundation

Risk makes the world go around. Or, at least many of the social science studies about the management and public acceptance of new technologies seem to focus on risk. Social science literature is replete with books, articles, and monographs trying to define, analyze, measure, and predict various kinds of technological risk and to track popular perceptions about them.

Take virtually any word in the English language and place it before or after “risk.” The result is the identification or creation of a whole field of social science study and thought—risk assessment, risk behavior, risk management, risk tradeoffs, affordable risk, ethical risk, involuntary risk, relative risk, zero risk, etc.

My point is that the study of nanotechnology offers government decision-makers and social scientists an opportunity to move beyond risk. New social science methodologies, tools, and approaches provide the capacity to develop a new framework for determining and tracking authentic public attitudes and complex societal dynamics beyond traditional qualitative and quantitative analysis.

More important, they can help researchers to lay out an effective road map for building the public "trust and community" that Harvard University scholar Sheila Jasanoff describes as necessary for democratic participation in the management and acceptance of the modern day risks and benefits brought on by 21st century scientific progress and technological change [1]. "Trust and community" is a complex and dynamic set of factors that go beyond what constitutes or builds trust in risk management institutions, or what shapes individual risk perceptions.

Ever since the revelations of early 1900s muckrakers, government has tried to protect its citizens from risk—from workplace hazards, public health threats, and modern lifestyle choices, to natural and man-made disasters. As former Environmental Protection Agency Administrator Lee Thomas once described, in order to carry out this responsibility, government regulators used experts to study a risk, and to assess how much is tolerable for vulnerable individuals or for at-risk societies. Then these experts judged a particular situation safe or unsafe, and the government puts in place mechanisms to best manage or mitigate risk.

In these judgments and regulatory schemes, experts allowed or acknowledged some residual risk. But politicians and regulatory agency heads didn't like to advertise or broadcast it. With initially little technical information available to the general public, most people were satisfied to rely on the experts—at least until something went terribly wrong, or until an unavoidable disaster was horribly mismanaged [2].

But nanotechnology is coming of age in a changed context for risk and public attitudes toward risk. First, the kind of risks we are concerned about in the 21st century often are more subtle, hard to quantify, and involuntarily assumed. Research also indicates that the public's level of aversion to risk is, in part, a function of how familiar the risk is and of how frightening its impact. So the fact that nanotechnology is being portrayed by science fiction writers and Hollywood as invisible, alien, and dangerous is not good news.

Second, the public is increasingly more educated about risk. Enabled by the Internet, citizen right-to-know laws, and regulatory transparency practices, people are better able to access and assess information about risk, and to communicate their views to government, industry, and the science community.

Third, the public at large generally mistrusts government. And more and more in Europe, citizens no longer trust scientists and technical experts as much as they once did. Science is viewed as increasingly commercialized. In disputes about risk in Europe, government and scientists—along with industry—often are seen by the public as pursuing their own institutional, political, or financial interests, and *not* as protecting the public good.

Much of what I describe is not news to anyone involved in social science research or active in the study of science, health, or environmental policy. But despite this new 21st century reality, many involved in government policy making continue to examine and manage the challenge of public technology acceptance or the phenomena of public rejection of a new technology—like nuclear power, or genetically modified food—as a risk–benefit equation. Authorities tend to view public resistance to a new technology as a case of inadequate public understanding of an expert-derived risk assessment, or as a lack of public science literacy. As a result, the remedy or solution to public concerns is simply to apply more sophisticated risk communication techniques on what's often characterized as a technologically naive and irrational citizenry [3].

In the specific case of the genetically modified food controversy in Europe—also known as the Frankenfood debate—this narrow, risk–benefit view of consumer and supermarket rejection of agricultural biotechnology resulted in the creation of what University of Lancaster social scientist Brian Wynne calls a series of 10 popularly accepted "myths" held by government decision-makers, industry, NGOs, and the media. Using significant qualitative and quantitative data from five European countries in a major comparative study funded by the European Commission, Wynne and his colleagues uncovered important lessons applicable to understanding public attitudes and acceptance (or rejection) of nanotechnology or of any other new technology appearing on our "brave new world" horizon. Wynne's group also developed an effective international research model that offers an exemplar for social science research on nanotechnology in the future [4].

The first "myth" Wynne and his group addressed was created with the help of 20 years of quantitative public opinion polling data. These surveys—supported by the biotech industry and governments—demonstrated that the public was generally ignorant of scientific facts and untutored in the fine points of genetic engineering. The false conclusion largely drawn from this research was that if you could put the adult population of Europe—along with any "biophobics" in the United States—through a basic biology course, then consumer concerns about "Frankenfood" would evaporate.

For scientists and public officials who embraced this myth (which British writer John Durant refers to as the "public deficit model"), the implication is that if you spend enough time and money educating people about the transformative benefits of nanotechnology, then the technology will be warmly embraced by the public.

A second myth from the GM food debate, which Wynne and his colleagues explored, was that public doubts about or rejection of agbiotechnology were all the fault of the British government's mishandling of mad cow disease (or BSE). If only the Thatcher-Major regimes hadn't made such a mess of managing the risk of mad cow disease from 1987 through 1996, and if only those modelers at Imperial College in London hadn't initially projected that over 30 years 500,000 Britons would die from the human form of mad cow disease because of apparent government and science ineptitude, then the Frankenfood controversy and European public rejection of agbiotech wouldn't have happened. (Current projections are a few hundred deaths in Britain from the human form of BSE rather than 500,000.)

Many in the nanotechnology arena take great comfort from this mad cow myth. Some because they believe nanoscientists and nanotechnologists are smarter than the scientists or companies that developed agbiotech. Others because it means that if they take enough precautions to avoid some kind of a nanotechnology train wreck—nanotechnology's version of Chernobyl or Three Mile Island—then nanotechnology will have smooth sailing among investors, regulators, and consumers.

Another myth from the agbiotech debate that Wynne examined centered around the growing belief that the European public is increasingly risk adverse and demands "zero risk" from any new technology or innovation. Or, as U.S. Trade Representative Robert Zoellick stated in January last year (2003), Europe's agbiotech policies—and by implication its people—were anti-science and "Luddite."

Wynne also looked at the myth that the anti-agbiotech public was the victim of a distorting, sensationalist media, especially in the U.K. This myth paints an image of citizens as dumb sheep who mindlessly follow wherever the popular press leads them.

By employing a combination of qualitative and quantitative research consisting of a sophisticated but manageable series of focus groups and validation polls, Wynne and his colleagues in four other EU countries developed data showing that none of the 10 commonly held "myths" about European public attitudes toward GM food held up. This research model is an important one for nanotechnology. It holds out the possibility of better understanding of public attitudes toward science in general and technological innovation in particular. And in an increasingly globalized world, it also takes such comparative work to a new depth and level.

More important for nanotechnologists, as I said earlier, this approach results in data that help develop a framework for determining and tracking authentic public attitudes. It also offers help in developing a kind of road map for building the public "trust and community" for the responsible science and technology that Jasanoff describes [1].

Yes, the public is ignorant about scientific facts. Every two years the National Science Foundation tests Americans' knowledge of science with questions like "What's DNA?" and "Does the earth orbit the sun, or vice versa?" In 1999, fewer than 30 percent of respondents could define DNA, and fewer than half knew that the earth goes around the sun once a year [5].

But despite a lack of science literacy, Wynne's research demonstrated that the public did not seem to hold entrenched or inalterable opinions "for" or "against" agbiotech. Like any technology, the public correctly surmised that genetically modified plants had pluses and minuses. And people mistrusted corporate or government officials who tried to mislead them by saying that any technology was risk free.

What the European public questioned was whether they, or their children, or their environment, would ever see the benefits of agbiotech. For them, the only winners in the agbiotech risk-benefit equation for the foreseeable future were bottom-line-driven U.S. corporations and American farmers. And in the case of GM food, the public objected to the fact that they seemed to have so little say in deciding who would benefit from the technology's commercialization.

As far as the British government's handling of mad cow disease in the 1990s (and on the Continent in late 2000), yes, it was awful. But people in Wynne's study did not regard government mishandling of the BSE case as an exception. Unlike in America—where the U.S. Food and Drug Administration (FDA) is held in very high regard—the UK government's handling of BSE was considered normal behavior. It was simply one more case of typical bureaucratic mismanagement. Worse, even with the establishment of a new European Food Safety Authority, the publics of these five European countries expect government institutions to behave in the same way with respect to other safety problems in the future.

Most surprising for Ambassador Zoellick, Brian Wynne's study showed that Europe's public found modern-day risk quite acceptable—but with several caveats. When subjecting themselves to risk, the public wants to receive some direct benefits before they or their children are subjected to what they perceive as a new kind of danger. For example, because of the benefits derived from cell phone usage, mobile phones are extremely popular in Europe despite some studies indicating that long-term use may cause brain cancer or tumors.

And when it comes to technology with potentially major societal or environmental implications and impacts, the public also wants to know that government authorities have a fallback plan, a "Plan B," just in case those expert risk assessments somehow go horribly awry. In addition, they want to have some confidence that the experts doing those risk assessments are not in the pocket of industry and are only acting in the public interest.

Participants in the Wynne study wanted product labeling on GM food, so that if there were unintended, harmful effects, the products could be recalled. And labeling would also make it possible for consumers to boycott products and financially penalize companies manufacturing what might be a dangerous product. Labeling also would help purchasers reward those corporations who seemed to care more about public health and environmental protection than their competitors.

Finally, publics aren't dumb enough to totally trust the media. But they trust their government officials even less. On complicated science questions, they tend to value the opinions of those they do trust. In Europe, the most trusted institutions on biotechnology are consumer (49 percent) and environmental groups (46 percent); whereas in America it's the FDA (41 percent), farmers (34 percent), and scientists (33 percent). No one on either continent trusts industry (United States, 5 percent; Europe, 8 percent).

What do Brian Wynne's results and his research tell us about studying public attitudes toward nanotechnology's risks and uncertainties? His key finding, in my view, is that it is not public knowledge or even the benefits of a new technology that sway public opinion; it's Jasanoff's "trust and community" backed up by concrete "confidence-building" measures and actions.

Any new technology—nanotechnology included—will have to pass the "trust and community" test. My recommendation is that research studies focused on nanotechnology move beyond risk and examine fully the many aspects of this complicated "trust and community" question. Most important, the research should be directed at trying to understand at a deeper level than has been studied before, what actions by government, industry, or scientists can either promote or discourage public acceptance or rejection of a new technology or science.

In June 2003, Britain's Science Minister Lord Sainsbury said in an interview that his government is determined that nanotechnology will not suffer the same fate as GM food. Rather, Sainsbury is committed to looking early at nanotechnology's possible ethical, health and safety, and societal implications, and to engaging in a genuine public dialogue over its future impacts [6].

Sainsbury's prescription for avoiding a GM-like scenario with nanotechnology draws on the hard lessons learned from Britain's mad cow disease crisis and Frankenfood debate. And it is built on the work of researchers like Wynne and others at Lancaster's Centre for the Study of Environmental Change.

Nanotechnology, biotechnology, and information and cognitive science are converging in laboratories with startling—and for some alarming—results. The pace and significance of discovery are dizzying. There has never been a more important time for social scientists to embrace—and to receive support for—a broader research approach that helps people and institutions better understand and manage the implications of such profound and rapid scientific and technological advancements.

References

1. S. Jasanoff, The dilemma of environmental democracy, *Issues in Science and Technology* (Fall 1996).
2. J. C. Davies, V. T. Covello, F. W. Allen, eds., *Risk Communication*, Washington, DC: The Conservation Foundation (1987).
3. J. Durant, M. W. Bauer, G. Gaskell, eds., *Biotechnology in the Public Sphere*, London: Science Museum (1998).
4. C. Marris, B. Wynne, P. Simmons, S. Weldon, *Public perceptions of agricultural biotechnologies in Europe, final report of the PABE research project*, funded by the Commission of European Communities, contract number: FAIR CT98-3844 (DG12 –SSMI) (December 2001).
5. National Science Board, Science and Engineering Indicators—2000, Chapter 8 in Publication NSP-00-1, Arlington, VA: National Science Foundation (2000).
6. A. Fazackerley, Nanotech "won't suffer GM fate," *Times Higher Education Supplement* (27 June 2003).

COMMUNICATION STREAMS AND NANOTECHNOLOGY: THE (RE)INTERPRETATION OF A NEW TECHNOLOGY

Toby A. Ten Eyck, Michigan State University

The trajectories of new technologies have long been influenced by their public presentations [1]. Growing sophistication in the manipulation of both information technology and content, along with the financial stakes involved in developing and marketing a new technology, points to a continuation of debates and polemics as different sides seek support among scientists, regulators, retailers, reporters, and consumers. As information about nanotechnology becomes more readily available within public spheres, those interested in the social dimensions of technology must take into consideration how this information is being developed, disseminated, received, and interpreted.

Numerous theories could be used to try to make sense of these processes. Luhman's systems theory, Beck's risk society, Latour's actor network theory could all be used to understand how nanotechnology has come to be developed and introduced to the public [2–4]. The approach presented here is based on a concern with the linkages between the history and current practices of journalism, as well as the actions, or lack thereof, of audience members. Studies by Zelizer and Gamson [5,6], as well as my personal experiences as a radio broadcaster and media researcher, provide the groundwork for this approach.

The current discourse surrounding nanotechnology contains many of the characteristics of a typical social movement, as argued by Rochon [7]. A small group of interested individuals (scientists) has found a solution to a problem (an efficient technology that has market potential and can help solve problems in various areas such as medicine and textiles), has made its stance known (nanotechnology is viable and the risks are minimal), recruited supporters (manufacturers who could take advantage of the marketability of nanotechnology), who, in turn, have tried to manipulate the public face of nanotechnology through control efforts aimed at reporters and regulators. Zelizer's work [5] on journalists as interpretive communities is appropriate in understanding the (re)actions of reporters to such control efforts. Zelizer contends that journalists approach their work through interpretations of both past and present journalistic practices. These individuals do not begin each news report with a blank slate, but with tools that they have borrowed from journalists who have gone before them, as well as their contemporaries (and, of course, themselves). Given the time proximity to biotechnology, journalists may find sources and metaphors that were used in the biotechnology debate helpful in making sense of nanotechnology, especially if biotechnology is viewed as a model topic. This style of journalism (de)legitimizes issues and sources, as news organizations continue to print or broadcast topics with specific characteristics (e.g., scientific or technological breakthroughs) with many of the same actors appearing across topics.

The making of a story is only part of the social landscape of information and communication. The other part, which is even more diffuse and nebulous, is the receiving and interpretation of information. According to Gamson [6], information is gathered three ways: through (1) experiential knowledge, (2) public wisdom, and (3) media discourse. Regarding nanotechnology, very little information is available from the former two methods, leaving media discourse to provide many of the details. The media, though, are not easily defined. While researchers often rely on news accounts of an issue to gauge public opinion, popular culture presentations—such as Michael Crichton's *Prey* or nanobots on the Web (e.g., http://www.hybridmedicalanimation.com/pages/nanobots.html) are also used to make sense of new issues, even if only in the form of metaphors.

As with studying the construction of stories, many theories exist for understanding how audiences receive and react to those stories. Morley [8] has offered an approach where various audience members have varying levels of interest in topics presented in the news media, ranging from taking some kind of action based on the information to completely ignoring it. Fiske [9] and Hoijer [10] have extended this approach to include the emotional states and biographies of the individuals who are watching a show, and how these factors play into interpretations of the information being presented. This same kind of approach could be used with regard to nanotechnology in the sense that researchers should be cognizant of their own interests in the topic and not apply those interests to their subjects. Yankelovich [11] has argued that too often researchers develop survey instruments based on their own knowledge of the topic, leading to measures of public opinion that is often vacillating due to the nascent state of the topic, how the questions are worded, or the state of the respondents (e.g., giving what they consider to be the socially acceptable answer to a question or trying to please the researcher). Instead, measurement instruments that capture a more stable set of opinions must be constructed—what Yankelovich refers to as judgments—especially if it is argued that public opinion will have some influence on the topic, such as decisions made in the marketplace.

The importance of taking into account what the public does or does not know is highlighted by two surveys. In one, nearly 20 percent of 1,200 respondents in a national Internet survey said they were familiar with *peptostreptococcus* as a food borne pathogen, which it is not. In a survey of 855 randomly selected individuals in the United States, more than half (52 percent) felt that nanotechnology would improve our way of life in the next 20 years, though it is difficult to believe that nearly half of the U.S. population has enough knowledge about nanotechnology to make this kind of judgment. The difficulty with such opinions comes when educational campaigns are developed. People who already feel they know about a subject may feel less inclined to learn more about it. If the information they have is wrong, their actions may have negative consequences—for themselves and others.

In summary, the construction, dissemination, and interpretation of information pertaining to nanotechnology must all be taken into consideration when trying to make sense of perceptions among the public. Understanding the control efforts of sources and gatekeepers to public arenas will offer insight into what information is available. Investigating the role of both traditional and new information technologies can shed light on the dissemination of this information. Finally, the study of audiences should be a gestalt or systemic approach to help understand what information is being received and interpreted, and the varying degrees to which audiences are paying attention to the information being made available to them.

References

1. T. A. Ten Eyck, M. Williment, The more things change . . . : Milk pasteurization, food irradiation, and biotechnology in the New York Times, *The Social Science Journal* **41** (in press).
2. N. Luhman, *The Differentiation of Society*, New York: Columbia University Press (1982).
3. U. Beck, *Risks Society*, Newbury Park, CA: Sage (1992).
4. B. Latour, *Science in Action*, Cambridge, MA: Harvard University Press (1987).
5. B. Zelizer, Journalists as interpretive communities, *Critical Studies in Mass Communication* **10**, 219-237 (1993).
6. W. A. Gamson, *Talking Politics*, New York: Cambridge University Press (1992).
7. T R. Rochon, *Culture Moves*, Princeton, NJ: Princeton University Press (1998).
8. D. Morley, C. Brunsdon, *The Nationwide Television Studies*, London, UK: Routledge (1999).
9. J. Fiske, Audiencing: A cultural studies approach to watching television, *Poetics* **21**, 345-359 (1992).
10. B. Hoijer, Reception of television narration as a socio-cognitive process: A schema-theoretical outline, *Poetics* **21**, 283-304 (1992).
11. D. Yankelovich, *Coming to Public Judgment*, Syracuse, NY: Syracuse (1991).

HISTORICAL COMPARISONS FOR ANTICIPATING PUBLIC REACTIONS TO NANOTECHNOLOGY*

Christopher Toumey, University of South Carolina

One way people try to envision the future of nanotechnology is to tell stories about the past, expecting that the future will continue certain features of nanotech's history. If one tells stories emphasizing that the founders of nanotechnology *past* were heroic geniuses, for example, that kind of emphasis would bless nanotechnology *present* and *future* as a noble effort in which heroic qualities endure. Or so the storyteller would hope.

Public reactions to nanotechnology in the United States are more difficult to envision this way because there has been practically no history of public awareness, let alone public reaction to it. We need to turn to past episodes of other technologies to anticipate public reactions to nanotechnology.

To do so, I'd like to employ an insight from cultural anthropology. I appreciate that numerous scholars in various disciplines have proposed theories about how a group's interests shape the ways its members recall the past. But instead of reviewing all of them, I focus on one that is especially germane. Bronislaw Malinowski proposed a relationship between social conditions in the present and the telling of stories about the past: people tell myths, not because they need to empirically reconstruct a true record of past events, but rather because they need to retroactively justify certain conditions in the present. The telling of myths gives legitimacy to current circumstances by tracing them to a "primeval reality," or by discovering precedent for the way things are now. And so myths seem to be a record of past events, but they are really a reflection of the present situation [1].

Malinowski drew his illustrations from his ethnographic work in the Trobriand Islands of the Western Pacific. The Trobrianders justify their geographical situations by reference to a First Principle of autochthony: it is right and proper that we live where we do because this is where our original ancestors emerged from underground. A group that is satisfied with its location will point out the exact spots at which its founders climbed up to the surface of the earth. But Trobriand clans and subclans are expansionary. They sometimes occupy lands beyond their rightful territory, which is to say that they subdue or displace other peoples. When this happens, they violate their First Principle by giving a moral justification. Our own first ancestors behaved virtuously, while the other peoples' ancestors behaved improperly. In still other circumstances, a group justifies its subjugation of another by marrying into the subjugated group, and then telling stories that exaggerate the rights that derive from marriage. Myth-telling for the

*The material in this document is based upon work supported by a grant from the National Science Foundation, NSF 01157, Nanotechnology Interdisciplinary Research Team [NIRT], Philosophical and Social Dimensions of Nanoscale Research: "From Laboratory to Society: Developing an Informed Approach to Nanoscale Science and Technology." All opinions expressed within are the author's and do not necessary reflect those of the National Science Foundation.

purpose of justifying the present situation is so open-ended that it is neither consistent nor reliable, even with respect to its own First Principle.

This kind of storytelling is most likely to arise when there are "certain inconsistencies created by historical events" [1, p.125]; or when there are some "specially unpleasant or negative truths" [1, p.136]; or when one group holds power over another; or when the credibility of a form of morality is less than secure. Then we can expect that myths will be told because the telling of myths enables people to resolve these anomalies and unpleasantries.

Here is a summary of Malinowski's theory of myth-telling:

- Myth-telling arises in certain circumstances of social or cultural tensions.
- Myth-telling need not be an accurate record of events in the past, even though it seems to be a convincing account of what happened before the present.
- Instead, myth-telling reflects conditions and problems in the present, which is to say that the past is reconfigured to serve the present.
- The result of myth-telling is to justify, legitimize, or rationalize the current circumstances in which people find themselves. Myth-telling is an exercise in coming to terms with present-day tensions.

That four-part formula is relevant to public reactions to nanotechnology if we imagine any of the following Malinowskian conditions:

- that the interests of the scientists who drive nanotechnology conflict with the interests of the public
- that one part of the public finds itself in serious conflict with another concerning nanotech
- that social or moral or political disagreements are rendered as controversies about nanotechnology, even if they have little or nothing to do with its scientific merits
- that large parts of the public find the consequences of nanotechnology to be puzzling, disturbing, or downright frightening

In other words, there are multiple possibilities for tension, unpleasantness, and social conflict that could bring nanotechnology into the kinds of conditions that generate storytelling in a Malinowskian style.

Now is a good time to think about this. Public awareness of nanotechnology has been negligible. While reports on nanotechnology appear regularly in *Scientific American*, *Wired*, *Small Times*, *Technology Review*, and the *New York Times*, and while several million people read these publications, there are several *hundred* million people who do *not* read them, nor do they read other newspapers, magazines, or Web sites that report on nanotechnology. This condition will probably not last much longer. For a short time, we have the luxury of anticipating the possible forms of public reactions to nanotechnology by selecting past cases that share important features with nanotechnology.

Landscapes of Nanohyperbole

One feature seems to me to be especially salient, namely, the climate of hyperbole that surrounds discussions of nanotech. Vivid predictions begin with the ur-text of nanotech, Richard Feynman's 1959 speech, "There's Plenty of Room at the Bottom" [2]. To cite but two examples, Feynman predicted an information technology in which "all of the information that man has carefully accumulated in all the books in the world can be written in this form in a cube of material one two-hundredth of an inch wide"[2, p.61]—and, there could be a "mechanical surgeon" so small that it can be swallowed, after which it would maneuver through to body to the site of a lesion, and then repair the lesion. If that was *nanoGenesis*, then *nanoDeuteronomy* was Feynman's 1983 speech, "Infinitesimal Machinery," which confidently reinforced the author's vision of a world transformed by nanotechnology [3].

If we stipulate that Feynman established the original tone of nanohyperbole, then the current ideological landscape includes several genres of thought about the value of nanotechnology. Four are particularly important. The first is extreme nanophilic hyperbole, an uncritical embrace of nanotechnology that looks ahead several decades to the arrival of nanotechnology's most amazing promises. We see this genre in some of the visionary predictions of K. Eric Drexler and his associates at the Foresight Institute [4–8]. Extreme nanophilic hyperbole is also represented in some works of science fiction, especially the novels of Kathleen Ann Goonan [9–11].

The second family of positions on nanotechnology is a somewhat less fantastic form of optimism. As the U.S. government gathered its various nanotechnology projects under the umbrella of the National Nanotechnology Initiative, it produced a series of enthusiastic documents, beginning with the government's colorful booklet on nanotechnology (Amato 1999). In the detailed blueprint for the NNI, it was said that "developments in … [nanotechnology] are likely to change the way almost everything—from vaccines to computers to automobile tires to objects not yet imagined—is designed and made" [12, p.13]. The next major NNI text told us that "The effect of nanotechnology on the health, wealth, and standard of living for people in this century could be at least as significant as the combined influences of microelectronics, medical imaging, computer-aided engineering, and man-made polymers developed in the past century" [13, 14, p.2]. Note that there are some important distinctions between the U.S. government's enthusiasm and the more dramatic style of visionary nanophilia. In particular, the government's optimism is much more concerned with immediate and near-future events, especially in materials science, medicine, information technology, and other areas in which commercial products can be delivered fairly soon.

My next category is that of measured skepticism, which comes from a group of science writers who recognize that important work is being done at the nanoscale, but who also express disdain, almost contempt, for the hyperbole of extreme nanophilia. *Scientific American* is their principal venue, and the epitome of this kind of writing is Gary Stix's 1996 profile of Eric Drexler, wherein Drexler and his followers are comic eccentrics [15]. Stix's next article on nanotechnology was slightly kinder to Drexler, but still found ways to diminish him [16]. When *Scientific American* reported on carbon nanotubes [17] and molecular computing [18], it found it necessary to suggest that stories of "microscopic robots rearranging atoms on command" might be "moonshine."[17, 18] "The hype," said John Rennie, "outruns the reality" [19, p.8]. In the September 2001 special issue on nanotechnology, a facetious opinion piece by Michael Shermer ridiculed the idea that "nanocryonics" will banish death [20]. Elsewhere, Adam Keiper bifurcated all the talk of a "nanotechnology revolution." On the one hand there are solid advances, incrementally achieved by hard-working scientists, and on the other there are the vivid fantasies of Drexler and such [21].

The fourth and final stance is an extreme nanophobic counter-hyperbole, approximately as intense as that of the visionary nanophiles. This last position follows the general outlines of the Frankenstein story to emphasize gloom-and-doom predictions that science is dangerous and that scientists are arrogant [22, 23]. Its rhetorical style has several features: (1) considering that nanotechnology has yet to kill humans or devour the earth, its evils are projected into the future with the words *would*, *might*, *possible*, and *possibly* appearing regularly, in lieu of empirical experience of danger caused by nanotech; (2) scientists, usually unnamed, are routinely depicted as being both irresponsible and undemocratic; (3) the hypothetical horrors of nanotechnology are assumed to greatly exceed any possible benefits; and (4) the proper response is a moratorium on research at the nanoscale.

Various combinations of these features are evident in recent articles by J. Smith and T. Wakeford [24] and by L. Broadhead and S. Howard [25], plus the comments by Prince Charles [26]. The most sustained commentary in this genre comes from the ETC Group of Winnipeg, Manitoba. Following several angry denunciations of the dangers of nanotechnology [27–29], this organization called for a moratorium on commercial development of nanotechnology [30–33], after which it published additional denunciations of nanotechnology [34–36]. The Greenpeace report on nanotechnology relied very heavily on the ETC Group's position papers but, after briefly flirting with the idea of a moratorium, it recommended instead a balance of industrial self-restraint and government oversight [37]. The *Chemical Market Reporter* expressed a sense of alarm in the business community that popular hostility to nanotech, regardless whether it had its basis in fact or in fiction, could poison the future of this kind of research [38].

The dark view of nanotechnology is also represented in science fiction films and some novels, the best known of which is Michael Crichton's *Prey* [39], which present visions of a world radically altered for the worse by nanotechnology [see also 40]. Yet another form of dramatic nanophobia comes from Bill Joy [41, 42] and Bill McKibben [43]. This subgenre indicates that nanotechnology is the centerpiece of a so-called convergence of technologies that will diminish human nature so much, in relation to high-performance machines, that our human qualities will become irrelevant: the end of humanity, so to speak.

In reviewing extreme nanophobia, I do not suggest that concern about this technology is categorically equivalent to paranoia. Vicki Colvin and others have instigated good questions about nanorisk [44, 45] while Doug Brown, Barnaby J. Feder, and Candace Stuart have chronicled these discourses [33, 46–48, 49–53]. My point, rather, is that some of this concern, e.g., that of the ETC Group, is so shrill that it polarizes discussions of nanotechnology between extreme nanophilic and extreme nanophobic hyperbole, and thereby erases the more nuanced ideologies in between. *The Economist* has noted that, unfortunately, common images of nanotechnology tend to arrange themselves into a bipolar division of love-nanotechnology-or-hate-nanotechnology positions [54].

The Case of Cold Fusion

If the public is going to be subjected to extreme forms of nanophilic and nanophobic hyperbole, we can look to past episodes of scientific or technological change that exhibited similar characteristics. I'd like to present two such cases: (1) one without Malinowskian conditions, and (2) one with. This contrast helps us see how hyperbole intersects with such conditions.

My case of techno-hyperbole without Malinowskian conditions is the story of cold fusion. Initial reports and speculations described a technological solution to our energy problems that would deliver abundant power at miniscule cost. Furthermore, this solution celebrated the simplicity of old-time technology. We would have all the energy we wanted by virtue of a plain gadget, a simple electrolytic cell that anyone could manage. No longer would we need legions of engineers, oil-producers, bureaucrats, and policymakers to make our electricity hum. Instead, we could do it ourselves with batteries, beakers, and liquids from the neighborhood hardware store, like a teen-age Thomas Edison. A quick fix, a cheap fix, and the simplicity of kitchen-table technology [55,56]. The story of *Exxon Valdez* oil spill dramatically amplified excitement about cold fusion. As NOVA put it, "most of the time when we think about such disasters, we're reduced to despair. But perhaps this time, from the deserts of Utah [where Stanley Pons taught at the University of Utah], somebody was offering a real answer" [71, p.1].

Thus the press had a story with "drama, heroes, wizardry, and the promise of unlimited energy," said Marcel LaFollette [58]. It was said that "a single cubic foot of sea water could produce as much energy as 10 tons of coal" [59, p.13], which is to say that "the top few feet of water in the world's oceans contain enough [cold fusion] energy to supply the world for 30 million years" [60, p.13]. The epitome of this technophilic hyperbole came from Norman H. Bangerter, governor of Utah, who said, "Knowing nothing about it, I am highly optimistic" [61, p.115].

These were not Malinowskian conditions. The process of getting energy from cold fusion was believed to be so simple and so inexpensive that everyone would benefit in approximately equal proportion. And, when the Pons-Fleischmann hypothesis was discredited, it embarrassed some people and ruined the careers of a few, but it did not give any particular class of people great power over another class. Cold fusion was a fascinating story about science and technology, but it was no great rearrangement of our society, our economy, or our values.

The Case of Recombinant DNA

The hyperbole surrounding recombinant DNA demonstrates a very different situation. An article in *Scientific American* announced that "Research with recombinant DNA may provide major new social benefits of uncertain magnitude: more effective and cheaper pharmaceutical products; better understanding of the causes of cancer; more abundant food crops; even new approaches to the energy problem" [61, p.22]. Jeremy Rifkin wrote, "With the unlocking of the secrets of DNA, we will eventually be able to change the cellular structure of living beings and to create entirely new species... The Nuclear Age was the age of the physicist; the Organic Age is the age of the biologist" [62, p.23].

Language like that was not always wise. Optimism about rDNA was matched with fears that "Old bugs might learn dangerous new tricks and might, if they escaped from a laboratory, demolish the intricate genetic balance that keeps all our chips in play" [64, p.44]. Jonathan King reminded others that at "the best microbiological containment facility ever built in the United States, the Army Biological Warfare facility at Fort Detrick, Maryland ... over a period of 20 years there were more than 400 cases of lab workers getting serious infections from the organisms with which they worked" [64, p.635]. Recombinant DNA might create an "Armageddon virus" [66] or an "Andromeda-type virus" [63]. Rifkin warned that such an organism could "spread a deadly epidemic across the planet, killing hundreds of millions of people. They [i.e., certain scientists] also fear that a new, highly resistant plant might be developed that could wipe out all other vegetation and animal life in its path" [62, p.24].

Much of this feeling stemmed from the use of *E. coli* to reproduce new genetic combinations. Units of DNA from other sources were implanted in *E. coli* because that bacterium multiplied rapidly. In one instance from 1971, a cancer researcher isolated viral DNA that was believed to be carcinogenic, and then recombined that genetic information with the genome of a strain of *E. coli* [67]. Many varieties of *E. coli* live within the human intestinal tract. And so there was a tangible concern that evil new forms of *E. coli* would move from genetic labs to humans' bodies [62, 65, 68]. "The worst that could be imagined was a cancer plague spread by *E. coli*" [63, p.46].

When these various individual concerns were summarized in general statements about the dangers of rDNA, the language could be extraordinarily dramatic:

- "The recombinant technology circumvents all the normal barriers to exchange of genetic material between species" [64, p.635].
- Some people imagined "worldwide epidemics caused by newly created pathogens; the triggering of catastrophic ecological imbalances; the power to dominate and control the human spirit" [62, p.22].
- "There is a class of technologies that can do great, perhaps irreversible harm. Recombinant DNA is a member of that class" [67, p.140].
- "Only one accident is needed to endanger the future of mankind." "The potential dangers [of rDNA] ... pose perhaps the single greatest challenge to

life that humankind has ever faced"; "science fiction's most horrible scenarios become fact" [62, p.26].

The original guidelines for minimizing risk were composed with no participation from public health experts, lab workers, or environmentalists [62, 65, 68]. This enabled Rifkin to frame the rDNA debate as "a question of the public interest groups versus the scientists" [66, p.21], and to capitalize on situations in which local officials were unaware of "secret research into recombinant DNA going on in laboratories in their communities" [62, p.25]. When it became known that some scientists had urged a moratorium on some forms of rDNA work in 1974, the popular interpretation was simplistic: "If scientists were banning some research, they [the public] reasoned, then all of it must be extremely dangerous" [63, p.49].

Public fears led to unpleasantness for working scientists. At Stanford Medical Center, Paul Berg had to terminate his experiment for inserting carcinogenic viral DNA into *E. coli* [67]. From that event came a brief moratorium on some kinds of rDNA experiments [62], followed by the Asilomar Conference of February 1975, which ranked rDNA experiments according to their potential dangers. The Asilomar document then became the basis for the NIH *Guidelines for Research on Recombinant DNA* [65, 69].

This did not satisfy all laypersons. In Cambridge, Massachusetts, a City Councilwoman learned that Harvard was building a P3 lab for rDNA. (P3 describes moderately risky experiments.) There had long been a "fragile relation" between Harvard and the residents of Cambridge, which played out in real estate values, tax bases, and other acrimonious disagreements [66]. The mayor of Cambridge initiated a series of hearings and investigations that emphasized the arrogance of the Harvard scientists in their dealing with the residents of Cambridge. "Who the hell do the scientists think they are," asked Mayor Alfred Vellucci, "that they can take federal tax dollars that are coming out of our tax returns and do research work that we then cannot come in and question?" [67, p.145].

During a long process of ritually humiliating the Harvard scientists, the Cambridge City Council temporarily banned "all recombinant research within the city limits" [66, p.21]. Later it eased that ban, and permitted rDNA work with certain specific safeguards. By 1981, there were similar laws regulating rDNA research in six cities across three states [68].

You might think that finally the scientists clearly understood the public's concerns and the consequences thereof. But Harvard found one more way to embarrass itself. NIH's Guidelines for rDNA research included a procedure for NIH to certify the safety of biological vectors before an rDNA experiment could employ them. Charles A. Thomas, who had been on the NIH committee that composed the guidelines, and thus ought to have known better, had proceeded with not-yet-certified vectors for producing insulin at Harvard Medical School. He was required to terminate his experiments, and his research team was very publicly embarrassed [70, 71].

Lessons from the Case of rDNA

When various elements of the public make sense of nanotechnology, will that process include the telling of lurid horror stories about evil scientists and their dangerous technology? Will public reactions to nanotechnology be as unpleasant as some of the reactions to rDNA? I suggest that the story of recombinant DNA will be relevant to nanotechnology when the following three conditions come together:

1. *Techno-hyperbole backfire*: When some people praise nanotechnology in words and images of unrestrained nanophilic hyperbole, we should remember one of the ironic lessons from the experience of rDNA: techno*philic* hyperbole inspires the opposite reaction too, namely, techno*phobic* hyperbole. How would you feel if someone announced that a small group of elite experts, unknown to you, will control an extraordinary powerful method for manipulating the material world, using knowledge and skills that you yourself cannot possibly control? This is what nanotechnology might sound like to many people.
2. *Malinowskian conditions*: Nanotechnology, like rDNA, is likely to affect different people in different ways; it may create profound historical changes; it might cause people to feel that they cannot understand the new situations in which they find themselves. In any of those conditions, stories about nanotechnology will bear a burden of helping people come to terms with anomaly, conflict, and change. These pressures are not likely to engender a dispassionate appreciation of nanotechnology.
3. *Disdain for public health and safety*: If those who make nanotechnology real are as arrogant and inconsiderate as some of the people who brought us rDNA, then we can expect nanotechnology to be humanized as a stirring drama of virtuous laypersons versus dangerous scientists. This is especially true if the makers of nanotechnology ignore its risks to the public, or if they know those risks but underestimate them, or if they know those risks but dissemble when they ought to be candid.

If all three conditions come together, I anticipate that many public reactions to nanotechnology will be at least as ugly as the initial public reaction to rDNA in Cambridge, Massachusetts. The first, techno-hyperbole backlash, is well underway. Indeed, frightening speculations are the bread-and-butter of the rhetoric of the ETC Group, Bill Joy, and Bill McKibben. Furthermore, nanotechnology is custom-made for Malinowskian conditions. It is likely to create profound historical changes. And, even if it benefits everyone to some degree, its political economy of patents, copyrights, and venture capital will give us a situation in which a limited number of people control those profound historical changes.

The third condition is yet undetermined. There has been too little public awareness of nanotechnology to craft a believable narrative of virtuous laypersons versus dangerous scientists. There have been a few extremely general warnings about the evils of nanotechnology, but no specific episodes of the makers of nanotechnology creating terrible risks to the public and then ignoring or concealing those risks.

Given that the first two conditions are here now, I suggest that the task of anticipating public reactions to nanotechnology should be focused on the last element. A little bit of recklessness or disdain will be easily magnified and transmuted into a compelling story about amoral scientists arrogantly producing terribly dangerous threats to our health and our environment. Perhaps the relevant scientific knowledge will be distorted, ignored, exaggerated, or manipulated, thereby leaving scientists feeling exasperated and powerless. But the important lesson is that hyperbole and Malinowskian conditions have already intensified the values, hopes, and fears that will be shaped into public reactions to nanotechnology.

Conclusions

Stories set in the past are a way of arranging people and values into a moral order. They enable us to say that one hero is better than another; or that some features are good, while others features are evil, and so on. Such stories compete with one another for credibility and historical authenticity: that nanotechnology is a blessing or a curse, that scientists can be trusted or should be feared, that all will enjoy its benefits, or that a few will control its powers. These kinds of pre-existing feelings about science will be as influential as the scientific merits of the research. The same was true in the earlier cases of fluoridation, cold fusion, creationism-versus-evolution, embryonic stem cell research, and many more forms of science and technology.

Nanotechnology is important enough to have its own collection of histories, tales, legends, myths, and anecdotes, but it is also new enough that it has to borrow information from comparisons and analogies until its own record of public reactions is established. As we anticipate those public reactions, let us recognize how they will be shaped by values and lessons that arise repeatedly in democratic societies, particularly if nanotechnology delivers Malinowskian conditions like inequalities of power and profound historical changes.

References

1. B. Malinowski, *Magic, Science, and Religion, and other Essays*, Garden City, NY: Doubleday Anchor (1954).
2. R. Feynman, There's plenty of room at the bottom: An invitation to enter a new field of physics. Talk given at the American Physical Society annual meeting, December 29th, California Institute of Technology (1959).
3. R. Feynman, Infinitesimal machinery, *Journal of Microelectromechanical Systems* **2**(1), 4-14 (1993 [1983]).
4. K. E. Drexler, *Engines of Creation: The coming era of nanotechnology*, Garden City, N.Y.: Anchor Press/Doubleday (1986).
5. K. E. Drexler, Machine-phase nanotechnology, *Scientific American*, 74-75 (September 2001).
6. R. E. Smalley, Of chemistry, love, and nanobots, *Scientific American*, 68-69 (September 2001).

7. K. E. Drexler, R. E. Smalley, Point, counterpoint: Nanotechnology, *Chemical & Engineering News* **1**, 37-42, (December 2003).

8. K. Change, Yes they can! No, they can't! Charges fly in nanobot debate, *New York Times* (9 December 2003).

9. K. A. Goonan, *Queen City Jazz*, New York: Tor (1994).

10. K. A. Gooonan, *Mississippi Blues*, New York: Tor (1997).

11. K. A. Goonan, *Crescent City Rhapsody*, New York: Avon Eos (2000).

12. NSTC (National Science and Technology Council), *National Nanotechnology Initiative: The Initiative and its Implementation Plan*, Washington, DC: National Science and Technology Council (2000).

13. M. C. Roco, W. S. Bainbridge, *Societal Implications of Nanoscience and Nanotechnology*, Dordrecht, Netherlands: Kluwer (2001).

14. B. C. Crandall, ed., Nanotechnology: Molecular speculations on global abundance, Cambridge, MA: MIT Press (1996).

15. G. Stix, Waiting for breakthroughs, *Scientific American*, 94-99 (April 1996).

16. G. Stix, Little big science, *Scientific American*, 26-31 (September 2001).

17. S. Mirsky, Tantalizing tubes, *Scientific American*, 40-42 (June 2000).

18. M. A. Reed, J. M. Tour, Computing with molecules, *Scientific American*, 86-93 (June 2000).

19. J. Rennie, Nanotech reality, *Scientific American*, 8 (June 2000).

20. M. Shermer, Nano nonsense and cryonics, *Scientific American*, 29 (September 2001).

21. A. Keiper, The nanotechnology revolution, *The New Atlantis*, 17-34 (Summer 2003).

22. B. J. Feder, Nanotechnology has arrived; A serious opposition is forming, *New York Times*, C3 (August 2002).

23. L. Mills, Brave new science might change the world—or, some say, end it, *(London) Sunday Telegraph* (20 June 2003).

24. J. Smith, T. Wakeford, Who's in control? *The Ecologist* **33**(4), 40-41 (May 2003).

25. L-A. Broadhead, S. Howard, The heart of darkness, *Resurgence* (221) (Nov.-Dec. 2003) (available at http://resurgence.gn.apc.org).

26. T. Radford, Brave new world or miniature menace? Why Charles fears grey goo nightmare, *The Guardian* (29 April 2003) (available at www.guardian.co.uk).

27. ETC Group, Nanotech takes a giant step down [news release] (6 March 2002) (available at www.ETCgroup.org).

28. ETC Group, Green goo: Nanotechnology comes alive!, *Communique* (77), (January/February 2003) (available at www.ETCgroup.org).

29. ETC Group, The little BANG theory, *Communique*, (78), (March/April 2003) (available at www.ETCgroup.org).

30. ETC Group, No small matter II: The case for a global moratorium. Size matters!, *Occasional Paper Series* **7**(1), (April 2003) (available at www.ETCgroup.org).

31. ETC Group, Size matters! [press release] (14 April 2003) (available at www.ETCgroup.org).

32. ETC Group, No small matter! Nanoparticles penetrate living cells and accumulate in animal organs, *Communique* (May/June 2002) (available at www.ETCgroup.org).

33. D. Brown, Activists: No more nanomaterials until we know whether it pollutes, Small Times (8 August 2002) (available at www.smalltimes.com).

34. ETC Group, Nanotech un-gooed! *Communique* (80), (September/Octobter 2003) (available at www.ETCgroup.org).

35. ETC Group, Much ado about nothing? *Communique* (81), (September/October 2003) (available at www.ETCgroup.org).

36. J. Thomas, Future perfect? *The Ecologis*t, 32–35 (May 2003).

37. A. H. Arnall, *Future technologies, today's choices* [a report for the Greenpeace Environmental Trust], London: Greenpeace Environmental Trust (July 2003).

38. I. Lerner, Nanotechnology faces recent problems in public image, *Chemical Market Reporter*, 10 (19 May 2003).

39. M. Crichton, *Prey*, New York: HarperCollins (2002).

40. G. P. Collins, Shamans of small, *Scientific American*, 86 (September 2001).

41. B. Joy, Why the future doesn't need us, *Wired*, 238-262 (April 2000).

42. B. Joy, Forfeiting the future, *Resurgence* (208), (September–October 2001).

43. B. McKibben, *Enough*, New York: Henry Holt (2003).

44. D. Rotman, Measuring the risks of nanotechnology, *Technology Review* **106**(3), 71-73 (April 1, 2003).

45. E. Tenner, Nanotechnology and unintended consequences. In *Societal Implications of Nanoscience and Nanotechnology,* ed. M. Roco & W.S. Bainbridge, Dordrecht, Netherlands: Kluwer (2001).

46. D. Brown, Drexler warns terror symposium: Nanotech has "extreme downsides," *Small Times* (19 December 2001) (available at www.smalltimes.com).

47. D. Brown, U.S. regulators want to know whether nanotech can pollute, *Small Times* (15 March 2002) (available at www.smalltimes.com).

48. D. Brown, Nanolitterbugs? Experts see potential pollution problems, *Small Times* (15 March 2002) (available at www.smalltimes.com).

49. B. J. Feder, Research shows hazards in tiny particles, *New York Times*, C8 (14 April 2003).

50. B. J. Feder, As uses grow, tiny materials' safety is hard to pin down, *New York Times* **C1, C4**, (3 November 2003).

51. C. Stuart, Nano should regulate itself before the government does, report says, *Small Times* (20 November 2002) (available at www.smalltimes.com).

52. C. Stuart, Watchdogs say stop nanotech, start worldwide dialogue, *Small Times* (31 Jan. 2003) (available at www.smalltimes.com).

53. C. Stuart, Survey finds the smaller the size, the bigger the possible risks, *Small Times* (17 April 2003) (available at www.smalltimes.com).

54. Trouble in nanoland, *The Economist*, 75-76 (7 December 2002).

55. C. P. Toumey, *Conjuring Science*, New Brunswick, N.J.: Rutgers Univ. Press (1996).

56. C. P. Toumey, Conjuring science in the case of cold fusion, *Public Understanding of Science* **5**, 121-133 (1996).

57. M. Heylin, Sociology of cold fusion examined, *Chemical & Engineering News*, 24-25 (19 March 1990).

58. R. Pool, How cold fusion happened–Twice! *Science* **244**, 420-423 (28 April 1989).

59. F. D. Peat, Cold fusion: The making of a scientific controversy, Chicago: Contemporary Books (1989).

60. G. Taubes, Bad science: The short life and weird times of cold fusion, New York: Random House (1993).

61. C. Grobstein, The recombinant-DNA debate, *Scientific American*, 22-33 (July 1977).

62. J. Rifkin, DNA, *Mother Jones* (February/March 1977).

63. W. Bennett, J. Gurin, Science that frightens scientists: The great debate over DNA, *Atlantic Monthly*, 43-62 (February 1977).

64. J. King, A science for the people, *New Scientist*, 634-636 (16 June 1977).

65. S. Krimsky, *Genetic alchemy: The social history of the recombinant DNA controversy*, Cambridge, MA: MIT Press (1982).

66. A. Budrys, The politics of deoxyribonucleic acid, *New Republic*, 18-21 (16 April 1977).

67. C. Nader, Technology and democratic control: The case of recombinant DNA. In *The gene-splicing wars*, ed. R.A. Zilinskas & B.K. Zimmerman, New York: Macmillan (1986).

68. M. Singer, Scientists and the control of science, *New Scientist*, 631-634 (16 June 1977).

69. N. Wade, Recombinant DNA: NIH rules broken in insulin gene project, *Science* **197**, 1342-1345 (30 September 1977).

70. N. Wade, Harvard gene splicer told to halt, *Science* **199**, 31 (6 January 1978).

71. NOVA, ConFusion in a Jar (transcript of show #1802).

7. EDUCATION

HUMAN RESOURCE IMPLICATIONS OF NANOTECHNOLOGY ON NATIONAL SECURITY AND SPACE EXPLORATION

Grant C. Black, Georgia State University

The emergence of nanotechnology and its expected increasing prevalence in society prompt careful examination of nanotechnology's potential impact on society, particularly national security and significant public enterprises, such as the space program. The United States lacks unchallenged dominance in nanotechnology. Several nations are poised to provide strong competition, given the global rise in skilled labor and technological infrastructure.

Human resource issues must be addressed in assessing the United States' ability to achieve its objective of being the leading nation in nanotechnology. Several of these issues may have significant implications for national security as well as for space exploration. Unfortunately, little existing research can shed light on these issues, highlighting the need for more direct investigation of questions related to the societal impact of nanotechnology.

At this early stage, the future labor market impact of nanotechnology is unclear. While job growth is expected, it remains to be seen whether a temporary shortage of skilled workers will arise. Based on recent job listings as a measure of labor demand, there does not appear to be such a labor shortage in the current nanotechnology environment [1]. This, however, could change based on the rate of future technological breakthroughs and adoption rates of nanotechnology products and processes. To achieve international dominance in nanotechnology, the United States must ensure an adequate pipeline of skilled labor sufficient to meet predicted nanotechnology needs over time. The ability to accurately forecast demand for a scientific workforce is suspect [2], yet it is pertinent to seek a reasonable indication of likely changes in nano-related labor markets. This, however, requires better information than currently exists on the estimated time frames, needs, and impacts of nanotechnology throughout society.

High priority must be placed on ensuring a sufficient supply of skilled labor, which requires maintaining and improving the educational infrastructure at all levels. Few nanotechnology academic programs currently exist; however, nanotech-related courses are offered at many institutions [3]. A major hurdle to establishing nanotechnology programs is the inherently interdisciplinary nature of nanoscience and nanotechnology, and formal interdisciplinary programs can be difficult to create and sustain under the current academic system [4]. Moreover, many U.S. students are weak in the training and competencies needed for careers in nanotechnology. For instance, in 2000, 20 percent of freshman reported needing remedial work in computer science, 14 percent in physical sciences and engineering, and 23 percent in biological sciences; interestingly, only 4 percent reported needing remedial work in mathematics [5]. Public policy initiatives could

M.C. Roco and W.S. Bainbridge (eds.),
Nanotechnology: Societal Implications — Individual Perspectives, 297–335.

be beneficial in stimulating the growth of nanotechnology programs and scientific training at U.S. institutions.

Government should consider incentives to attract students into nanotech-related fields. This has long been a concern in science and engineering (S&E) as a whole. The number of associate degrees earned in the natural sciences, math and statistics, and engineering represented only 4 percent of all associate degrees in 1998; of those 21,730, 3 percent of the total were earned by foreign citizens [5]. This small number may be inadequate if speculation on the usefulness of two-year training in nanotechnology proves correct [3]. In 1998, 205,355 bachelor's degrees were awarded in the natural sciences, math and computer science, and engineering—17 percent of all bachelor's degrees; of these, 4.7 percent went to foreign citizens [5]. As a benchmark, the proportion of white college students interested in majoring in math and statistics has declined since the early 1980s, stabilizing in the 1990s at approximately 0.7 percent. Those interested in majoring in computer science were approximately 3 percent in 2000, while approximately 8 percent were interested in majoring in engineering that same year after highs over 11 percent in the early 1980s. Government could attract students through financial support that targeted nanotechnology fields and research. The proportion of full-time graduate students receiving federal support in 1999 was approximately 35 percent in the physical sciences, 9 percent in mathematics, 15 percent in computer science, 25 percent in the life sciences, and 24 percent in engineering [5]. Of the students with federal support, NASA funded only 3.5- to 4.2 percent of them in 1996–1999.

A master's-level education may be one avenue for training a professional workforce in nanotechnology. This concept has recently taken root in the United States with the emergence of professional science degrees, due in large part to the support of the Alfred P. Sloan Foundation and the National Science Foundation. At this time, only one of the Sloan professional master's programs is specifically related to nanotechnology. Such programs, however, take time to implement and can face considerable obstacles [4]. For overall master's training, 56,660 master's degrees were awarded in the natural sciences, math and computer science, and engineering in 1998, representing 13 percent of all master's degrees; of these, almost 32 percent were earned by foreign citizens [5].

The development of nanotechnology will depend to a great extent on a doctorally trained workforce involved in nano-related R&D. In 1999, 44 percent (18,226) of all doctorates were awarded in the natural sciences, math and computer science, and engineering; of these, 7,472 (41 percent) were received by foreign citizens [5]. The increasing number of foreign students has driven the majority of the sizeable growth in Ph.D. production in the last 20 years [6, 7]. In 1981, 20 percent of science and engineering Ph.D. recipients were temporary residents, while more than 32 percent were temporary residents in 1999. By the late 1990s, approximately 40 percent of Ph.D. recipients in mathematics, computer science, and engineering were temporary residents. Most foreign students come from a handful of Asian countries. During the 1990s, 60 percent of temporary-resident Ph.D. recipients came from the People's Republic of China, Taiwan, India, and South Korea.

The issue of foreign students receiving S&E training in the United States is a legitimate concern for national security and well being [8–11]. Considerably more information and research are needed to adequately examine the potential impact of foreign students in nanotechnology. In the 1990s, 7,110 Ph.D. recipients were temporary residents from the 26 countries included on the U.S. Department of State's watch list developed after the terrorist activities on September 11, 2001 [6]. These Ph.D.s represent 3.6 percent of all S&E doctorates earned during this period but 11 percent of doctorates awarded to temporary residents during this period. Approximately 10 percent of the 7,110 Ph.D.s were in "sensitive" fields, defined as nuclear and organic chemistry; chemical and nuclear engineering; bacteriology; biochemistry; biotechnology research; microbiology; molecular biology and neuroscience; and atomic, chemical, molecular, and nuclear physics. Most of the Ph.D.s in sensitive fields came from Turkey and Iran, followed at a considerable distance by Pakistan, Malaysia, Egypt, and Jordan.

From self-reported information collected by the National Science Foundation through the Survey of Earned Doctorates, an annual census of all new Ph.D.s in the United States, some knowledge can be learned about foreign students' immediate postdoctoral plans—at least about whether or not they plan to stay in the United States [7, 12]. Of the 7,110 from watch-list countries earning S&E Ph.D.s in the 1990s, 43 percent reported plans to stay; of the 732 in sensitive fields, 56 percent planned to stay [7]. Plans to stay were quite low for graduates from a number of Persian Gulf and North African countries. For all temporary-resident S&E Ph.D. recipients, the proportion planning to stay in the United States has increased considerably, from approximately 40 percent in the early 1980s to more than 70 percent by 1999. Plans to stay, however, vary considerably by country of citizenship. For instance, approximately 80 percent of S&E Ph.D. recipients from China plan to stay, while only 15 percent of those from Brazil plan to stay. The location decision for these students is complex. Significant factors influencing the likelihood of staying in the United States include a student's age, marital status, prior ties to the United States (such as a U.S. undergraduate education and work experience), nationality, field of training, and quality of training.

Beyond the issue of foreign students, the ease with which information is transferred, and whether such transfers should be more carefully monitored, should also be considered for their implications on national security. At the present time, little effective monitoring occurs. In the past 20 years, however, the transfer of information has become less costly and more frequent—due in large part to technological innovations such as the Internet and e-mail. Moreover, collaborative research networks in the sciences have expanded in size and grown increasingly international [13]. These trends have increased the likelihood of information transfers with potential security risks.

In summary, considerably more information and research are needed on human resource issues related to nanotechnology. The human resource implications of nanotechnology on national security and significant public enterprises, including space exploration, are particularly unclear and must be addressed. Pertinent issues concern (1) ensuring a sufficiently skilled workforce to meet future demand; (2)

the ease with which information of national interest is transferred globally; and (3) the prevalence of foreign students trained in the United States.

References

1. P. Stephan, Human resources for nanotechnology. Paper presented at the National Science Foundation Workshop on Societal Implications of Nanoscience and Nanotechnology, 3-5 December, Washington, DC (2003).
2. National Research Council. Forecasting demand and supply of doctoral scientists and engineers: Report of a workshop on methodology, Washington, DC: National Academy Press (2000).
3. S. Fonash, Education and training of the nanotechnology workforce, *Journal of Nanoparticle Research* **3**, 79-82 (2001).
4. P. Stephan, G. Black, Bioinformatics: Does the U.S. system lead to missed opportunities in emerging fields? A case study, *Science and Public Policy* **26**(6), 382-392 (1999).
5. National Science Board. *Science and Engineering Indicators–2002*, Arlington, VA: National Science Foundation (2002).
6. G. Black, P. Stephan, Importance of foreign Ph.D. recipients to U.S. science. In *Science and the University*, ed. R. Ehrenberg, P. Stephan, Madison, WI: University of Wisconsin Press (forthcoming).
7. P. Stephan, G. Black, J. Adams, S. Levin, Survey of foreign recipients of U.S. Ph.D.s, *Science* **295**(22), 2211-2212 (2002).
8. G. Borjas, The impact of foreign students on native enrollment in graduate programs, unpublished manuscript (2003).
9. G. Borjas, An evaluation of the foreign student program, Center for Immigration Studies, Washington, DC (2002).
10. S. Levin, A. Winkler, G. Black, P. Stephan, Differential employment patterns for citizens and non-citizens in science and engineering in the United States: Minting and competitive effects, *Growth and Change* (forthcoming).
11. P. Stephan, S. Levin, Exceptional contributions to U.S. science by the foreign-born and foreign educated, *Population Research and Policy Review* **70**(1-2), 59-79 (2001).
12. M. Finn, Stay rates of foreign doctorate recipients from U.S. universities, 1997, mimeograph, Oak Ridge, TN: Oak Ridge Institute for Science and Education (2000).
13. J. Adams, R. Clemmons, G. Black, P. Stephan, Patterns of research collaborations in U.S. universities, 1981–1999, *Research Policy* (forthcoming).

TRAINING STUDENTS TO BE INTERACTIONAL EXPERTS

Michael E. Gorman and James Groves, University of Virginia

> Productive work on societal implications needs to be engaged with the research from the start. Ethicists need to go into the lab to understand what's possible. Scientists and engineers need to engage with humanists to start thinking about this aspect of their work. Students need training now that will take their understanding of nanotechnology from laboratory to society. These students today, trained in the right interdisciplinary setting, will become a cadre of scientists, engineers, and scholars used to working together, thinking about the societal and technical problems side-by-side. Only thus, working together in dialog, will we make genuine progress on the societal and ethical issues that nanotechnology poses.
>
> Davis Baird, in testimony before the Senate Committee on Commerce, Science and Transportation, May 1, 2003

Note the emphasis in the above quote on interdisciplinary collaboration and training of the most radical sort, including engineering, science, ethics, and social science. Such collaborations depend not only on disciplinary depth, but also on what Collins and Evans refer to as *interactional expertise*, which includes "the ability to pass interesting scientific information from one scientist to another. One can see that this does need a degree of painfully acquired scientific expertise, but not enough to qualify its owner as a contributor to the field. Another thing you can do with interactional expertise is to argue a devil's advocate position, but not well enough to take scientific decisions based upon it" [1, p.447].

Collins is a sociologist of science who developed the concept of interactional expertise from his own experience studying scientific processes like the attempt to replicate a new form of laser [2]. He did not have sufficient contributory expertise to suggest improvements in the laser design, though he was able to play a kind of devil's advocate role when it came to troubleshooting problems.

The kind of collaboration advocated by Baird involves more than social scientists studying nanotechnology. The social scientists will have to become partners in the enterprise, and in order to do that, they will have to acquire interactional expertise in areas of nanotechnology. Similarly, nanoscientists will have to acquire interactional expertise in ethics and social sciences. In effect, they will have to form a trading zone and develop a common language, or creole, in order to exchange knowledge [3].

To see if this sort of trading zone is really possible, the authors conducted a pilot project at the University of Virginia, supported by a small ($100,000) NSF award. We co-supervised a master's student. This trading zone included NSF funding of the graduate student's stipend and a small portion of the time of the two advisors. Access to laboratory facilities and equipment was provided by an NSF-sponsored Center for Nanoscopic Materials Design.

Each of us began the trading zone with different goals. The student wanted a master's degree in Materials Science and wanted to do a laboratory thesis, though she was strongly interested in the societal implications as well. Groves had expertise in metal oxides and wanted to extend this expertise into nanotechnology. Gorman wanted to see if he could create a small, interdisciplinary trading zone where we would first become interactional experts, then perhaps contributing ones.

Groves liked to use matrices to plot research strategy, something Gorman often did in his psychology work. Gorman suggested we might extend this method to combine social and technical dimensions. Groves suggested a matrix that combined world ills with possible engineered devices that could mitigate such ills, and the graduate student got to work on it. While mitigation of certain world ills could involve development of a newly engineered sensor system (e.g., to detect a chemical, biological, or radiation hazard), other world ills research might seek to develop systems for purification (e.g., of groundwater for drinking, of manufacturing plant effluent, or of fluids used in medical treatments). Still other world ills research might investigate newly engineered medications and delivery systems for improved human and biosphere health. Still other problems might be made worse by nanotechnology. For example, differences between the developed and developing world might be exacerbated by nanotechnology.

An implicit constraint on this kind of research is local expertise and knowledge. The Center for Nanoscopic Materials Design was established in 2000 to investigate the directed self-assembly of materials onto patterned surfaces, primarily in the silicon-germanium material system. Groves wanted to extend this sort of work to metal oxides, a relatively understudied area. With regard to this particular project, the literature suggests that metal oxides might be useful as a foundation for a bio-nano scaffold [4], the type of development that could mitigate world ills like terrorism, disease, and pollution.

So far, our limited creole involved Gorman learning the meaning of terms like *deposition*, *nanodot*, and *focused ion beam*, and Groves and the graduate student learning terms like *trading zone* and *problem space*. In order to explain what we were doing at a strategy level, Groves—a hiker like Gorman—proposed a metaphor. The world ill we were trying to mitigate was a distant mountain. The graduate student's research would be a bridge toward that mountain. Perhaps she could focus on finding two candidate metal oxides that might be useful in a biomedical application, and focus on depositing one on the other.

Therefore, our project team added a biomedical engineer associated with the Center to our growing trading zone. We are currently exploring how metal oxides could impact biomedical research and development in a positive way by providing a bio-nano scaffold that assists in understanding wound healing, progression of atherosclerosis in arteries, and tumor cell invasion. Each of these applications could serve as a mountain in our metaphor, and the graduate student's work would build a bridge that could potentially lead to any peak in this whole range.

Each of us acquired interactional expertise with respect to social and technical aspects of our joint project, evidenced by posters, papers, and presentations we had to create together. For example, Gorman had to acquire interactive expertise across several types of knowledge [5], including basic information about aspects of nanotechnology, including terms needed as part of our creole. Gorman also had to acquire judgment concerning what constituted an appropriate research project, essential if Gorman were going to give the graduate student good advice about the scope and direction of her thesis. Gorman tried to acquire this kind of knowledge by constantly asking for justification of decisions like our preference for metal oxide systems, which led us to construct more matrices of possible combinations and search the literature for relevant articles.

This small pilot study indicates that a social scientist and a materials scientist can share a graduate student and collaborate on a nanotechnology research project—acquiring interactional expertise in each others' areas, and beginning to make joint contributions. The result is a different research process. Is the final product any different? We discussed that issue at length, and agreed that to be sure, we would have to have a control group—somehow run a clone of the graduate student through a project with the same advisor, minus the interaction with Gorman. This kind of study is both unethical and impossible.

We are now in the process of recruiting others to try similar trading zones, to see if our approach is generalizable to other trading zones in the Social and Ethical Implications of Nanotechnology (SEIN).

Implications for Education

To fulfill Baird's dream of "a cadre of scientists, engineers, and scholars used to working together," students will have to be trained in interactional expertise as well as disciplinary depth. But once outside elementary school, students are taught science, technology, and social sciences as separate domains, with very little interaction across these areas. The best way to teach interactional expertise is to model it—to have teachers working together in teams to introduce nanotechnology and its impacts. The students should see the social science, math, science, English, and history teachers struggling to understand one another as they present this new frontier. This kind of teacher collaboration will require the development of special scaffolding that includes project ideas, materials, hints on how to collaborate, and ways to assess progress and problems.

A challenge will be fitting such an idea into the disciplinary testing structures mandated by states and by the Advanced Placement system. The only way a teaching project on nanotechnology will succeed is if it also permits coverage of disciplinary skills and knowledge. The scaffolding should contain suggestions on how to cover or reinforce specific disciplinary objectives in the course of a multidisciplinary module on nanotechnology. Let us consider an example.

Nanotechnology Policy Simulation

Students at the secondary and post-secondary levels could be involved in an interactive trading zone simulating nanotechnology policy. Students would be placed into groups representing various decision-making bodies in the government, thereby creating a trading zone that simulates how policy is determined. Roles might include the following:

1. The Office of Science and Technology Policy (OSTP) at the White House that directs research policy and develops budget recommendations.
2. The House Committee on Science, which holds hearings and makes budget recommendations. The Congress and the White House often have different agendas.
3. The National Science Foundation, which makes a significant investment in nanotechnology at the program level.
4. The Woodrow Wilson Policy Institute, which has a Foresight project that looks ahead at technology directions and informs Congress and the White House.
5. The non-government group ETC that has recommended a moratorium on nanoscience research until its societal impacts are thoroughly studied.
6. Groups representing multidisciplinary nanotechnology research efforts, like the NSEC at Rice, which is focused on environmental implications of nanotechnology.

Students would study the groups that play these roles in the real world and interact from these perspectives to make decisions about where nanotechnology research ought to go, including whether there ought to be a moratorium.

It is easy to see how such a simulation would fulfill the goals of a government class. But properly constructed, this simulation would also achieve science and technology objectives. Students would have to research and understand actual nanotechnology research frontiers, aided by scaffolding created for the simulation that includes articles, links to Web sites, even interviews with experts. Teaching moments occur frequently in such simulations, creating the opportunity for brief lectures on basic scientific principles that students need to understand in order to make intelligent decisions [6].

This simulation is only a rough example of what might be done. The point is to design educational environments that encourage students to complement their growing disciplinary knowledge with the ability to interact across specialties.

References

1. H. M. Collins, R. Evans, King Canute meets the Beach Boys: Responses to the third wave, *Social Studies of Science* **33**(3), 435-452 (2003).
2. H. M. Collins, R. Harrison, Building a TEA laser: The caprices of communication, *Social Studies of Science* **5**, 441-445 (1975).
3. M. E. Gorman, Expanding the trading zones for convergent technologies. Pp.424–428 in Converging Technologies for Improving Human Performance: Nanotechnology, biotechnology, information technology, and cognitive science, ed. M. C. Roco, W. S. Bainbridge. Dordrecht, Netherlands: Kluwer (2003).

4. R. Michel, I. Reviakine, D. Sutherland, C. Fokas, G. Csucs, G. Danuser, et al., A novel approach to produce biologically relevant chemical patterns at the nanometer scale: Selective molecular assembly patterning combined with colloidal lithography, *Langmuir* **18**, 8580-8586 (2002).
5. M. E. Gorman, Turning students into professionals: Types of knowledge and ABET engineering criteria. *Journal of Engineering Education* **91**(3), 339-344 (2002).
6. M. E. Gorman, J. A. Plucker, C. M. Callahan, Turning students into inventors: Active learning modules for secondary students. *Phi Delta Kappan,* **79**(7), 530-535 (1998).

EDUCATING UNDERGRADUATE NANOENGINEERS

Tanwin Chang, National Bureau of Economic Research

Nanotechnology is understood to sit at the intersection of established scientific and engineering disciplines. But is there something more? Will a coherent body of knowledge emerge that is considered the "native" domain of nanotechnology? The answer to this question has important implications for the education and productivity of the presumed nanotechnology workforce of the future. That is, how is the nation to train our students of science and engineering so that they can be best utilized in an age of nanotechnology?

In particular, how do we prepare our undergraduate students for careers in nanotechnology? This article takes the position that, yes, it is possible for a distinct field of study—a "stand-alone discipline"—to coalesce, thereby forming an epistemological base for nanotechnology. Further, it is possible to construct nanoengineering baccalaureate curricula that teach students valuable skills and useful techniques.

This article presents an outline for a nanoengineering curriculum that is consistent with the typical requirements of a university in the United States for an ambitious undergraduate. But such degree programs should not be implemented simply because they are possible. I argue that the practice of awarding nanoengineering degrees will hasten the transfer of nanoscale discoveries from academic institutions to industry. Another reason to allow students to earn the "nanoengineer" label is to enhance workforce versatility during the development of novel products and processes enabled by nanotechnology.

Nanotechnology Undergraduate Educational

Americans categorize themselves by their undergraduate degrees. When someone declares at a cocktail party, "I was a math major," the likely intention is to convey information regarding that person's quantitative skills, idealism, and perhaps taste in literature. An undergraduate degree symbolizes a level of basic competency in a specific discipline of knowledge and also implies membership in a culture of like-minded individuals. Societal progress has resulted in the establishment of new degree programs in the past 50 years: witness Environmental Science. However, not every advancement is accompanied by a baccalaureate degree program. For example, with the rise of the Internet, we have no "Internet engineers." Instead, workers who support the Internet may hold bachelor's degrees in electrical engineering or computer science. Presently, there are no four-year nanotechnology undergraduate degree programs in the United States. However, the NSF has initiated a number of funding opportunities for "nano" education in conjunction with the National Nanotechnology Initiative (NNI). Will the phrase "I was a nanotechnology major" be frequently overheard at cocktail parties of the future?

The problem of creating undergraduate curricula for nanotechnology is embodied in its inter- or multidisciplinary nature. Consider an excerpt from the report of the NSF's workshop on Nanotechnology Undergraduate Education (NUE) in September 2002:

> ...there was no support at the workshop for a stand-alone degree at the B.S. level in nanotechnology at this time... In the future, there may be opportunities in pure engineering, engineering technology, and the sciences for independent degree programs in nanotechnology... Nanotechnology should not be considered a single discipline but should permeate the undergraduate science and engineering curriculum. This can be accomplished through modifications of existing courses and creation of new courses, which could lead to concentrations and/or minors in nanotechnology or nanoscience. [1, p.8]

The message here is that nanotechnology is not the basis for a stand-alone discipline at this time. Still, the excerpt seems to have been written with the future in mind. For instance, the workshop recommends the creation of new nanotechnology courses and concentrations. This may be interpreted as a kind of preparation for the anticipated "independent degree programs" that will presumably be established when nanotechnology does comes of age as its own discipline.

If nanotechnology may be a stand-alone discipline in the future, then the question is, "Why not now?" If it can be done, shouldn't educators embark on baccalaureate degree programs immediately? The slow prudent approach may reflect uncertainty on the ability of industry to absorb graduates with a degree titled "nanotechnology." Better to allow students to earn a traditional degree that is easier to place in the job market. The efforts at the University of Washington to create a "nano" Ph.D. program are informative:

> In a new field, it is critical to insure a student's viability upon completing the program: brand-new degrees are met with incomprehension, or active distrust, in both academia and industry. The field is too young to develop a stand-alone 'Nanotechnology' Ph.D. program. [2, p.500]

If educators are not ready to embark on doctoral programs, they are probably no more eager to offer baccalaureate degrees in nanotechnology.

Note, however, that bachelor's degrees in nanotechnology *do exist*, but not in the United States. At the time of this writing (January 2004), several programs are offered internationally, particularly in Australia. Flinders University's nanotechnology undergraduates first matriculated in 2000 [3], and recently the University of Queensland and Curtin University began to offer nanotechnology specializations. In Canada, the University of Toronto now offers a nanoengineering option for its Engineering Science program. Are these universities too hasty in their decision to offer such programs? When will nanotechnology be ready to fit into an American undergraduate curriculum?

Nanoengineering vs. Nanoscience

A common problem with forward-looking statements on this topic is the ambiguity arising from the fluid usage of the word "nanotechnology" and associated terms. Let's adopt the following clarifications, which are adapted from various sources in the literature:

- "Nanoscience" is a convergence of the natural sciences—physics, biology and chemistry—that deals with the behavior and properties of matter on length scales between the molecular and the micron size [4].
- "Nanoengineering" refers to the use of nanoscience methods and knowledge to manipulate matter at the nanoscale in order to create products.
- "Nanotechnology" is a catch-all term that refers to the entire enterprise of nanoscience, nanoengineering, and the nano-enabled products themselves.

Assuming that baccalaureate programs in nanotechnology will arrive eventually, which are more likely to be offered by universities: degrees in nanoscience or nanoengineering? Anecdotally, nanotechnology is often said to be mostly an engineering effort. That is, much is already known about the way things work at the nanoscale, which is the "nanoscience." What remains is the larger challenge of how to *manipulate* matter in this regime, which is the "nanoengineering." Evidence of this viewpoint can be inferred from the distribution of disciplines of faculty members at research institutes in the United States. Figure 7.1 shows data gathered from the Web sites of 12 "nano" institutes.

Each faculty member was categorized into one of the disciplines listed in the National Opinion Research Center's 2001 Doctorate Records File (DRF). In most cases, the "home" department of the faculty member coincided with a category in the DRF. However, many of the faculty belonged to more than one department, or otherwise did not fit neatly into the DRF's listed disciplines. In those cases, the categorization was based on research interests, the publication record, and/or the member's Ph.D. discipline. The data includes a total of 448 faculty researchers in 51 disciplines, although Fig. 7.1 breaks down only the top 10 disciplines. If all 51 disciplines are examined, then we find that 52 percent of the researchers are categorized as engineers, 44 percent are natural scientists, and the remainder are social scientists or in the humanities. However, the pie chart shows that engineers are more dramatically favored in the more important disciplines. For instance, the top three disciplines represented at nanotechology research institutes are engineering disciplines. (Materials Science is traditionally taught in engineering schools, and it is listed as an engineering discipline in the DRF). Additionally, 68 percent of the researchers in the top 10 disciplines are categorized as engineers, while 32 percent are natural scientists. If universities place a similar emphasis on engineering vs. science into undergraduate nanotechnology degrees, then we may expect that those graduating with bachelor's degree in nanoengineering will outnumber those holding nanoscience degrees.

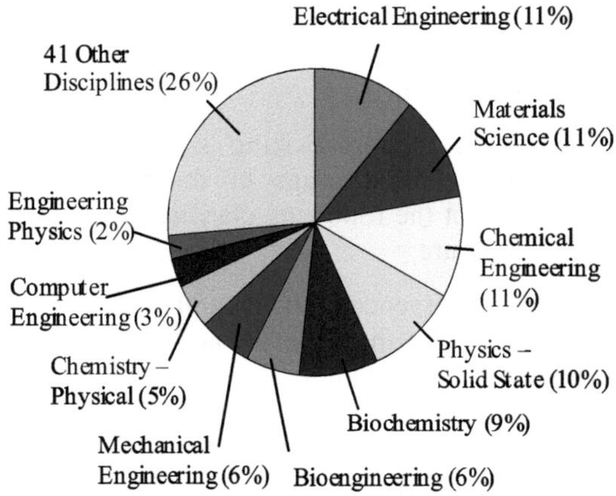

The surveyed institutes
California Nanosystems Institute
Cornell Center for Nanoscale Systems
Columbia Center for Elec. Trans. in Molecular Nanostructures
Harvard Nanoscale Science and Engineering Center
MIT Institute for Soldier Nanotechnology
Northwestern Nanoscale Science and Engineering Center
Purdue Center for Nanoscale Devices

Figure 7.1. Top 10 Faculty Disciplines at "Nano" Institutes in the United States.

A B.S. in Nanoengineering. Is It Just BS?

In order to limit ambiguity, let's focus on *nanoengineering*, specifically, the wisdom of implementing baccalaureate programs for *nanoengineers*, although portions of the discussion may be applicable to notional nanoscience programs also. Here is a skeptical viewpoint:

> Nanoengineering covers too wide a range of potential applications and requires a knowledge base so disparate that it cannot be codified into a stand-alone discipline.

This perspective implies that even a lifetime of learning is insufficient to produce the proverbial nanoengineer, and that there is no nano-Esperanto that can be developed to bridge the cultural gap between, say, bioengineering, chemical engineering, and materials science. Under this interpretation, the ambitious technologist who sets out to master all the disciplines of nanotechnology ends up assembling a plethora of metaphorical jigsaw puzzle pieces, but finds it impossible to build a unified puzzle since the pieces belong to several distinct and separate puzzles. In this case, it seems a dubious enterprise to offer a bachelor's in nanoengineering. The best preparation might instead be to offer double majors for students wishing to become nanotechnology workers. Here is a more hopeful viewpoint:

> Nanoengineering *can* be integrated into a single discipline; however, it is so broad and deep that an undergraduate career is simply not enough time to properly train a student in this subject.

This statement has traction among researchers and educators involved with nanotechnology who lament, "How can a student be trained to be an expert at *all* the natural sciences and associated engineering disciplines?" That is, in a four-year program, there's simply not enough time to gain sufficient depth of knowledge to compete against those who hold degrees in traditional disciplines. This position implies that any attempt to implement a bachelor's nanoengineering curriculum would inevitably produce skills-poor, unemployable, technological generalists. However, there is another path implied here: if students were willing to extend their undergraduate careers beyond the traditional 4 years, or perhaps were committed to graduate careers at the outset, there could be hope of training a cadre of nanoengineers. But now consider an even more optimistic perspective:

> Nanoengineering *can* (1) be integrated into a single discipline that *can* (2) be consolidated into a curriculum effectively taught to undergraduates, and (3) the implementation of such degree programs would be a good thing.

This last viewpoint does have proponents in the nanotechnology community. For these believers, the question is "How do we implement a nanoengineering major?" rather than "Is it possible or prudent?" The remainder of the document explores the validity and ramifications of this position.

Nanoengineers Need Not Know All

While the commercial applications for nanotechnology are very broad, the underlying nanoscience may have a unified story. This is the position of the government's IWGN Report "Nanotechnology Research Directions" (published in 2000) regarding the various scientific disciplines involved in nanoscale research:

> Each of these disciplines has evolved its own separate view of nanoscience; the opportunities for integrating these views and for sharing tools and techniques developed separately by each field are today among the most attractive in all of science. [5, p.2]

This declaration implies that at the nanoscale, there exist unifying principles, tools, and techniques that may serve as a pedagogical foundation for a nanoengineering stand-alone discipline. The integrated nanoscience, in turn, can support a consolidated nanoengineering education. This unification of viewpoints is more like a blending of cultures (the way that tap dancing was a blend of African-American, Irish-American, and other immigrant traditions), rather than some grand theoretical synthesis (like the unification of electricity and magnetism).

Here is an important point: a nanoengineer is not required to be an expert at all of the natural sciences, only the aspects of the natural sciences that are concerned with the nanoscale. This is a key realization that brings an undergraduate nanoengineering curriculum into the realm of possibility. Yes, the nanoengineer must gain a basic level of expertise in molecular biology, but does not need to know about ecology, or mammalian anatomy, or population genetics. These latter subjects, which may be essential for a bioscience or bioengineering degree, can be omitted from a nanotechnology curriculum. Likewise, a nanoengineer must fathom many aspects of modern physics, but the entire mathematical apparatus of classical electromagnetic wave theory, which is an essential aspect of a physicist's education, is unnecessary for a nanoengineer. EM wave theory is a framework to explain "action at a distance"—the interaction of charged particles at relatively long distances. But for nanoengineers, "interactions of close proximity," such as van der Waals forces, are dominant.

A decisive issue in the design of an effective bachelor's nanoengineering program is the assessment of which ideas and techniques can be omitted from the traditional disciplines, and which ideas must remain to be repackaged as a coherent "nano" narrative. Indeed, construction of a nanoengineering curriculum will require substantial reorganization of existing programs. However, it may not actually require a comprehensive overhaul of the way that engineering is currently taught at universities in the United States. Even so, some advocates suggest that the time is right for a just such an overhaul. This is called "reversing the pyramid."

> Reversing the pyramid of learning would provide a holistic view, deductive understanding, and motivation to students in physical, chemical, biological, and engineering sciences at all levels. [6, p.1248]

While a nanoengineering curriculum does not depend on "reversing the pyramid," it may serve as an effective stepping stone towards that goal. That is because intermediate-level science courses will have to be rewritten in order to emphasize the cross-disciplinary concepts that are relevant to nanotechnology, thereby forming a "nanoscience core" that will be a unique amalgamation of subject matter from existing courses. The nanoscience core will likely embody a more holistic philosophy of the natural sciences, as the "reversing-the-pyramid" strategy advocates. However, it is also possible to fit the nanoscience core into the current undergraduate engineering paradigm that makes nanoengineering simply another discipline among many other engineering disciplines. A better understanding of how the nanoscience core fits into a conceptual nanoengineering program can be obtained by examining the course structure of an ambitious sort of undergraduate: the pre-med.

Nanoengineering Curriculum: A Proposal

Pre-meds, i.e., undergraduate students who plan to attend medical school, are required to have a well-rounded background in the sciences in addition to their regular major or concentration. They are the existing embodiment of an undergraduate natural science "generalist." A typical pre-med course load consists of four to five classes per semester over four years, for a total of about 36 courses. Medical schools all impose a core science requirement that roughly comprises of one year each of calculus, biology, inorganic chemistry, organic chemistry, and physics. On top of that, most colleges have a general education requirement along with courses that count towards the major.

To build our conceptual nanoengineering degree program, we take first the pre-med's basic science core, as shown in Table 7.1. The general education requirement found at most universities is filled by the addition of 10 humanities or social science courses. Thus far, the courses for this program are taken from ones already existing at the university.

Next add the nanoscience core that comprises eight courses. The two math/compsci courses may be drawn from already existing intermediate-level courses. The remaining six courses will be courses specifically designed to emphasize the concepts of physics, chemistry, and biology that are relevant to nanotechnology. The challenge for curriculum developers is to create these "nanoscale" courses so that they complement and build upon each other, while minimizing any unnecessary redundancy among themselves.

Table 7.1
National Undergraduate Nanoengineering Program

36 Courses (each one semester)		
Basic Science Core 12 courses designed for freshmen and sophomores. Note that this sequence more than satisfies premed science requirements at most medical schools.	Math/ CompSci	Calculus I
		Calculus II
		Statistics/data reduction
	Physics	Fundamental physics I
		Fundamental physics II
		Modern physics
	Biology	Introductory biology I
		Introductory biology II

Table 7.1
National Undergraduate Nanoengineering Program

36 Courses (each one semester)		
	Chemistry	Principles of chemistry I
		Principles of chemistry II
		Organic chemistry I
		Organic chemistry II
General Education 10 courses.	Language	(2 semesters)
	History	(2 semesters)
	Philosophy	(2 semesters)
	Arts	(2 semesters)
	Politics	(2 semesters)
Nanoscience Core 8 courses. The 6 "nanoscale" courses must be designed to emphasize nanoscale phenomenon.	Nanoscale Physics	2 semesters emphasizing fluids mechanics, Brownian motion, thermodynamics, surface effects, basic quantum mechanics, GMR, etc.
	Nanoscale Chemistry	2 semesters emphasizing macromolecular interaction, single molecule chemistry, catalysis, transduction, etc.
	Nanoscale Biology	2 semesters emphasizing biochemistry, cellular machinery, hierarchical construction of matter, etc.
	Math/ CompSci	Math for engineers
		Intro. computer modeling
Engineering and Techniques Core 6 courses that emphasize specific skills and know- how.	Students choose 6 courses from a menu that may include a spectrum of difficulty level. Examples in next column.	Introduction to Fullerenes
		Scanning probe microscopy
		Modeling molecules
		Lithographic techniques
		Self-assembly techniques
		NEMs devices
		Cellular signaling

Table 7.1
National Undergraduate Nanoengineering Program

36 Courses (each one semester)		
		Computational biology
		NMR spectroscopy
		Proteomics
		Molecular memory
		Colloids and nanoparticles

Finally, the engineering and techniques core enables nanoengineering students to gain specific skills and know-how relevant to nanotechnology. Students choose six courses from a menu that may include a spectrum of difficulty level. Some of these may be taken early in the undergraduate career, while others may have prerequisites that can only be fulfilled by upperclassmen. Judicious choice of the courses in the engineering sequence can add a certain emphasis to the student's expertise that mirrors his or her interest.

Note that a well-designed nanoengineering program may serve as an ideal education for a future MBA who plans to manage a high-tech company. By choosing nanoengineering as a major, the pre-MBA demonstrates dedication and scholarship while earning an exceptionally strong and well-rounded scientific background. This is especially useful for gaining the respect of the company's engineering staff, as well as the trust of investors, which are both essential ingredients to a successful venture.

Nanoengineers Complement Star Scientists

The CEO of a nanotechnology startup has many important decisions to make. Among the most important is the composition of the product development team. What does the chief executive officer prefer? On the one hand is a team with a chemical engineer, a biochemist, a condensed matter physicist, and an electrical engineer. The other team has four nanoengineers. Under what circumstances will the former group be favorable over the latter? This thought experiment is perhaps a trivialization of the hiring process. Hiring managers usually attempt to find the best fit for a specific need based on demonstrable skills and past experience, rather than simply based on the discipline of the applicant. Then the question is this: Are applicants who are the "right fit" more likely to be those trained as nanoengineers or those trained in traditional disciplines?

The recipe for success of a nanotechnology company is still unknown. It is, as yet, unclear where, when, and how successful commercialization will occur. Certainly, the operating strategy may vary, depending on the specific type of nanotechnology that is being developed at any given venture. Still, the recent history of high-tech commercialization, specifically in the biotechnology industry, may yield some applicable clues. Zucker and Darby found that in this field of science, the time and location of active publication of so-called "star scientists" (exceptionally prolific academic researchers) was correlated with the emergence of commercialization efforts by existing firms or new startups [7]. Further, they found that an increased collaboration between star scientists and biotechnology firms resulted in a greater probability of commercial success, as measured by a larger number of marketed products and higher levels of employment. Star scientists in academe assist the product development process by transferring their knowledge through collaboration with industrial development teams or by founding their own startups.

> The conclusion is that star scientists play a key role in regional and national economic growth for advanced economies, at least for those science-based technologies where knowledge is tacit and requires hands-on experience. [7, p.12714]

A key concept in Zucker and Darby's argument is that of natural excludabili*ty*, which refers to the idea that newly discovered and highly technical scientific knowledge is effectively transferred between organizations only through the participation of the discoverers themselves. The period of time when natural excludability is most effective is before the knowledge is effectively distilled and recorded into educational materials (such as textbooks), or incorporated into commercially available tools.

If natural excludability is also applicable to the commercialization of nanoscience, then star scientists will have a pivotal role to play in nanotechnology, as in biotechnology. In a recent publication that takes this position, Darby and Zucker have, in fact, reported preliminary evidence "of a close coincidence in time and place of firms engaging in nanotechnology and the location of academic scientists and engineers making major breakthroughs" [8, p.26].

Research at academic nanotechnology institutes is regarded as a highly interdisciplinary effort. In interdisciplinary academic teams, friction among the investigators may arise when compensation levels are revealed in the grant-writing process [9]. For instance, in collaborations between post-doc biologists and computer scientists on bioinformatics projects, a junior-level computer scientist will often command a higher salary than a better educated and more-credentialed biologist. This is problematic if an academic team decides to found a startup company, the success of which is sensitive to the harmonious cooperation of the founding team.

Be that as it may, the issue more relevant to this discussion does not involve the founders per se, but the composition of the industrial product development team of engineers and/or scientists to whom the academics are transferring their know-how. If the commercial team is a group educated at nanoengineering programs, will it be more or less successful at apprehending knowledge from academic scientists than an interdisciplinary industry team?

Undoubtedly, there is great value in a multidisciplinary collaboration in the academic setting, where diversity of thought can lead to new ideas. In the commercialization effort, however, the basic technology is already established, and the challenge is to effectively translate the base technology into a consumable product or process. For this task, there is an advantage to the more homogenous team of nanoengineers because (a) each nanoengineering team member is equally equipped to understand the star scientists of the academic team, and (b) the nanoengineers can communicate very effectively with each other. This is opposed to an interdisciplinary product development team, where only, say, the company biochemist may understand the star biochemist, and then have difficulty imparting that knowledge to his or her company coworkers. Diffusion of knowledge will be more efficient in the homogeneous nanoengineering team, leading to greater understanding and consensus of engineering priorities, and faster execution of the final product.

Workforce Versatility

Scaling up from the needs of a single company to the perspective of a national initiative, another reason to educate a more homogeneous legion of nanoengineers (rather than a nanotechnology workforce of many disciplines), is to maintain a versatile and easily retrainable workforce. That is because there will be no "nano-technology" industry in the same sense as there is a "fiber-optics" industry or "automobile" industry.

> There is not one nanotechnology industry, comparable for instance with computers, but many different applications for the growing ability to manipulate materials on the nanoscale, most of which will be in combination with non-nanotechnologies. [4, p.19]

That is, technological applications for nanoscience are far reaching, and nanotechnology will serve as an underlying infrastructure for manufacturing in a wide variety of industries. Therefore, commercial return from nanotechnology is not going to happen in a single economic boom, whether prolonged or short-lived. Instead, one particular sub-nanotechnology will first garner investment capital (anti-cancer nanoprobes, for example), followed several years later by another nanotech-related innovation (for instance, molecular electronics). At the tail end of each of these sub-booms, a portion of the nanotechnology workforce must be ready to move quickly to the next big thing. The architects of Penn State's Associate Degree in Nano-Fabrication (a two-year program) advocate creating "a workforce that can move from industry to industry as the breakthroughs and job opportunities shift back and forth among nanotechnology fields" [10, p.80]. While they were referring specifically to the *technician* workforce, the same can be said for the nanotechnology *engineering* workforce.

The versatility of a workforce is limited by the ability of its workers to be retrained. Consider that in the wake of the telecommunications bust of the past two years, many optical engineers needed jobs. At the same time, biotechnology manufacturers are increasingly developing devices that use fiber optics, infrared lasers, and other photonic technologies to probe and provide therapy to the human body. Unfortunately, the growth of this "photonic therapy" industry is slow, and a possible impediment is the lack of properly trained engineers. While there is a large supply of optical engineers, very few of them have a background in the biological sciences, which may make biotech companies reluctant to hire them.

This type of scenario would not impede a workforce holding nanoengineering degrees. They will be competent in the fundamentals of the natural sciences and yet have the engineering skills useful to the development of nano-enabled products. If nanoengineers lack a particular skill due to a shift in the focus in the nanotechnology marketplace, they can be credibly and quickly retrained due to the common language and understanding of nanoscale issues. Taking another cue from the medical community, medical students are required to perform rotations in order to acquaint them with specialties. Similarly, nanotechnology centers-of-excellence may provide workers with access to rotations in nanoscience laboratories, either before their first job, or for retraining between employment. Such programs might be appropriately called "Research Experience for Professionals."

Conclusion

The characteristic of the future nanotechnology workforce will have important consequences for the National Nanotechnology Initiative. It is true that the character of academic nanoscience teams may well benefit from interdisciplinary synergy, but since nanotechnology is largely an engineering effort, it is important for educators to focus on and prepare for an engineering workforce that is dedicated to commercialization. A more productive nanoengineering workforce will lower the cost of manufacturing as a whole, thereby allowing the population-at-large to enjoy cheaper goods while encumbering fewer people in boring or repetitive jobs. The key to maximizing productivity is to train a versatile cadre of nanoengineers who understand the principles of natural science and also have engineering skills and expertise specific to nanoscience. Such a workforce could easily be retrained as fads in nanotechnology shift from one industry to another. Further, commercialization of academic discoveries will be more efficient when nano-product development teams have been all trained to "speak the same language," as opposed to having a workforce that comprises many different disciplines.

It is a great challenge to create baccalaureate programs in nanoengineering for four-year undergraduate students. Skeptics may argue that it can't be done because nanotechnology requires competence in too many fields of knowledge for a student to gain meaningful mastery in an undergraduate career. But the future nanoengineer doesn't need to know everything about all the sciences, only those aspects of the traditional scientific disciplines that are relevant to the nanoscale. Therefore, one of the main challenges of curriculum writers is to understand which concepts may be omitted from a nanoengineering education, and which concepts must be retained and then integrated into a coherent "nanoscale story."

References

1. Nanotechnology undergraduate education: A report and recommendations based upon a National Science Foundation workshop held 11–12 September in Arlington, VA (2002).

2. V. Vogel, C. T. Cambell, Education in nanotechnology: Launching the first Ph.D. program, *International Journal of Engineering Education* **18**(5), 498-505 (2002).

3. J. G. Shapter, M. J. Ford, L. M. Maddox, E. R. Waclawik, Teaching undergraduates nanotechnology, *International Journal of Engineering Education* **18**(5), 512-518 (2002).

4. S. Wood, R. Jones, A. Geldart, The social and economic challenges of nanotechnology, Swindon, UK: Economic & Social Research Council (ESRC) (2003) (available at http://www.esrc.ac.uk/esrccontent/DownloadDocs/Nanotechnology.pdf).

5. M. C. Roco, R. S. Williams, P. Alivisatos, eds., *Nanotechnology Research Directions: IWGN workshop report*, Dordrecht, Netherlands: Kluwer (2000).

6. M. C. Roco, Converging science and technology at the nanoscale: Opportunities for education and training, *Nature Biotechnology* **21**(10), 1247-1249 (October 2003).

7. L. G. Zucker, M. R. Darby, Star scientists and institutional transformation: Patterns of innovation in the formation of the biotechnology industry, *Proceedings of the National Academy of Sciences* **93**(23), 12709-12716 (12 November 1996).

8. M. R. Darby, L. G. Zucker, Grilichesian breakthroughs: Inventions of methods of inventing and firm entry in nanotechnology, National Bureau of Economic Research Working Paper No. 9825 (July 2003).

9. B. Schachter, Partners in research, competitors in pay, *The Scientist* **16**(5), 44 (March 2002).

10. S. J. Fonash, Education and training of the nanotechnology workforce, *Journal of Nanoparticle Research* **3**, 79-82 (2001).

INTERACTIVE, ENTERTAINING, VIRTUAL LEARNING ENVIRONMENTS

Judith Klein-Seetharaman, University of Pittsburgh and Carnegie Mellon University

The convergence of Nanotechnology, Biotechnology, Information Technology, and Cognitive Science (NBIC) is believed to ultimately result in a "comprehensive understanding of the structure and behavior of matter from the nanoscale up to the most complex system" [1 p.1]. In biology, high-throughput methodology now allows the accumulation of unprecedented amounts of scientific data, such as genome sequences, gene expression profiles, and structural and functional proteomic data. These advances have stirred great hopes for improving human health and performance, but the quantity of data demands convergence with information technology to interpret and utilize these data to advance human performance and quality of life. This requires an understanding of the complex interactions between the components of biological systems by both domain and non-domain experts. This is particularly challenging in the biological domain because of the massive data and knowledge accumulation. Facilitating convergence of NBIC technologies therefore requires novel interactive, virtual learning environments that engage users from different backgrounds.

Such a virtual learning environment should be a computer system that provides all the computational facilities needed to immerse into a problem derived from any of the four NBIC domains. It should hide the intricacies of computer modeling of physical phenomena so that the user can concentrate on developing an approach to solve a given problem such as finding a cure for a disease. At the same time, the virtual learning environment must be scientifically accurate and include access to the state-of-the-art in available data, knowledge, and technology without requiring the user to bring domain expertise and extensive experience with the technical intricacies of the virtual learning environment that would present the user with tiresome activation barriers. In the above example, this requires that thermodynamic and kinetic processes of biological molecules and cells that control the dynamics of motion and morphological changes in a nanoscale should be translated to macro-scale changes in combating diseases. These in turn need to be interfaced graphically to respond to actions in real-time for the user and also allow for networked interaction between several users.

Progress to Date

Different aspects of virtual learning environments have been addressed to date. Modern biological education makes extensive use of visualization of biological processes and concepts. For example, the publication of the human genome sequence was accompanied by a CD-ROM that presented genome background as well as DNA sequencing techniques in animations [2]. However, these visualization tools are mostly designed to complement traditional teaching techniques and are not very interactive. More interaction is provided in the "Virtual Cell," a virtual environment in which question-based assignments are given to users in a simulated laboratory [3]. A submarine is launched that immerses the user in the virtual environment of the cell populated by sub-cellular components that the user can investigate. With a toolbox, various cellular processes can be investigated experimentally. The results of these investigations and experiments allow users to solve the assignments at their own pace and through their own motivation. It was shown that this approach significantly improves authentic learning, in particular for large-enrollment general biology classes [4]. At the other end of the spectrum, realistic, fully interactive virtual laboratories have been developed to simulate chemical [5], biological [6], and recently nanoscience [7] laboratory experiments.

For inexperienced users, specific virtual laboratories may not be appropriate, because such users may not yet have sufficient insight needed for discovery and for solving problems in a virtual laboratory without the background in the specific domain of the laboratory. Insight, i.e., the capability to make non-obvious connections between the complex interactions of the components of these systems, is the main requirement for solving any type of problem [8]. Such insightful solutions can often be found in an interactive and visual virtual learning environment, as demonstrated, for example, by the fact that despite the modern numerical computing technologies, biophysicists today still use Gedanken experiments for concept development [9]. Although there are many virtual reality three-dimensional molecular models available, biochemists still use hand-made models for intuitive reasoning. Intuitive simulation is one of the most powerful approaches to creative problem solving.

Challenges—and Some Novel Approaches

In principle, a virtual learning environment could teach insight into a particular problem to users who lack background in the given problem area. Major challenges in designing such an environment are how to leverage existing but dormant learning capabilities and how to avoid frustration with the novelty and quantity of the data and concepts required to achieve the insight. It is generally believed that human capabilities for learning can be leveraged through enjoyment [10]. Novel approaches are being developed to this end; for example, a storytelling system has been presented to fertilize multidisciplinary biological problem solving [11]. More generally, the concept of "edutainment"—the marriage between education, software, and entertainment—is currently booming (see, for example, the Edutainment Section at amazon.com). Recently, the concept of edutainment was taken a step further towards developing a software that has the potential to be a virtual problem-solving environment targeted at children and professionals alike: an interactive and visual problem-solving environment for the biological domain, BioSim, has been proposed based on game design principles [12, 13]. BioSim is described in more detail below to illustrate these design principles. The first version of BioSim describes a simple world model of the human vascular system and a biological problem that involves an infection by *Neisseria meningitidis* where the biological characters are white and red blood cells and *Neisseria* cells. The users can explore biological interactions in the biological world model by three mechanisms.

1. Role Play

The system allows the user to be a biological character in the game. Cognition science shows that role play is an important way to stimulate creative ideas. It enables the user to have an intimate connection to the character. Also, personalization of a biological character makes a game more interactive.

2. Voyage

The user can navigate through the biological system in the game, either as a character, or by using a "ship," supporting different view angles, e.g., traveling through capillaries and tissues. Voyage allows exploration at the user's chosen leisure, accommodating users with various backgrounds.

Figure 7.2 shows an example for a biological system. The user plays a game in which the human body is infected with Neisseria bacteria and the goal of the game is to cure the infection. A ship provides transportation through the body (left). A control panel inside the ship has sensing and action capabilities. Bacteria can be marked by the user with histamines, introducing the concepts of molecular recognition. Bacteria that have not been marked can divide undisturbed (gray bacteria in 1,2 on the right). When the ship approaches the bacteria (3), the histamine sensor identifies the bacterial infection and the user can mark them (color change). After marking, the user can attract macrophages that will "eat" the bacteria (4).

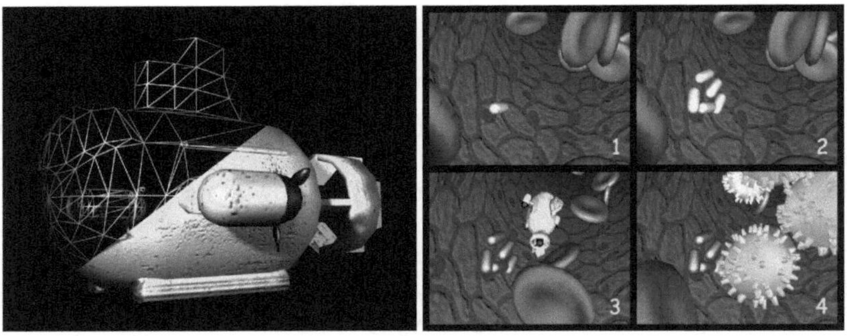

Figure 7.2. Immersion of the user in virtual learning environments. (The figure is modified from Figure 7 in ref. [13] and is produced with permission from Elsevier.)

3. Distributed Problem Solving

The game engine allows users to play the game over the Internet so that large problems can be solved collaboratively or antagonistically, i.e., some users can play macrophages and others can play bacteria. The distributed problem solving enables diverse game strategies and more excitement of the game. The user can also choose between two aims, rather than playing the role of a single biological component only. The user can assume the roles of multiple biological characters, thus studying their individual influence on a particular aim. These aims are to induce an infection with *Neisseria* and ensure its successful propagation in the human body, or to fight the *Neisseria* infection.

The system architecture of these "characters" contains four main components:

1. Bio-behavior is modeled using cellular automata.
2. Bio-morphing uses vision-based shape tracking techniques to learn from recordings of real biological dynamics.
3. Bio-dynamics implements mathematical models of cell growth and fluid-dynamic properties of biological solutions.
4. Bio-sensing is based on molecular principles of recognition to identify objects, environmental conditions, and progression in a process. For example, each activity is determined by availability of "energy points," which have to be carefully balanced to minimize consumption and maximize effectiveness. The user knows the status of energy points via a control panel, which also provides for the various possibilities of action. For example, in a state of high energy, the user can afford to travel actively with the ship to a point of infection. However, in a state of low energy, the user would choose to travel passively with the blood stream. This will allow the user to further develop

decision making skills in a biological virtual learning environment. In the future, there will also be additional bio-sensing capabilities available through the control panel, for example, a histamine sensor (Fig. 7.2,) and mechanisms of the immune system to distinguish self from non-self. This will allow introduction of molecular-level information. For example, the user will need to use molecular docking of the immune system's antibody structures to those of the bacterial surface structures. This will train users to view protein structures and understand the mechanisms of complementarities of two structures. The player seeking to evade the immune system would need to develop strategies to evade antibody marking, e.g., through surface mutation. Thinking about possible strategies from each point of view will allow the user to gain deep insight into the factors controlling the health of the organism, from the molecular to the macroscopic level, ultimately aiding in the development of novel solutions for biological problems such as the *Neisseria* infection.

The specific example for a virtual learning environment described here for a problem in the biological domain (*Neisseria* infection) only represents an example for the principles of design of such systems. Different types of environments can be envisioned, encompassing any area within NBIC, implementing any specific type of problem. One of the major advantages of a virtual learning environment over traditional learning experiences with special relevance to NBIC is the ability to visualize concepts and matter that cannot be seen by the normal eye. This is particularly important for teaching nanotechnology concepts that involve the atomic scale. For example, users without scientific background can learn the concept of "smallness" and the relative range of size scales involved.

Virtual Learning Environments Enhance the Learning Experience

Virtual learning environments are meant to support strategic and creative thinking and professional innovative problem solving, and to help users to learn complex concepts and interactions. The main principle is that education is directly linked to enjoyment, thereby enhancing the learning experience as proposed by Leonard [10]. In the virtual learning environment, cross-disciplinary education will be on-demand, entertaining, and interactive, allowing focus on discovery and creativity rather than on one-way tutoring. The users can be from any background or age. Virtual learning environments support equally well the education of pre-school children, professionals with diverse backgrounds and experiences, and the general public.

References

1. M. C. Roco, W. S. Bainbridge, Overview: Converging technologies for improving human performance: Nanotechnology, biotechnology, information technology, and cognitive science (NBIC). In *Converging Technologies for Improving Human Performance*, ed. M. C. Roco and W. S. Bainbridge, Arlington, VA: National Science Foundation (June 2002) (available at http://www.wtec.org/ConvergingTechnologies/Report/NBIC_report.pdf).

2. ***Science*, 291** (5507) 16 February 2001, http://www.sciencemag.org/content/vol291/issue5507/ and http://www.genome.gov/Pages/EducatonKit/download.html
3. R. White, P. E. McClean, and B. M. Slator. The virtual cell: An interactive, virtual environment for cell biology. Paper presented at World Conference on Educational Media, Hypermedia and Telecommunications, Seattle, WA (1999) (available at http://wwwic.ndsu.edu/abstracts/ed-media.htm).
4. P. E. McClean, B. Saini-Eidukat, D. Schwert, B. M. Slator, and A. R. White. Virtual worlds in large enrollment science classes significantly improve authentic learning, In *12th International Conference on College Teaching and Learning*, Jacksonville, FL: Center for the Advancement of Teaching and Learning (2001) (available at http://wwwic.ndsu.edu/publications/McClean-et-al-ICCTL-2001.pdf).
5. The irYdium Project, Carnegie Mellon University, http://ir.chem.cmu.edu/irproject/applets/virtuallab/
6. Biointeractive Home Page, Howard Hughes Medical Institute, http://www.biointeractive.org/
7. M. Guggisberg, P. Fornaro, T. Gyalog, and H. Burkhart, An interdisciplinary virtual laboratory on nanoscience, *Future Generation Computer Systems* **19**, 133-141, (2003).
8. Y. Cai, Pictorial thinking, *Journal of Cognition Science* (1986).
9. T. Hayden, The inner Einstein, *U.S. News*, 60, (9 December 2002).
10. G. Leonard, *Education and Ecstasy*, Mill Valley: ITP (1968).
11. Kuchinsky, K. Graham, D. Moh, A. Adler, K. Babaria, M.L. Creech. Biological storytelling: A software tool for biological information organization based upon narrative structure, *ACM SIGGROUP Bulletin* **23**(2), 4-5 (August 2002).
12. Y. Cai, I. Snel, S. Bharathi, C. Klein, J. Klein-Seetharaman, Towards biomedical problem solving in a game environment. In *Lecture Notes in Computer Science* 2659, 1005-1014 (2003).
13. Y. Cai, I. Snel, B. Y. M. Cheng, S. Bharathi, C. Klein, J. Klein-Seetharaman, BioSim—A biomedical character-based problem-solving environment, *Future Generation Computer Systems* 21(7), 1145-1156 (2005).

INCORPORATING NANOTECHNOLOGY INTO K–12 EDUCATION

Kristen M. Kulinowski, Rice University

Nanotechnology is hip. If it were a person, nanotechnology would be today's "it-girl": desirable, ubiquitous, a little mysterious, the one you want to be photographed with on the red carpet, the JLO of R&D. Evidence of this can be found throughout popular culture. Stereotypical geeks —such as the one portrayed in the General Electric ad (Fig. 7.3), who is paired with a supermodel to tout a refrigerator's marriage of brains and beauty—used to be computer programmers or genetic engineers; today, they're nanotechnologists. Nanotechnology is a key plot device or casual throwaway reference in movies and television shows such as *Spiderman*, *Hulk*, *Agent Cody Banks*, and *Jake 2.0*, not to mention that 800-pound gorilla of nanotechnology pop culture, the novel *Prey* [1]. If nanotechnology stands to enjoy more than the average it-girl's fifteen minutes of fame—which seems likely given the recent passage of multi-year nanotechnology legislation [2]—then its cachet presents us with an extraordinary opportunity to draw young people into science and engineering.

Figure 7.3. GE ad featuring nanotechnology professor and supermodel (courtesy BBDO/New York).

The problem is a perennial one: the near-continuous decline in enrollments of U.S.-born students in science, technology, engineering, and mathematics (STEM) programs at the undergraduate and graduate levels threatens to jeopardize the U.S. position as global R&D leader in cutting-edge technologies. According to a report by the National Science Board, the share of doctorates awarded to U.S. citizens in the natural sciences and engineering decreased from 70 percent to 56 percent over the last 25 years [3]. Foreign students, who make up an increasing fraction of the Ph.D. degrees awarded in STEM disciplines, have allowed academic research in the United States to continue apace. However, many choose to return to their native countries after their studies, thus exporting both their intellectual firepower and the U.S. investment in their training. And, as other countries build up their R&D capacity, fewer of these talented students will come to the United States for their training, thereby reducing the numbers who ultimately enter the U.S. workforce [3]. Policymakers and legislators agree: we need to train more of our own.

So why aren't U.S. students choosing STEM disciplines? Blame it on the (stereo) typical white-male STEM practitioner, whose slice of the U.S. demographic pie grows smaller with every census reading. Without a concomitant increase in STEM enrollments by women and members of historically underrepresented groups, such as Hispanics and African-Americans, the whole U.S. STEM pie shrinks. Blame it on the K–12 educational system, No Child Left Behind, the soft bigotry of low expectations, outdated science curricula, poorly trained teachers, administrators hamstrung by rigid bureaucracies. Blame it on TV, video games, text messaging, and other symbols of the short-attention-span culture that is at odds with the deliberative and measured pace of scientific discovery. Blame it on whatever you like. The question is, what are you going to do about it? Many of the pressing challenges to learning faced by our teachers and students do not have easy solutions and show every sign of persisting. It is our responsibility to find creative ways of communicating the excitement of our enterprise to the next generation of producers, consumers, and policymakers despite these challenges.

Anyone who has spoken about nanotechnology to a group of middle school students can tell you that it fascinates children. This interest should be exploited to engender further interest in scientific inquiry. When I do this kind of outreach activity, I begin by polling the crowd to get an understanding of their preconceived ideas about nanotech. Invariably, most of these involve self-replicating nanobots of the type first imagined by Drexler [4] and recently popularized in *Prey* [1]. Rather than being scared by the prospect of little machines that swarm around and eat people, they seem positively enthralled, in the way that children are by insects and slime before they learn to be disgusted. Indeed, one could argue that the highest social value of the concept of self-replicating nanobots is its ability to excite the imagination of the average 12-year-old. Having gained children's attention, I then discuss what nanotechnology is now and probably will be in the future. I conclude with some guidance on the course of action they should take if they want to grow up to be nanotechnologists. At the top of my list of recommendations is—no surprise—further study in science and mathematics. It's hard to gauge the impact of these "one-offs," but it's certain to be limited. If long-term impact could be achieved this easily or quickly, we wouldn't have dozens of reports and workshops on the challenges of recruiting students into science and engineering. To effect real change, we need follow-through and greater exposure. We need to infiltrate the curriculum.

Effective incorporation of nanotechnology into K–12 curricula presents new obstacles. Chief among these, perhaps, is the difficulty of representing and conceptualizing the nanometer length scale. Research shows that even graduate students in chemistry have difficulty visualizing and comprehending objects and processes at the nanoscale [5]. Another is the issue of readiness. Nanotechnology concepts and processes can't readily be understood until students have been introduced to the concept of atoms, molecules, and chemical interactions. This occurs at different age levels in different states, ruling out a one-size-fits-all strategy for K–12 nanotechnology education and making the youngest students especially hard to reach. Before students can learn, their teachers must gain mastery at working with these concepts, or at least achieve a threshold comfort level. Yet, nanotechnology is only beginning to infiltrate the curricula of undergraduate students, and the ones targeted are science and engineering students, not pre-service teachers. Then there is that pesky interdisciplinarity. Nanotechnology research occurs at the interfaces of traditional disciplines; it's not unusual in a single project for researchers to rely upon fundamental concepts, theories, or tools from physics, chemistry, biology, and engineering. It's one thing for an incoming graduate student to enter the nanotechnology research enterprise as a baccalaureate subject-matter expert in a single discipline. That student's discipline expertise can be leveraged and applied to the new activity with little difficulty. It's quite another matter to integrate nanotechnology concepts into rigidly discipline-bound, K–12 state curricula, where such a foundation has not yet been built. Finally, many state curricula are not only rigidly discipline-bound, but rigidly content-bound. Teachers will tell you the pressures to cover the required material are so high that there's no room for new material.

The approaches taken by the Center for Biological and Environmental Nanotechnology (CBEN), an NSF Nanoscale Science and Engineering Center at Rice University, address many of these challenges. The underlying principle relies upon a concept taken from R&D: that nanotechnology is an enabling technology, not the thing itself. In the applications arena, nanotechnology already is an enabler of existing, commonplace technologies such as those resulting in consumer products and drug delivery devices. Nanotechnology devices and materials are not expected to be stand-alone products or technologies themselves for a long time, if ever. Think "Nano Inside," an analogy to the Intel marketing slogan. You don't buy the chip, you buy the computer that contains the chip. You're not going to buy nanomachines, you're going to buy a car whose parts contain nanomaterials. As it is with applications, so, I argue, it is in education. Nanotechnology can enable STEM education but may not be appropriate as its own subject in the K–12 curriculum. Distilled down to its essence, nanotechnology really relies on nothing more than the subjects of standard school science curricula: biology, chemistry, and physics. Throw in some computer science and engineering and you have the basis for all of nanotech.

CBEN's approach is to work within the existing curriculum and show teachers how nanotechnology can be used to illustrate concepts in their required curriculum. We target teachers for several reasons. First, teachers are in the classroom with students every day; therefore, their impact on a single student is far greater than that of a researcher who may make a couple of visits to the school per year. Second, a single teacher can impact hundreds to thousands of students over the course of a career. Finally, teachers can disseminate new content and methods to their peers, thus multiplying the effect even more. CBEN's strategy for infiltrating the science curriculum is simple: link cutting-edge nanotechnology research to underlying concepts that are taught in school science classes. For a variety of reasons, we focus on 9th–12th grades. The mechanism is a semester-long evening course that meets once per week for three hours. The first hour is a review of a concept taught in chemistry or physics class. For example, we discuss the concepts and vocabulary of chemical solubility such as "like dissolves like," intermolecular forces, and so on. In the second hour, a nanotechnology researcher presents his or her work, which relies upon the concept discussed in the previous hour. Building upon the previous example, a review of solubility is followed by a nanotechnology researcher who is working to solubilize single-walled carbon nanotubes (SWNTs). The researcher can describe why SWNTs are insoluble in so many solvents using the same vocabulary that we used in the first hour, thus helping establish the connection between the science curriculum and nanotechnology research. This can motivate the research by explaining the technologies that could be developed around solubilized SWNTs such as new drug delivery devices or diagnostic agents. The third hour is dedicated toward making those connections explicit and brainstorming about how best to bring these new ideas back to the classroom. The hope is that the next time the teacher discusses solubility with the class, a little nanotechnology might sneak its way into the traditional lesson.

The teacher-training course is only the first part of a three-phase program designed to immerse science teachers in the concepts, principles, and practices of today's nanotechnology researchers. The second phase is a summer research program akin to the Research Experiences for Teachers (RET) program funded by NSF. This phase is a more in-depth exposure to a particular research program to which they were introduced in the spring course. Ours is a four-week program where the teachers spend most of the time working in a research lab, either assisting on a graduate student's project or conducting their own projects. One half-day per week is spent developing classroom-ready curricular materials that build upon their research experience.

The final phase is an intense residency program in which the high school science teachers spend a one-year sabbatical in an inner-city school studying and class-testing discovery-based approaches to science education that more closely replicate the process of discovery that occurs in the research lab. Teachers chosen for this experience leave their home schools for a year and enter a special classroom called the Partners for Houston (pH) Model Science Lab at Lee High School to refresh and update both their science content knowledge and their teaching methods. Working with a reduced teaching load, the teachers in the pH lab spend half their time teaching students and the other half on learning new content, such as nanotechnology, and student-centered approaches to teaching. Dr. Elnora Harcombe, Associate Director of the Center for Education at Rice University, modeled the pH lab after the very successful Model Science Lab she implemented for middle school teachers, which has been in operation for more than 12 years and has been praised as a model for urban education. The MSL's tangible achievements include an almost perfect teacher retention rate (95 percent after 11 years), the professionalization of a cadre of teachers who act as peer mentors, and the improvement in student test scores [6]. There are early indications that the CBEN-funded high-school pH lab will replicate the successes of its precursor. Taken as a whole, the three-phase teacher-training program offers the teacher a comprehensive package of nanotechnology research, connection of that research to existing science curricula, and discovery-based approaches to science teaching that better replicate the process of scientific discovery in the research lab.

While the CBEN approach seems to be having an impact in its educational setting, the number, breadth, and diversity of the challenges faced by teachers and students around the country suggest that new education research is needed to determine the best ways to incorporate nanotechnology into pre-graduate curricula. The National Science Foundation has launched a focused program in Nanoscale Science and Engineering Education designed to deal head-on with these issues. The new Nanoscale Centers for Learning and Teaching bring together education researchers, nanotechnology researchers and teachers to answer fundamental questions on the most effective way to communicate nanotechnology to students in grades 7–16. These centers will produce a new generation of education experts with a specialty in nanotechnology education who will serve as a resource for everyone doing educational outreach in this area.

References

1. M. Crichton, *Prey*, New York: HarperCollins (2002).
2. 21st Century Nanotechnology Research and Development Act, 108th Congres, 1st sess., Dec. 3, 2003, Public Law 108–153, 15 USC 7501, Page 117 Stat. 1923 (Washington, D.C.: United States Government Printing Office, 2003).
3. National Science Board, *Science and engineering indicators–2002*, Arlington, VA: National Science Foundation (2002).
4. K. E. Drexler, *Engines of Creation: The coming era of nanotechnology*, New York: Anchor Press/Doubleday (1986).
5. G. M. Bodner, I have found you an argument: The conceptual knowledge of beginning chemistry graduate students, *Journal of Chemistry Education* **68**, 385-388 (1991).
6. E. Harcombe, Science teaching/science learning: Constructivist learning in urban classrooms, New York: Teachers College Press (2001).

EDUCATION OPPORTUNITIES RELATED TO THE SOCIETAL IMPLICATIONS OF NANOTECHNOLOGY

Bruce E. Seely, Michigan Tech

I wish to suggest two areas where educational initiatives related to the societal implications of nanotechnology could be pursued. The first relates to the education of undergraduate engineers, the second to efforts to advance what in the past has been labeled "technological literacy." In both areas, education programs focused on developing broader understandings of the consequences, positive and detrimental, of emerging sciences and technologies might have far-reaching consequences.

In engineering, opportunities exist to better educate beginning professionals to understand not only the technical side of nanotechnology, but also the ethical and societal implications of this still-emerging field. The task is not easy, but several factors are encouraging changes in engineering education as we move into the 21st century. After concentrating primarily on undergraduate instruction before 1940, engineering colleges in American universities changed significantly after the Second World War. Most schools witnessed the emergence of graduate education and externally funded research programs, the placement of greater emphasis upon fundamental and theoretical (i.e., scientific and mathematical) studies, and the appearance of hybrid science-engineering fields (e.g., materials). This change was so pronounced that by the 1980s significant pressure emerged to return the displaced problem-solving and other real-world elements to undergraduate curricula

[1–4]. A pivotal element in this latest swing of the engineering education pendulum: the accreditation requirements adopted by the Accreditation Board for Engineering and Technology (ABET) in the late 1990s. The new evaluation criteria replace a pattern of counting inputs (*x* courses in the humanities, *x* courses in the social sciences, etc.) with efforts to assess the learning of engineering students in a range of areas. Significantly, several criteria established for evaluating engineering education are obviously non-technical (a knowledge of contemporary issues, an ability to communicate effectively, the broad education necessary to understand the impact of engineering solutions in a global and societal context). Others with a technical dimension also include significant non-technical components (e.g., an understanding of professional and ethical responsibility; an ability to function on multidisciplinary teams; recognition of the need for, and an ability to engage in, lifelong learning).

This change in accreditation criteria presents an opportunity for advancing educational efforts intended to present engineering students with opportunities to examine the societal and ethical implications of all emerging technologies, including nanotechnology. While incorporating such material into already-jammed curricula is not easy, the shift in evaluating engineering curricula opens an opportunity to use creatively the limited number of non-technical courses to broaden the education of future engineers. Engineering students can learn not only how to manipulate nature at the nanoscale, but also how to think about the possible consequences of doing so. At Michigan Tech, a Nanotechnology Undergraduate Education (NUE) award from NSF is currently supporting efforts to introduce a significant number of first-year engineering and science students to these ideas, using modules in the required first-year engineering sequence and in selected math and physics courses. A one-credit seminar on the Fundamentals of Nanoscale Science and Engineering places equal emphasis on the science, the applications, and the societal implications. And a limited number of Research Experiences for Undergraduates (REU) will introduce a few students to research in nanoscale science and engineering. In line with the goals of the NUE program solicitation, the focus at Michigan Tech is on introductory efforts, and students (not just first-year students) are responding with enthusiasm. Such activities also conform to the National Nanotechnology Initiative (NNI) recommendations not to develop specialized undergraduate majors in the distinctly multidisciplinary domain of nanotechnology.

Since the directions for this workshop urged participants to think outside the box, let me suggest that one of the most important changes that might emerge from attention to work at the nanoscale is the possibility of a fundamental reform of engineering education. For more than a century, engineers have attempted to find ways to include everything new while retaining a traditional core. At the same time, they need to prepare new professionals and do all this in only four years. Experimental efforts to extend the curriculum to five years failed at several colleges after World War II, although most engineering students now spend that long in college. A perhaps related problem has concerned ways to incorporate multidisciplinary fields such as materials, biomedical, and computer engineering into the departmental structure formed around the classic disciplines and their professional societies (civil, mechanical, and electrical).

Eric Walker, dean of engineering and president of Penn State in the 1950s and 1960s, proposed an interesting solution. Walker envisioned an education that taught engineering problem solving to undergraduates but left specialized professional study for graduate school. Such a general engineering approach would allow more time for a broader general education for students, on the assumption that the M.S. degree should be the first professional engineering degree. Walker argued that this reform could make engineering the liberal arts degree of the 21st century. Others have accepted Walker's rhetoric, but until recently only Dartmouth's engineering school embraced anything like this concept. Two new engineering schools, the Franklin W. Olin College of Engineering in Boston and Smith College's Picker Engineering Program, are taking steps in this direction. Picker, for example, has adopted as its motto "Engineering Redefined."

Such a curricular structure seems especially well suited to nanotechnology and other multidisciplinary fields of engineering and science. Undergraduates interested in nanotechnology could gain a stronger interdisciplinary foundation in science, math, and engineering, before developing a research specialization in graduate school. As importantly, there would be opportunities to provide stronger foundations in the liberal arts and social sciences, the foundations for examinations of the societal implications of emerging technologies. Such a structure might offer advantages in preparing engineering students for nanoscale work, in comparison with the departmental and discipline-driven majors found in current engineering programs.

Engineering students, however, are not the only college graduates who should learn about nanotechnology. Non-technical students, whether in business, the liberal arts, or social sciences, also need to understand some of the basic science and technology related to the nanoscale. One approach to this situation can be found in efforts to improve public understanding of science—a matter that other workshop participants are likely to discuss in detail. On the education side, however, it may be worth reviving an insight that is about as old as Eric Walker's concept of a broader engineering curriculum. This is the concept of the New Liberal Arts (NLA), which the Sloan Foundation promoted during the 1980s. Believing that liberal arts students needed to encounter more science and technology in their curricula, Sloan awarded 23 liberal arts colleges nearly $20 million to develop academic programs that strengthened quantitative reasoning and technological literacy; historically black colleges received another $2 million. A number of individual courses emerged, and Sloan supported workshops and a newsletter to disseminate information. Some of the courses included *Episodes in American Invention* (Princeton University and Bryn Mawr College); *The Economics of Technology* (Montana State University); *Hyperacoustics: Digital Sound.* (Middlebury College); *Chemistry and Crime* (Williams College); *Bioengineering and Health Technology* (Davidson College); and *Management of Public Risk* (Brandeis University). Several scholars also published books similarly designed to connect liberal arts students to science. Among the titles published jointly by MIT Press and McGraw-Hill were Joseph D. Bronzino, Mourice L. Wade, and Vincent H. Smith, *Medical Technology and Society;* Robert Mark, *Light, Wind, and Structure: The Technology of Historic Architecture;* Frank Wattenberg, *Personal Mathematics and Computing: Tools for the Liberal Arts;* and Newton H. Copp and Andrew W. Zanella, *Discovery, Innovation, and Risk: Case Studies in Science and Technology.*

When Sloan funding ended in 1992, evaluators found that some goals had been met, a few had been misplaced, and final results varied. But the idea of introducing liberal arts students in formal ways to the basic science, technology, and numeracy on which the modern world rests continues to have appeal. Occasional articles present the idea anew, and science, technology and society programs on a number of campuses embrace the idea of advancing technological literacy. The University of Texas in 1998 started a new liberal arts concentration entitled Technology, Literacy, and Culture (TLC) that "aims to produce graduates skilled in changing technologies." The logic of all of these efforts is unchanged since the Sloan NLA program took form: liberal arts students should have a better grasp of the potentially significant ways that society and technology interact to form our world.

Both of these educational approaches share a belief that students need to develop analytical tools and concepts for thinking about nanotechnology and other emerging technologies—especially their societal implications. To make sense of the increasing number of fictional accounts or the recent television advertising by General Electric and Hewlett-Packard, students need familiarity with both technical and non-technical information. Providing students from engineering, science, and the liberal arts alike with knowledge and tools might dampen the willingness to accept either the unrealistic hyperbole that surrounds much of the current discussion about nanotechnology, or the unfounded fears of novelty found in fictional scenarios of self-replicating nanobots running amok. Science fiction books and films such as Michael Crichton's *Prey* [5], and the 1966 film *Fantastic Voyage* [6], offer, in fact, useful educational tools for engaging both types of student audiences. What is needed is a range of educational experiments to examine the efficacy of various approaches to student encounters with the societal implications of nanotechnology.

References

1. B. E. Seely, The other re-engineering of engineering education, 1900-1965, *Journal of Engineering Education* **88**, 285-294 (July 1999).
2. B. E. Seely, Research, engineering, and science in American engineering colleges, 1900-1960, *Technology and Culture* **34**, 344-386 (April 1993).
3. S. W. Leslie, The cold war and American science: The military-industrial-academic complex at MIT and Stanford, New York: Columbia University Press (1993).
4. L. P. Grayson, The making of an engineer: An illustrated history of engineering education in the United States and Canada, New York: Wiley (1993).

5. M. Crichton, *Prey*, New York: HarperCollins Publishers (2002).

6. R. Fleischer (dir.), *Fantastic Voyage*, 100 min., U.S.: 20th Century Fox (1966).

HUMAN RESOURCES FOR NANOTECHNOLOGY

Paula E. Stephan, Georgia State University

In thinking about nanotechnology, it is instructive to use an analogy. Some use electricity as an appropriate analogy. Others use the Internet. I prefer to use the assembly line as an appropriate analogy.

The assembly line was a major innovation approximately 100 years ago. It transformed the way we make things. It was associated with job growth for assembly line work related to (1) the introduction of new products; (2) the development of a new way of producing traditional products, such as the sewing machine and typewriter; (3) increase in quantity of goods demanded by consumers brought about by lower costs of production associated with the assembly line. It was also associated with an increased demand for scientists and engineers.

The assembly line meant that muscle jobs became less important; so did being a "prime age male." The analogy is also useful in the sense that the assembly line wasn't one industry; rather it related to how we make things and although it began in autos it spread to many other industries. Where it was adopted was driven by economics.

Nanotechnology has the potential of fundamentally changing how we make things. It is not an industry; it spreads across industries. Nanotechnology has the potential to make muscle less important. The sectors where nanotechnology is adopted and implemented will depend upon cost considerations. It is too early to know the full labor market implications of nanotechnology, just as it would have been too early to know the labor market implications of the assembly line 100 years ago.

The analogy between the assembly line and nanotechnology suggests that jobs will be spread across a considerable spectrum of skill. It also suggests that the overall job impact will depend upon the extent to which nanotechnology results in the introduction of new products, the extent to which it lowers the cost of existing products, the extent to which nanotechnology products are substituted for existing products, and the degree to which these industries are labor intensive.

Implications of Nanotechnology for the Highly Trained Workforce

My goal here is to limit the analysis, focusing not on the broad spectrum of jobs but instead on the implications of nanotechnology for the highly trained workforce. An initial question that I wish to address is whether we have a labor market problem.

Market economies function reasonably well in transmitting signals concerning new opportunities. Labor markets for highly trained individuals have special peculiarities, however, due to the length of time associated with the pipeline. The economy can't create more highly skilled workers overnight. So, when I ask if we have a labor market problem, what I really want to ask is, "Do we have a bottleneck or a shortage that could slow down growth in nanotechnology in the future?" I rely on three indicators to assess the bottleneck issue: (1) position ads, (2) salary data, and (3) anecdotes from recruiters.

Position Announcements

The first piece of evidence comes from looking at position announcements in *Science* for 2001 that mention either the work "nano" or MEMS. Searching through all position announcements placed, we find 36 announcements, related to 37 positions. Thirty-two of these are in universities; three in government; two in firms, one of which is a firm–university collaboration. All but two of the announcements specify a Ph.D. as a requirement. When we examine the ads in *Science* one year later, for the period 2002, we find only a small amount of growth: 32 announcements for 41 positions. Twenty-six are in universities, six in government (two of these are government–university collaborations), and two are in a firm. If a degree is mentioned, it is always a Ph.D.

We also searched for ads on Web sites. For example, on Nanogugsy.com for the period of January–October 2003, we found 44 positions listed. On Monster.com, for the period September–October 2003, we found 16 nanotechnology-related positions listed. Seven of these were science-engineering specific and one was in academe. On ScienceJobs.com, for the period September–November 2003, we found four positions, three in academe. We found five positions, all in academe, in Naturejobs.com, for the period October–November 2003. From Nanotechweb.org (October–November 2003), we found 44 positions, most in Europe. On Careerbuilder.com (October–November 2003), we found seven positions, one in academe.

Salary

None of the ads mentioned anything related to salary.

Recruiters' Anecdotes

Nano Circuit [1] asked top recruiters about current and future employment opportunities in nanotechnology. According to Jason Finkelstein, a partner of Glocap Tech, "There will *never* be a nano hiring frenzy." He went on to say that startup companies in nanotechnology hire directly out of university and government labs. Herman Collins, CEO of NanoOptic Networks, stated that current recruiting activity is "pretty stable, with small but steady growth." He went on to say that he expects a hiring frenzy down the road, but based his comment on the NSF figures predicting two million workers in nanotechnology by 2015, rather than on anything he had observed.

We conclude that there are no signs of a bottleneck now in the field of nanotechnology. This conclusion is cautioned by the "sniff test." It is difficult to know exactly who is taking a job in nanotechnology; the term "nano" is not always in the position announcement.

By way of a benchmark, we consider the situation in bioinformatics that we observed seven years ago. In 1996 articles appeared in the press concerning the high salaries that were being earned in bioinformatics. Concern was expressed that the field was eating its "seed corn" by luring faculty away from universities to high-paying jobs in industries. When we analyzed position announcements in *Science* [2, 3] we found that they doubled over the course of a year. In response to these strong signals, the number of university degree programs in bioinformatics/computational biology grew from 20 to 70.

Supply of Nanotechnology Talent

Here we examine the supply of nanotechnology talent by enumerating the new programs and courses in nanotechnology as well as an estimate of the flow of talent trained in complementary areas.

To date, there appear to be only two graduate programs in nanotechnology. The University of Washington has an "optional Ph.D. in nanotechnology." Rice University has a professional master's of science in nanoscale physics. While these are the only formal programs, numerous training opportunities exist around the 26 (or more) nanotechnology centers that have been funded at universities.

The answer as to why there are so few formal graduate programs involves several parts. First, there is the obvious issue that programs must span disciplines. This means that they must go across university departments and often colleges in a university. There is also the issue that engineers lack training in biology and physical sciences, and students in the physical sciences and biology lack training in engineering. The requirements for nanotechnology programs can also be daunting from a student's perspective since the coursework is usually of an "add on" nature, not a substitute for other, required courses. Concern is also expressed that the curricula in a nanotechnology program will be too general, depriving students of in-depth expertise. And then there is the concern that the nanotechnology field is "too young" to develop into a stand-alone program [4].

We examined the catalogs of 11 universities to see the extent to which courses are offered in nanotechnology. With the exception of the University of Wisconsin, the 11 are either members of the National Nanofabrication Users Network or are NSF Centers of Excellence. Wisconsin is included because it is known for being active in the nanotechnology field.

Our findings are displayed in Fig. 7.1. We see that Stanford has the most courses with "nano" in the prefix or course description for the period 2003–2004, followed by Cornell and Rice universities. Overall, we find that 70 of the courses are in engineering, while 33 are in the sciences. Two are in "other" areas. With regard to specific disciplines (see Fig. 7.4) the largest number of offerings comes from materials/mechanical engineering (33), followed by physics, applied physics, and engineering physics (27). We find only a handful (8) listed in the biological sciences/engineering.

We also examined the training and occupations of people working in nanotechnology institutes to get an idea of the number of people already trained who have the potential to work in nanotechnology. More precisely, we searched

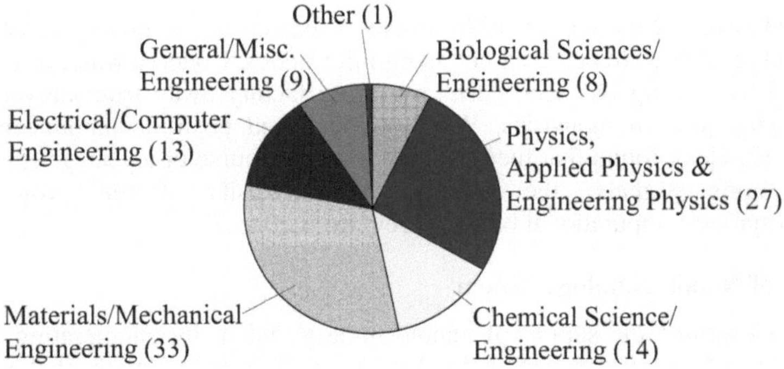

Figure 7.4. Nano-courses by pedagogical discipline.

the Web pages of 12 nanotechnology institutes in the United States and examined the profile and research interests of each listed faculty member. The institutes are California Nanosystems Institute, Cornell Center for Nanoscale Systems, Columbia Center for Electron Transport in Molecular Nanostructures, Harvard Nanoscale Science and Engineering Center, MIT Institute for Soldier Nanotechnology, Northwestern Nanoscale Science and Engineering Center, Purdue Center for Nanoscale Devices, Notre Dame Center for NanoScience and Technology, Rensselaer Center for Directed Assembly of Nanostructures, Rice Center for Nanoscale Science and Technology, SUNY Albany Nanotech, and University of Wisconsin Center for Nanotechnology. When possible, we then matched them to disciplines listed in the Survey of Earned Doctorates, an annual census of all new Ph.D.s in the United States. The top 10 disciplines were Electrical Engineering (11 percent); Materials Science (11 percent); Chemical Engineering (11 percent); Physics—Solid State (10 percent); Biochemistry (9 percent); Bioengineering (6 percent); Mechanical Engineering (6 percent); Chemistry—Physical (5 percent); Computer Engineering (3 percent); Engineering Physics (2 percent). Other disciplines accounted for the remaining 26 percent.

We then used the Survey of Earned Doctorates, administered by NSF, to estimate the flow of Ph.D.s in these fields over time. We estimate that in 1980, 2,395 Ph.D.s were awarded in these fields, representing 14.4 percent of all Ph.D.s in S&E. By 1990 the number had grown to 4,169 or 18.2 percent of all S&E Ph.D.s. By the year 2000, it was 5,017 or 19.3 percent. We know that some individuals with training in these (as well as other fields) have transitioned to working in nanotechnology. We assume that others could as well. We believe that the transition is facilitated by working in interdisciplinary teams where individuals learn from each other. We also know that learning the "nano" language facilitates working in the area. These "language skills" can be acquired by working in nanotechnology labs. This suggests that there is a demand for short-term post-doc positions to help individuals trained in related fields to make the transition to nano. It also suggests that training opportunities be created for individuals with master's and doctoral degrees who don't want to take off a year or more to study but would like to transition into nanotechnology.

Conclusion

We conclude that there is no indication that a bottleneck exists for highly trained scientists and engineers in nanotechnology at the present time. We also conclude that a large number of individuals have training that predisposes them to being able to pick up nanotechnology skills. Providing training opportunities to facilitate the transition could be important. We see few new graduate programs starting at this stage.

We close with a cautionary tale from bioinformatics. The hype surrounding bioinformatics in the late 1990s resulted in the creation of a large number of new academic programs. Indeed, as noted above, in a period of fewer than five years, the number of formal programs grew from 20 to 70. Now there are few jobs in bioinformatics. Positions announcements in *Science* declined by 43 percent from 2000 to 2002 and most that are placed are for university positions related to hiring needs of new programs. There is strong evidence that there are a large number of disappointed students.

The lesson for nanotechnology is clear. There is no sign at the present time that there is a shortage of highly skilled individuals in nanotechnology. There are also no disappointed students. A large number of Ph.D.s who trained in related fields could, if and when there is a shortage, transition into nanotechnology. The conclusion: Grow programs in nanotechnology, but at a rate that the market can absorb.

References

1. Sandra Helsel, The Nano Circuit: Where Nanobiotech and Electronics Meet, Shalom Equity Fund Newsletter, Monday, June 10, 2002, Volume 1 Issue 17 (available at http://www.imakenews.com/shalomequityfund/e_article000075542.cfm)
2. P. E. Stephan, G. Black, Bioinformatics: Does the U.S. system lead to missed opportunities in emerging fields? A case study, *Science and Public Policy* **26**(6), 382-392 (1999).
3. P. E. Stephan, G. Black, Bioinformatics: Emerging opportunities and emerging gaps, Capitalizing on new needs and new opportunities: Government—industry partnerships in biotechnology and information technologies, Washington, DC: National Academy of Sciences (2002).
4. V. Vogel, C. T. Campbell, Education in nanotechnology: Launching the first Ph.D. program, *International Journal of Engineering Education* **18**(5), 498-505 (2002).

APPENDIX

GLOSSARY: TERMS RELATED TO THE SOCIETAL IMPLICATIONS OF NANOSCIENCE AND NANOTECHNOLOGY

By William Sims Bainbridge

Adaptive System - A complex set of interacting entities that adjusts in some way to internal pressures or environmental influences, often with the assumption that it acts in some way to preserve its own equilibrium or to evolve in a particular direction.

Alienation - A personal sense of estrangement from society or powerlessness, usually associated with lack of trust for societal leaders.

Amino Acid - The molecular building blocks that make up proteins; twenty naturally-occurring amino acids combine in various ways to form protein molecules, which are often much more complex.

Angstrom Unit - One tenth of a nanometer, a unit of distance commonly used in atomic physics and related fields, describing the scale just slightly smaller than the nanoscale.

Anomie - An unhealthy condition of society in which people lack shared values, stable standards of judgment for their own lives, and limits that might prevent harmful behavior.

Aquaporins - Proteins that serve as water channels across membranes, thus regulating water transport inside living cells.

Assembler - A hypothetical nanoscale device that could assemble other nanostructures from raw materials consisting of small molecules or even individual atoms.

Atomic Force Microscope - A scientific instrument that can make images of nanoscale details on a physical surface by scanning a tiny, flexible ceramic or semiconductor probe just above the surface, where it will be attracted or repelled slightly by features on the surface, and the deflection can be detected with a laser.

Biomimetic - Engineered structures or devices that imitate biology in their functions or methods of manufacture.

Bionic Technology - Engineered devices and systems that combine biology with electronics, including prosthetic devices and biosensors.

Boundary Work - Effort invested in defining fields of activity, delineating the differences between concepts, and locating things in a socio-cultural context.

Buckyball - A familiar, non-technical synonym for *fullerene.*

Bureaucracy - A system for organizing work and other group activities according to a set of formal rules and procedures that define how participants should behave, usually in the form of a hierarchy of authority.

Capitalism - An economic system characterized by private enterprise, relatively free markets, and a regime that provides legal protection for investment in corporations; widely regarded as especially effective in promoting technological progress, while its connections to social progress are debated.

Co-evolution - The simultaneous development of two different phenomena in interaction with each other, for example the emergence of nanotechnology coupled with the emergence of research on its societal implications.

Common Law - Strictly speaking, an English tradition of unwritten law based on judicial decisions and popular customs; more generally, a set of widely shared but informal legal or ethical principles.

Complexity - The property of being composed of many interrelated parts, or a system of high entropy that cannot be reduced accurately to a very simple model.

Consilience - Convergence of the natural sciences into one, unified body of knowledge.

Constituency - The set of people who are potential supporters for a particular policy or political leader.

Construction - In social science and the law, the social process by which people interpret or construe a particular set of facts or phenomena, at least to some degree arbitrary and influenced by people's self-interest.

Contextualize - To analyze something in terms of a defined set of conditions rather than abstractly, recognizing for example that the meaning and impact of nanotechnology depends upon the context in which it is developed and applied.

Converging Technologies - Unification of nanotechnology, biotechnology, information technology and cognitive technologies, based largely on the unity of nature at the nanoscale and on transforming tools for research and production.

Coordination - With respect to government-funded science and engineering initiatives, coordination involves communication and sometimes cooperative efforts between two or more agencies, to avoid duplication of effort and to make sure that each agency is working at the cutting edge with the highest quality technical guidance appropriate to its mission.

Cultural Lag - A maladjustment in the culture that comes about when rapid change in one area (such as technological innovation) is not immediately matched by concomitant change in other areas that are closely related to it.

Demographic Collapse - The impending loss of population by most advanced industrial nations, as birth rates drop below the level required to replenish the number of people in the society who die during a representative period of time.

Devolution - Transfer of political, economic, and industrial power from a central point, such as the national government or dominant commercial and manufacturing cities, to many locations spread widely across society.

Digital Divide - A supposed social problem consisting of a great inequality between rich and poor (or between ethnic or gender groups) in their access to information, especially to Internet and the Web.

Diffusion - The spread of a technical idea or other cultural element from one group, place, or application to another.

Disruptive technology - New technologies that have the effect of changing the relations between existing organizations and institutions in society, often providing a net benefit even as some people may be harmed at least temporarily.

DNA - Deoxyribonucleic acid, the double-helix molecule that provides the basis of genetic heredity, about 2 nanometers in diameter but often several millimeters in length.

Dual Use - Technologies that may be used for both military and civilian purposes.

Dystopia - The opposite of utopia: An imagined or anticipated state of society that is extremely undesirable, perhaps even implausibly so.

Elasticity - In economics, the tendency of one variable to respond to changes in another, such as a rise in prices responding to an increase in demand.

Entropy - A measure of disorder or of unavailable energy in a closed thermodynamic system; the information content of a message or system, or the irreducible complexity of a scientific model.

Epitaxial Film - A thin crystal layer, perhaps of nanoscale thickness, deposited on the surface of another substance by processes such as vapor deposition.

Ethics - A branch of philosophy that seeks to identify principles defining right or proper behavior of human beings.

Externalities - In economics, costs or benefits that accrue to people other than those making the investment or decision.

Focus group - A method of exploratory public opinion research that brings together a group of people and encourages them to talk more-or-less freely about a topic of interest.

Fullerene - A category of roughly spherical carbon nanoscale structures named after Buckminster Fuller's geodesic spheres.

Futures Market - A market in which prices are established for commodities that will not be delivered until some time in the future.

Gap Analysis - A method of policy analysis that begins with a clear picture of a desired future state, compares it with the current state, and identifies actions that need to be taken to bridge the gap between them.

GDP - Gross Domestic Product, a measure of a nation's annual output of goods and services.

Gedanken Experiment - The mental equivalent of a laboratory experiment, a "thought experiment" in which one thinks through the logical implications of selected assumptions.

Gene Expression - The process by which genetic information is biologically decoded from the genome and gives rise to the cellular structures and genetically-determined characteristics of an organism.

Genome - A single set of chromosomes and their genes; the genetic code for a particular individual.

Genome Sequence - The ordered set of base-pairs that constitute the genetic code in a particular sample of DNA or RNA.

Grey Goo - A fictional nano-engineered substance that can take on any desired form, stimulating the nightmare that it might escape from human control.

Hermeneutics - Interpretation of the variable meaning of something, such as nanotechnology, across time or cultural backgrounds, or within a particular historical context.

Hyperbole ("Hype") - Extravagant exaggeration of claims, for example about the benefits or risks of nanotechnology, sometimes necessary to communicate past a wall of indifference, like shouting so the deaf may hear.

Inelasticity - In economics, the tendency of one variable to remain approximately fixed despite changes in another, such as Americans' tendency to continue to buy fuel-inefficient vehicles despite increases in the price of gasoline.

Initiative - A generic term for a special, time-limited funding emphasis in one or more government agencies, supporting research and development in an area of great scientific promise and potential human benefit, such as the National Nanotechnology Initiative.

Institution - A significant standard practice or kind of organization in public life.

Intellectual Property - A right recognized in copyrights and patents that benefits the creator or current owner of a particular expression of an idea; notably our society grants intellectual property rights for technological inventions but not for scientific discoveries.

Interactional Expertise - The skill and experience required to pass information from one scientist to another, especially the ability to act as a cultural broker between significantly different fields.

Interdisciplinary - A field of endeavor that integrates principles from two or more disciplines.

Invention - The process by which new forms of technology are created; sometimes described as the first stage in the process of technological innovation, followed by development and application of the invented idea.

Labor Market - The system of supply, demand, and communication that connects workers with various skills to available jobs.

Laser Tweezers - A popular term for methods employing lasers to manipulate individual molecules.

Luddites - A protest movement among underpaid English textile workers around 1815 who smashed machines, probably because they did not have the legal option of organizing a labor movement peacefully rather than because of innate hostility to technology; today the term implies stupid opposition to technology.

Melting Pot - The quality of a society or group that readily absorbs immigrants and induces them to give up distinguishing characteristics.

Micron - A distance unit representing one thousand nanometers, one thousandth of a millimeter, or one millionth of a meter.

Molecular Electronics - Electronic circuitry, including computer hardware, in which the separate components are individual molecules or consist of small assemblages of molecules; the hoped-for advantages include cost reduction, increased speed, higher bit density, reduced power requirements and less waste heat.

Monolayer - A layer, film or coating that is only one atom or molecule thick.

Moore's Law - Proposed by Intel founder Gordon Moore in the 1960s, this is a variously stated observation that the density of transistors on an integrated circuit chip has been doubling every 18 to 24 months, or the cost of a transistor has been dropping by half, or that the general capabilities of microelectronics have been improving at an exponential rate.

Multidisciplinary - Activity in which people with two or more different kinds of disciplinary expertise cooperate.

Myth - In anthropology, a myth is not merely a fanciful story but a traditional way of expressing or dealing with a significant social reality, through techniques of metaphor, rhetoric, and narrative.

Nanobot - Fictional nanoscale robot capable of functioning or perhaps even reproducing in some part of the natural environment, such as inside the human body.

Nanocomposite - A material composed of two or more substances of which at least one has a nanoscale dimension, such as nanoparticles dispersed throughout another solid material.

Nanofabrication - A general term for methods to create, assemble, or otherwise form nanoscale structures.

Nanofluidics - Science or engineering involving the flow of liquid or gas through nanoscale spaces.

Nanomanipulator - A tool for moving individual molecules or nanoscale objects, such as an atomic force microscope.

Nanometer - A distance unit representing one billionth of a meter, one millionth of a millimeter, or roughly one millionth the thickness of an American dime.

Nanophobia - An unreasonable fear of nanotechnology, often resulting from false definitions and excessive claims about the power of the technology promulgated by the mass media or special interest groups.

Nanoporous - Substances that have holes or pores on the nanoscale, used for example to separate particles or molecules by size.

Nanoscale - The size range of roughly 1 to 100 nanometers, where many of the fundamental structures of biology are formed, composite materials may take on their distinctive characteristics, and many important physical phenomena are found.

Nanoscience - The study of the unique properties of matter at the nanoscale; an interdisciplinary field of science combining physics, materials science, the chemistry of complex molecules, and related disciplines.

Nanosensor - A device for sensing radiation, forces, chemicals, or biological agents, in which some portion of the device operates at the nanoscale, for example by having receptors into which the particular molecules to be sensed fit.

Nanotechnology - The ability to engineer at the molecular level, atom by atom, to build new structures, materials, and machines.

Nanotube - Hollow, cylindrical structures, often but not necessarily composed of carbon, with walls as thin as a single atom, having remarkable strength and electrical properties.

NBIC - The convergence of four engineering realms: Nanotechnology, Biotechnology, Information technology and new technologies based on Cognitive science.

NEMS - Nanoelectromechanical systems are nanoscale devices - including motors, vibrators, and flex devices such as stearable mirrors - in which mechanical movement is electrically induced.

Niche Market - A finely defined segment of a market requiring specialized products or services; these tend to be relatively small, but in the modern global economy serving their unique needs can be profitable.

Opinion Leader - An influential person who tends to adopt innovations and then convinces other people to adopt them as well.

Organization - A collectivity oriented to the pursuit of relatively specific goals and exhibiting a relatively formal social structure.

Qualia - The fundamental quality of a particular sensory phenomenon, as directly experienced by a human being.

Quality of Life - The standard of living, well being, or satisfaction of people living under particular conditions.

Quantum Dot - A very small clump of material capable of holding a single electron, for example, as part of an electronic or spintronic quantum computer, having a very wide range of potential scientific and industrial applications.

Photolithography - A technique to manufacture objects like microelectronic circuits from optical patterns; contrary to the pessimistic predictions of a few years ago, it has been able to produce circuits with components into the nanoscale such as 50-nanometer wide transistors.

Polymer - A chemical compound, typically formed by connecting smaller molecules together, that consists of repeating structures, often arranged in a chain.

Post hoc analysis - Studying the societal implications of something (such as nanotechnology) after it has already been produced, rather than studying the processes of its creation or being directly involved in them.

Post-industrial Society - The most modern form of society, in which service professions are more important than manufacturing, the professional and technical class dominates other elites, and theoretical knowledge is central for the formulation of policy.

Private Goods - Economic valuables that can be owned by individuals or organizations, can be traded in markets, and are such that for one person to enjoy one is likely to mean that another person cannot.

Protein - Complex combinations of amino acids that serve as the building blocks of life; a very large number exist and play crucial roles in biological systems; their structures, folding, and interactions take place at the nanoscale and present challenging but tractable problems for contemporary science.

Proteomics - The scientific study of proteins, their structure and functions, especially how genetic patterns are expressed through the production of particular proteins in particular parts of a living body at particular points in its life cycle.

Prudential Principle - A proposed ethical principle discouraging the taking of unnecessary risks or requiring debate about whether an innovation is necessary and whether it presents a risk.

Public Engagement - More than mere public awareness about technology issues, but a full citizen involvement in the deliberations about the proper course for technology policy or investment.

Public Goods - Economic valuables that cannot be appropriated or owned by an individual, are not traded in markets, and that are often difficult to support by voluntary contributions because free riders may benefit from them.

Risk Assessment - A set of analytic approaches in the decision, risk and management sciences, typically considering both the probability and the potential cost of a particular type of event.

Risk Aversion - A measurable tendency to avoid taking risks in general, varying across individuals and groups in society.

Scale economies - Differing costs of an activity, depending upon whether it is carried out at a larger or smaller scale; for example, economies of scale achieved by increasing production when many costs are fixed.

Scanning Probe Lithography - A technique in which an atomic force microscope or scanning tunneling microscope scratches, indents, or heats to produce nanoscale features on a surface.

Scanning Tunneling Microscope - A scientific instrument that can make images of nanoscale details on an electrically conductive surface by moving a sharp metal probe very close to that surface, passing a low voltage electric current across, and measuring tiny fluctuations in the current as the probe is scanned across the surface.

Self-Assembly - A process in which a given nanostructure spontaneously constructs itself, generally limited to very specific structures in chemical environments precisely defined in order to promote the self-assembly.

Social Judgment - Evaluation or decision making in which individuals significantly affect each others' perceptions and thoughts, thereby producing an outcome different from what the people might have arrived at separately.

Social Movement - A relatively organized attempt to change some significant aspect of society or to prevent such change.

Speech Recognition - Automatic machine transcription of the sound of a human voice into written words or computer-stored concepts, by means of hardware and software systems based on some mixture of research in signal processing, machine learning, or statistical analysis of language.

Spintronics - A novel form of electronics in which it is the "spin" of the electron rather than its charge that carries information; it holds great promise for further miniaturization.

Stakeholder - An organization, person, or category of people that has a material interest in a pending policy decision and thus arguably should be involved in some way in the decision process.

Sustainability - The characteristic of a technology or economic system that allows it to be used indefinitely without fear of exhausting resources or causing irreparable harm.

Technological Determinism - The theory that technology is the primary cause of social change, and that technological development itself is primarily self-generating.

Technology Assessment - Formalized methods for evaluating the impact of a new technology, including its unintended or second-order consequences.

Technology Transfer - The expansion of a technical idea, device, or new material from one area of application to another.

Therapeutic Cloning - Cloning of human cells not to create a human being but merely to produce tissues or organs that could provide medical benefit to the donor of the cells.

Thermodynamics - The laws governing heat-related processes in matter, notably the energy given off by some chemical reactions and required by other reactions.

Tort - In law a wrongful act that may become the focus of a civil action, rather than a criminal action or breach of contract.

Toxicity - The extent to which a chemical substance is poisonous or through chemical action destroys living tissue.

Trading Zone - A metaphoric place, consisting of shared assumptions and terminology, where specialists from different fields can come together, communicate, and cooperate.

Transilience - In technology, a revolutionary leap that renders older technology obsolete, often involving convergence of technologies and a new relationship between producers and customers.

Transistor - A solid-state electronic device based on a semi-conductor material that regulates the flow of current between two of its electrodes and therefore is capable of amplifying a signal; since its invention in 1948, the transistor has become constantly smaller (recently entering the nanoscale) and more efficient, serving as the basis of most electronic devices including computers.

Tribology - The study and design of interacting surfaces, for example lubrication to reduce friction and wear, that often involves surface features and other phenomena on the nanoscale.

Two Cultures - The theory propounded by C. P. Snow that the science and humanities in modern society are distinct cultures separated by a gulf of ignorance and misunderstanding.

Utopia - An imagined or anticipated state of society that is extremely desirable, perhaps even implausibly so; see also *dystopia*.

Values - The most general principles that guide social action, in the form of abstract goals that people feel should be achieved.

Vapor Deposition - A chemical process commonly used in the semiconductor industry to apply thin films of one substance on a surface composed of another substance.

Virtual Laboratory - A computer-simulation of scientific experiments or observations, perhaps employing full virtual reality display techniques, either for educational or research-planning purpose.

INDEX

MIX
Papier aus verantwortungsvollen Quellen
Paper from responsible sources
FSC® C105338

If you have any concerns about our products,
you can contact us on
ProductSafety@springernature.com

In case Publisher is established outside the EU,
the EU authorized representative is:
Springer Nature Customer Service Center GmbH
Europaplatz 3, 69115 Heidelberg, Germany

Printed by Libri Plureos GmbH
in Hamburg, Germany